SIXTH EDITION

Experiments in Physiology

Gerald D. Tharp
University of Nebraska

MACMILLAN PUBLISHING COMPANY
NEW YORK
MAXWELL MACMILLAN CANADA
TORONTO
MAXWELL MACMILLAN INTERNATIONAL
NEW YORK OXFORD SINGAPORE SYDNEY

Editor: Robert Pirtle
Production Supervisor: Publication Services, Inc.
Production Manager: Aliza Greenblatt
Text Designer: Publication Services, Inc.
Cover Designer: Proof Positive/Farrowlyne Associates

This book was set in Melior by Publication Services, Inc. and printed and bound by Quebecor.
The cover was printed by Phoenix Color Corp.

Printed in the United States of America

Macmillan Publishing Company
866 Third Avenue, New York, New York 10022

Macmillan Publishing Company is
part of the Maxwell Communication
Group of Companies.

Maxwell Macmillan Canada, Inc.
1200 Eglinton Avenue East
Suite 200
Don Mills, Ontario M3C 3N1

Library of Congress Cataloging-in-Publication Data

Tharp, Gerald D.
Experiments in physiology / Gerald D. Tharp. — 6th ed.
p. cm.
Includes index.
ISBN 0-02-419853-6
1. Physiology, Experimental—Laboratory manuals. I. Title.
QP44.T53 1993
591.1'078—dc20 92-19266
CIP

Printing: 1 2 3 4 5 6 7 Year: 3 4 5 6 7 8 9

Preface

TO THE INSTRUCTOR

A laboratory manual is never the work of one author alone; it represents a blend of ideas from other lab manuals, other teachers, and personal experience in the laboratory. I have selected the experiments in this manual because they fulfill two key criteria:

1. They produce consistently successful results—students need not be trained scientists to get meaningful data.
2. They teach significant physiological concepts.

The manual is written in a "generic" format so that it can be used with any text and with a variety of laboratory equipment. Detailed directions for operating specialized equipment are not included, so the manual is more versatile and less cluttered with excessive directions that are seldom read by students.

The following features of the manual have made it an effective tool for student learning and efficient teaching by instructors.

1. *Experiments are grouped into 23 "teaching units."* Each teaching unit consists of a group of related experiments suitable for a 3-hour laboratory period. By eliminating some of these experiments, an instructor can also provide an effective 2-hour lab. The teaching units presented are those that I have found can be successfully performed and discussed by students during a typical 3-hour lab. This grouping of experiments was initiated in the fourth edition and has been well received by instructors and students because it helps them better organize the learning that takes place each week. It also helps students focus on related concepts in physiology, thus maximizing their learning.
2. *A lab report for each teaching unit is provided after each exercise.* Lab reports consist of data tables, graphs, and questions designed to stimulate students' thinking on what they have seen and done in lab. These reports can be removed from the manual and turned in for grading. The questions posed are not meant to be comprehensive but are to accent the major concepts explored experimentally in each unit. Most of the questions can be answered by short statements that can be easily graded by the lab instructor. Comments by instructors indicate that these lab reports are a major feature of the manual. Students have commented that the reports help them understand what they are doing in lab and make it easier to relate their findings to the theoretical concepts studied in lecture.

   Some questions are more complex than others; they require an application of knowledge to new situations. In the elementary physiology course, the Laboratory Report questions are used as a basis for discussion at the end of the lab period, and the answers to complex questions are provided after discussion. In upper-level courses, students are expected to think through these questions on their own.
3. *The* Instructor's Guide *provides sample data, graphs, and answers to questions in the lab reports.* These experimental results represent average values obtained in our teaching labs; they provide guidelines the instructor may use to compare with his or her own results. Instructors are encouraged to devise additional questions for lab quizzes that probe students' knowledge of other facets of the lab experience. The *Instructor's Guide* also provides lists of materials and equipment needed for each teaching unit and the quantities needed for a lab of 20 to 24 students. This simplifies the ordering of supplies and preparation of solutions.
4. *An adequate number of experiments is included* so that the manual can be used for one-semester introductory courses in physiology or for upper-level courses one or two semesters in length. Instructors teaching upper-level courses may wish to use this manual for the foundation core of experiments and to add a few extra experiments that use additional equipment or techniques available in their laboratory.

The revision in this sixth edition of *Experiments in Physiology* represent microevolutionary, rather than macroevolutionary, changes. No new teaching units have been added, but several have been drastically reorganized to enhance learning and streamline the experimental procedures.

1. The AIDS epidemic has caused much concern among physiology instructors because human blood has traditionally been used in several laboratories. Four teaching units in this manual use blood (Units 2, 12, 20, 21). Each of these units now has an introductory paragraph of caution on the use of human blood, referring the students to a new "Precautions for Handling Blood" section (Appendix A).
2. The experiments in the Blood Physiology labs (Experiments 20 and 21) have been reorganized so that Experiment 20 focuses solely on erythrocyte functions and can be run entirely with animal blood, which is available to many instructors through veterinarians or animal science departments. Animal blood also works well for the tonicity and cell permeability experiments in Experiment 2. Experiment 12, "Insulin Regulation of Blood Glucose," and Experiment 21, "Leukocytes, Blood Types, Hemostasis," are designed to test human blood, and precautions for handling blood must be restrictly followed if these units are used in the course. Experiment 12 now contains detailed instructions for using the Glucometer II (Ames Division, Miles Labs) for measuring blood glucose more accurately. Instruments such as the Glucometer II are readily available and are commonly used by diabetics to monitor their blood glucose.
3. A new "Nerve-Muscle Activity" unit (Experiment 15) combines experiments from the fifth edition's "Muscle Contractility" and "Neuromuscular Activity" units. Directions for dissecting either an *in vitro* or an *in vivo* muscle preparation are provided so that the instructor can follow his or her own preference. The sequence of experiments will work a little better with the *in vivo* preparation because the sciatic nerve remains protected in the thigh and blood flow is maintained to the muscle. The major advantage of combining the units is that it helps students learn not only the characteristics of isolated muscle contractions but also the importance of motor units and neuromuscular junctions in regulating muscle activity.
4. Two experiments have been removed from the manual because students have perceived them as being cruel to animals: the spinal frog reflex demonstration (from Experiment 5) and insulin shock (from Experiment 12).
5. Sections that have been completely rewritten include the tuning fork tests (Experiment 7), near point of accommodation (Experiment 8) and the immunologic pregnancy test (Experiment 9). New sections that have been added to the manual are the sample concentration problem (Experiment 1), and the astigmatism (Experiment 8) and arterial pulse experiments (Experiment 17). New Laboratory Report questions have been written for Experiments 3, 10, 12, 17, 20, 21.
6. The unit on "Cardiovascular Principles" (Experiment 11 in the fifth edition) has been deleted. The microcirculation experiment has been added to the blood physiology lab (Experiment 20) because it provides a good demonstration of erythrocytes in action, and the heart diagram has been added to Experiment 16 ("Cardiac Function").
7. The sequence of teaching units has been completely reorganized to correspond with the lecture sequence I have found most effective for student comprehension of physiology concepts. The sequence begins with fundamental principles and membrane transport mechanisms, followed by renal physiology, which provides good examples of how these basic processes are used in the kidney to produce homeostasis. Then the neural and endocrine integrative system units are presented, which provides the foundations needed to understand the organ systems: digestion, metabolism, muscle, and cardiovascular, respiratory, and blood physiology. The physical fitness and exercise units conclude the manual because they require an understanding of all physiological systems.

TO THE STUDENT

The study of physiology is only half accomplished if you never enter the laboratory. It is one thing to hear a concept explained in lecture, but quite another to see the concept unfold before your eyes in a laboratory experiment. The study of physiology is both fascinating and practical—fascinating for its examination of the awesome complexity of body processes, and practical for its future usefulness in our lives.

The experiments presented in this manual are designed to illustrate the basic principles of physiology. They are also meant to develop your ability to carry out measurements, make observations, and formulate reasonable deductions—characteristics of the scientific process.

Physiology might appear at first to be an easy science to master, especially to students whose prior schooling has included some study of the heart, brain, eye, and ear. As the complexity of the subject becomes apparent, however, some students may become discouraged. I urge you to accept the difficulties as a challenge; you will find that the more effort you expend, the more interesting physiology will become.

The following suggestions will make the laboratory experience more valuable to you.

1. Study the lab experiment and lab report *before* coming to the lab. Usually the instructor will give a short introduction to the lab, but this introduction is to help you organize your work and not to give all the details for conducting the experiments.

2. Arrive promptly for lab and become acquainted with the location of equipment and supplies. Use the instructor's introductory comments to help you plan how your team can accomplish the lab work most effectively.

3. Participate actively in the lab. Don't expect to listen passively and let others do the work. Research has shown that true learning occurs best when a person learns actively, and that active learning is stored longer in the memory systems of the brain than is passive learning.

4. Be prepared to work closely with others. Physiology lab work is a team effort that requires an exchange of information and interaction with your classmates and with your instructor. Working closely with others is an important feature of the lab experience, one that will provide benefits to you beyond the mere acquisition of knowledge.

5. As you conduct the experiments, try to relate the theoretical information presented in lecture and textbook with your lab observations. Don't just perform the work mechanically. There is always the danger that, as you struggle with technical difficulties, you will lose sight of the purpose of the experiment. You should continually ask yourself, "What is this experiment trying to show us?"

6. Get in the habit of promptly recording all data as soon as they become available. Much information is lost because a person is too lazy to write things down immediately. If a recording is made of some parameter, write on the record the date, experiment, experimental conditions, and results so that you have complete data when you examine the record later.

7. Use the lab report as a guide for recording the data from the experiments and studying the major concepts explored in each lab. If used properly (that is, to engage your mind in active, critical thought about what you have seen in lab), the lab report can be a useful learning device. Don't expect to gain much if you copy someone else's answers so that you can turn in the report and hurry off to do something "more important." If you don't do your own reports, you will be the loser because you will have missed a valuable opportunity to put your mind to work and learn something.

8. Ask for help when you don't understand how something works. The instructor will gladly help you get your experiment working, but you should also attempt to solve the minor problems yourself. Don't be too discouraged when an experiment fails or you obtain data that do not agree with the expected results. Because of "biological variation" there will be times when things don't work out exactly right (this also happens in real life), but try not to get discouraged, and keep doing your best.

9. When the lab is completed, clean your lab table and discard any waste in the wastebasket. Return all equipment and supplies to their appropriate places.

10. **Caution!** Some experiments in human physiology in this manual, such as the step tests and the exercises to be done in Experiments 22 and 23, induce some degree of cardiovascular stress. *Students who have cardiovascular difficulties, such as cardiovascular insufficiency or hypertension, should not take part in any experiment that causes cardiovascular stress unless they have permission from their physician.* If you feel that you should not participate in any experiment for personal health reasons, be sure to tell your instructor. If you suffer any ill effects while in an experiment, stop and inform your instructor.

If you keep these suggestions in mind and approach the physiology lab with a positive attitude, you will be richly rewarded with a positive learning experience.

Good luck. I hope your study of physiology is as exciting as mine has been.

ACKNOWLEDGMENTS

I would like to express my appreciation to many of my students for their constructive criticisms, which have improved the effectiveness of this manual. I am most indebted to my teaching assistants, who have worked closely with me in preparing, teaching, and criticizing the labs.

A major factor in any endeavor such as this is the support of those who make all of this effort worthwhile: my wife Dee, who has loved me in spite of the long hours I spend with my nose in the books; my children, Danny, Kathy, Tim, and Jeanine, who have given my life much meaning and excitement; and my parents, Evae and Glenn, who instilled in me the idea that each person is valuable in God's eyes and gave me the confidence needed to accomplish my goals in life.

Also, I would like to thank my first and best physiology professor, Paul Landolt, who made physiology so interesting that I decided to enter the field and follow in his footsteps, a decision I have never regretted.

G.D.T.

Contents

Preface v

To the Instructor v

To the Student vi

Acknowledgments viii

1 Fundamental Physiological Principles 1

Units of Measurement 1
Concentration of Solutions 2
Acid-Base Balance 4

2 Movement Through Membranes 9

Diffusion 9
Osmosis 10
Tonicity 10
Cell Permeability 12

3 Renal Physiology 21

Kidney Regulation of Osmolarity 22
Urinalysis 22

4 Neuroanatomy 33

Organization of the Nervous System 33
Spinal Nerves and Spinal Cord 33
Cranial Nerves 35
External Structures and Landmarks of the Brain 36
Sectioning of the Brain 37

5 Reflex Functions 41

Human Reflexes 41
Reaction Times 44

6 Membrane Action Potentials 49

Resting and Action Potentials 49
Stimulation of Tissues 49
Oscilloscope 50
Sciatic Nerve Compound Action Potential 51

7 Sensory Physiology I: Cutaneous, Hearing 63

Sensory Receptors 63
Cutaneous Receptors 63
Hearing 65

8 Sensory Physiology II: Vision 75

Functions of the Eye 75
Anatomy of the Eye 81
Ophthalmoscopy 83

9 Reproductive Physiology 89

Influence of Hormones on Reproduction 89
Testicular and Gonadotropic Hormones 91
Ovarian Hormones and Estrus Cycle 92
Pregnancy Tests 94

10 Digestion 99

Salivary Digestion of Carbohydrates 99
Gastric Digestion of Protein 100
Digestion of Fat with Pancreatic Lipase and Bile Salts 100

11 Smooth Muscle Motility 107

Responses of Intestinal and Uterine Segments 107

12 Insulin Regulation of Blood Glucose 111

Action of Glucose 111
Glucose Tolerance Test 112
Operation of the Glucometer II 113

13 Measurement of Metabolic Rate 121

Human Metabolism: Calorimetry 121
Relationships of Metabolism to Surface Area and Body Weight 124

14 Thyroid Function 135

Thyroid Effects on Metabolism 135
Thyroid Uptake of Iodine 137

15 Nerve-muscle Activity 143

Dissection of Nerve-muscle Preparation 143
Isolated Muscle Responses 147
Stimulation of Motor Points 149

16 Cardiac Function 163

Characteristics of Heart Contractility 163
Anatomy of Amphibian or Reptilian Heart 163
Physiology of Amphibian or Reptilian Heart 164

17 Human Cardiovascular Function 175

Auscultation of Heart Sounds 175
Measurement of Blood Pressure 176
Arterial Pulse Wave 178
Valves in the Veins 180
Electrocardiogram 180
Electrical Axis of the Heart 184

18 Respiratory Function 189

Respiratory Movements 189
Respiratory Volumes 191
Pulmonary Function Tests 193

19 Regulation of Circulation and Respiration 199

Instrumentation 199
Anesthetics for Rabbits 200

20 Blood Physiology I: Erythrocyte Functions 205

Functions of Blood 205
Blood Hematocrit 205
Hemoglobin Determination 206
Blood Cell Counting 208
Microcirculation 212

21 Blood Physiology II: Leukocytes, Blood Types, Hemostasis 219

Identification of White Blood Cells 219
Blood Typing 222
Blood Coagulation (Hemostasis) 224

22 Physical Fitness 229

Muscular Strength and Endurance 229
Flexibility 230
Body Composition 230
Cardiorespiratory Endurance (Aerobic Fitness) 234

23 Physiology of Exercise 243

Parameters Modified by Exercise 243

Appendix A: Precautions for Handling Blood 251

Appendix B: Solutions 253

Appendix C: Tables and Nomograms 255

Index 263

1 Fundamental Physiological Principles

Physiology is a quantitative science; physiologists are constantly trying to measure changes occurring in living organisms. Experimental work in physiology therefore requires a knowledge of certain fundmental principles such as units of measurements, concentration of solutions, and acid-base balance. This exercise is designed to acquaint or reacquaint students with these principles, which we will use during our study of physiology. For students with previous courses in biology and chemistry, this will be old material. For others it will be a first encounter with these concepts. It is hoped that all students will be on more even ground after completing this lab.

UNITS OF MEASUREMENT

The following units are used fairly frequently in physiology:

Length

kilometer[1] (km) = 1000 m = 0.62 mi
1 mi = 1.61 km = 1760 yd

hectometer (hm) = 100 m

dekameter (dkm) = 10 m

meter (m) = 39.37 in. = 1.09 yd

decimeter (dm) = $\frac{1}{10}$ m = 10 cm

centimeter (cm) = $\frac{1}{100}$ m = 10 mm = 0.3937 in.
1 in. = 2.54 cm

millimeter (mm) = $\frac{1}{1000}$ m = 1000 μm (microns) = $\frac{1}{25}$ in.

micrometer (μm) = one-millionth m = 1000 nm = 1 micron

nanometer (nm) = one-billionth m = 1 mμm = 10 Å

angstrom (Å) = 0.1 nm

picometer (pm) = one-trillionth m

Volume

liter (L) = 1000 ml = 1000 cc = 1.05 qt = 0.264 gal

deciliter (dl) = 0.1 L = 100 ml

milliliter (ml) = $\frac{1}{1000}$ L = 1000 μl

ounce (oz) (fluid) = 8 fl drams = 29.57 ml

quart (qt) (fluid) = 32 oz = 946 ml = 0.946 L

Weight

metric ton = 1,000,000 g = 2204.62 lb

kilogram (kg) = 1000 g = 2.2 lb = 35.27 oz

gram (g) = 1000 mg

454 g = 1 lb = 16 oz

1oz = 28.35 g

milligram (mg) = $\frac{1}{1000}$ g = 1000 μg

microgram (μg) = $\frac{1}{1000}$ mg = 1 gamma

Temperature

0° centigrade (°C) = 32° Fahrenheit (°F) = 273 Kelvin (K)

$°C = \frac{5}{9} (°F - 32)$

$°F = \frac{9}{5} °C + 32$

Pressure

Pressure is force per unit of area.

1 atmosphere = 34.0 ft of water = 760 mm (or 29.92 in.) of mercury (Hg) = 14.7 lb/in.^{2}

Energy

One calorie (cal) is the amount of energy required to heat 1 g of water 1 °C (at 15 °C).

One kilocalorie (large calorie or kcal) is the amount of energy required to heat 1 kg of water 1 °C (at 15 °C).

1 kcal = 1000 cal = 3086 ft-lb = 426.4 kg-m

1 g of carbohydrate = 4.1 kcal

1 L of oxygen used in burning glycogen (RQ of 1) = 5.047 kcal = 15,575 ft-lb = 2153 kg-m

1 L of oxygen in a closed circuit system = 4.825 kcal in the postabsorptive state (RQ assumed to be 0.82) and 4.862 kcal on an ordinary mixed diet (RQ assumed to be 0.85).

Work

Work is force times the distance through which it acts.

1 ft-lb = 1 lb of force times 1 ft

1 kg-m = 7.23 ft-lb = 0.002 343 kcal = 2.343 g-kcal

1 kg-m = 1 kg of force times 1 m

Power

Power is work or energy per unit of time.

1 horsepower (HP) = 33,000 ft-lb/min = 550 ft-lb/sec = 4564 kg-M/min = 76.07 kg-m/sec = 746 watts (W) = 10.694 kcal/min = 0.178 kcal/sec

1 kilowatt (kw) = 1000 W = 1.341 HP = 0.239 kcal/sec

CONCENTRATION OF SOLUTIONS

Many of the physiological properties of solutions depend on the number of molecules, ions, or particles in the solution, and therefore it is important that you understand the various means of expressing concentrations that are used in physiology.

Percentage (%) Solutions

This is probably the simplest means of expressing concentration and one that is commonly used. Percent means "parts in 100." Percentage is the number of grams of solute dissolved in 100 ml (deciliter) of solution. It is calculated using this formula:

$$\text{Percentage} = \frac{\text{Grams of solute}}{\text{Volume of solution}} \times 100$$

Thus, a 12% solution **weight to volume (W/V)** of glucose would contain 12 g of glucose in each 100 ml of solution (12 g/dl), or 120 g/L. If 2 g of NaCl is dissolved in 25 ml of water, the percentage will be:

$$\text{Percentage} = \frac{2\text{ g}}{25\text{ ml}} \times 100 = 8\%$$

In living organisms, the concentration of many substances is so low it is more easily expressed as **milligrams percent (mg%).** For instance, the average blood glucose concentration is around 90 mg%. This simply means that in every 100 ml of blood there is 90 mg of glucose. If this concentration were expressed as percentage it would be 0.09%, which is a more awkward number to use.

Molar (M) Solutions

A 1 molar (M) solution contains 1 mole of solute in 1 L of solution. One mole of a substance contains 6.024×10^{23} molecules (Avogadro's number). Thus, solutions of equal molarity have the same number of molecules in solution, even though their molecular weights may be different. One mole is equal to the molecular weight (MW) or atomic weight of the solute in grams.

For example, the MW of glucose is 180. To prepare a 1 M solution of glucose, we would weigh out 180 g of glucose and dissolve it in a total volume of solution (solvent + solute) of 1 L. Eighteen grams of glucose in 100 ml of solution would also be a 1 M concentration and would be an 18% solution. Why?

To make a 1 M solution of NaCl (58.5 MW), we would dissolve 58.5 g of NaCl in a liter of solution. This would be the same as 5.85 g of NaCl in 100 ml, or a 5.85% solution. You can see from these examples that decreasing the amount of solute and solution by the same proportion does not change the concentration of the solution.

Because of the low concentrations of solutes in body fluids, we often use millimolar (mM) measurements in physiology. If 180 mg of glucose is dissolved in 1 L of solution, a 1 mM concentration is produced.

The 90 mg% glucose concentration in the blood = 90 mg/100 ml = 900 mg/L.

$$\frac{900 \text{ mg}}{180 \text{ mg/mM}} = 5 \text{ mM glucose}$$

Osmolar (Osm) Solutions

Osmolar concentrations are used mainly in the biological sciences to express the osmotic effect of a solution. To understand their use we will look at some examples. (Osmosis and osmotic effects are further explained in Experiment 2, "Movement Through Membranes.")

A membrane permeable only to water separates a container into two compartments (Figure 1.1a). Water molecules are on side A, and glucose molecules are trapped on side B. In an effort to reach equilibrium, the water molecules pass through the membrane into side B, moving from a higher to a lower water concentration. We call this water movement **osmosis.** The force of the water movement on the membrane is called the **osmotic pressure** and is determined by the number of molecules in compartment B that cannot penetrate the membrane. The 1 M glucose solution is 1 Osm in the osmotic pressure it produces on the membrane.

If the concentration of glucose (a nonelectrolyte) is doubled on side B to 2 M (Figure 1.1b), the water moves across with twice the osmotic pressure, because there are twice as many osmotically active particles in solution. Thus, the 2M glucose solution is 2 Osm in its osmotic effect.

Some electrolyte molecules, such as NaCl (table salt), do not remain as molecules in solution but dissociate into ions.

$$1 \text{ mole NaCl} \rightarrow 1 \text{ mole Na}^+ + 1 \text{ mole Cl}^-$$

1 M 1 M = 2 Osm

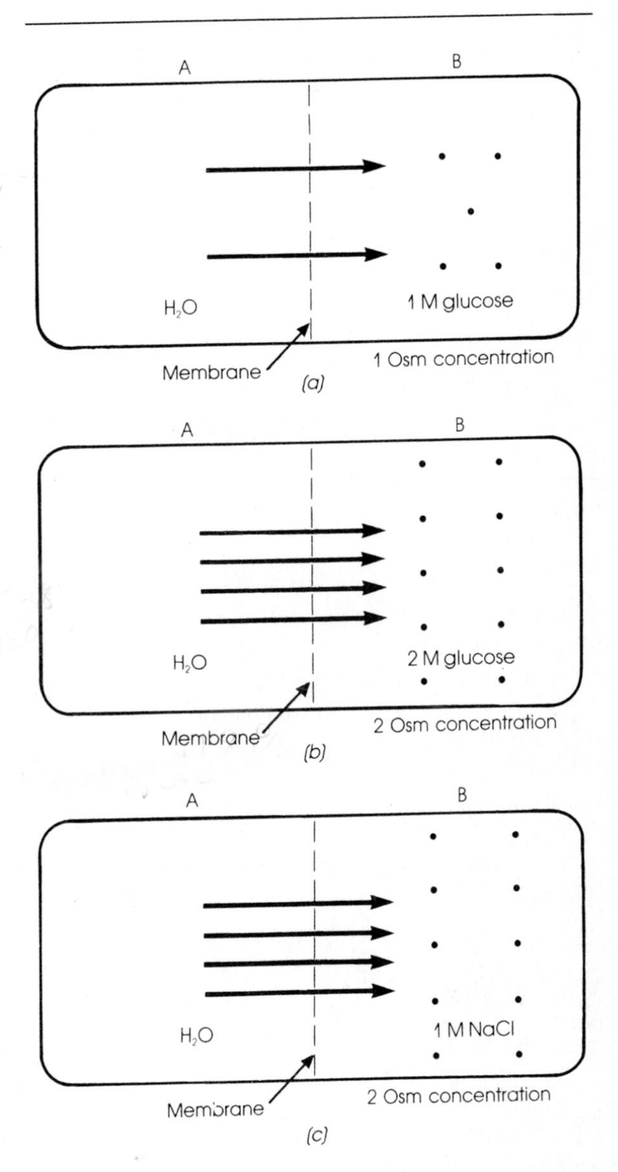

FIGURE 1.1. Examples of osmotic effects of solutions.

Each ion acts as an osmotically active particle to cause water movement into the region of higher solute concentration (osmosis). Thus, a 1 M NaCl solution produces the same osmotic pressure as a 2 M glucose solution (Figure 1.1c). Both solutions are therefore 2 osmolar (2 Osm) in their osmotic concentration. The osmole provides a measure of a solution's ability to produce osmosis or osmotic pressure.

The total number of osmotic particles in the cells of a mammal have a concentration of 0.3 Osm. If we bathe the cell in a solution having the same osmolar concentration (e.g., 0.3 M glucose or 0.15 M NaCl), there will be no net movement of water in or out of the cell, and the cell will retain its size and shape (Figure 1.2). Solutions

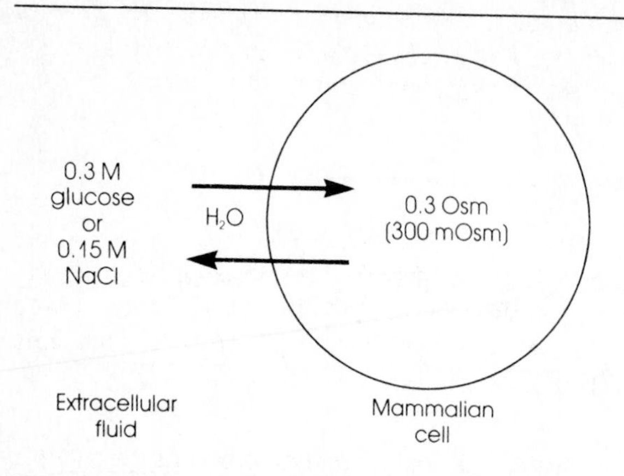

FIGURE 1.2. Example of osmotic equilibrium.

having the same osmolar concentration as the concentration inside the cell are said to be **isotonic** (same "tone" or "tension"). Solutions with a higher osmolar concentration are called **hypertonic** and those with a lower osmolar concentration are **hypotonic.** Which way will water move if a cell is placed in each of these solutions? Some examples of osmolar calculations are as follows:

To make a 1 Osm solution of NaCl (58.5 MW), dissolve 58.5 g/2 = 29.25 g in each liter of solution (two ions formed in solution).

To make a 1 Osm solution of $CaCl_2$ (110 MW), dissolve 110 g/3 = 36.6 g in each liter, because $CaCl_2$ dissociates into three ions in solution.

Equivalent (Eq) or Milliequivalent (mEq) Solutions

This means of expressing the concentration of ions is used extensively in chemistry and the various biological sciences. We will not use it in this manual, but you should be aware of it, because it appears often in textbooks and research papers. Equivalent weights reflect the combining power of substances during a chemical reaction, which in turn depends on the valence (charge) of the atoms involved. For univalent ions like Na^+, K^+, and Cl^-, the milliequivalents are equal to millimoles. Only when we deal with divalent (Ca^{++}) or trivalent (Fe^{+++}) ions will the milliequivalent and millimole concentrations differ.

Sample Concentration Problem

To solve concentration problems, we need to know the relationship between percentage, molarity, and osmolarity concentrations. If these are related in a stepwise fashion, it is easier to understand and solve such problems. As an example, let us use a series of steps to solve the following problem: A 2% solution of KCl is what osmolar concentration?

1. Percentage refers to the number of grams in 100 ml of solution (mg% refers to milligrams per 100 ml of solution)

   2% KCl = 2 g of KCl dissolved in 100 ml of solution

2. To go from percentage to molarity we must first determine the number of grams in one liter (1000 ml) of solution.

   2 g KCl in 100 ml = 20 g KCl in 1000 ml (1 L) (because 1000 is 10 ×100)

3. To determine molarity we first determine the number of g per L in a one molar solution of the substance.

   1 M KCl (74 MW) = 74 g per liter

   Therefore, the 20 g per liter of KCl we have is less than 1 M. Specifically, we have 20 g/74 g = 0.27 M of KCl.

4. To go from molarity of osmolarity we need to know if the substance is an electrolyte, which dissociates into ions in solution, or a nonelectrolyte, which remains as molecules in solution. KCl is an electrolyte that forms 2 ions (K^+ and Cl^-) in solution.

   Moles × Number of ions = Osmoles

   Therefore, 0.27 M × 2 ions = 0.54 Osm KCl.

ACID-BASE BALANCE

It is critical for the body's **homeostasis** (maintenance of a constant internal environment) that the concentration of hydrogen ions (H^+) in the blood be maintained within the narrow range of around pH 7.0 to 7.8. If the pH goes below or above these limits, death results, because most enzymes cannot operate properly if the pH is outside this range. In light of the great importance of H^+ regulation, let us review the concepts of pH and acid-base balance.

pH

The pH scale is simply a way of expressing small molar concentrations using whole numbers (Figure 1.3). It was devised by the Danish chemist Sorenson. Note in Figure 1.3 that a difference of one pH unit represents a tenfold change in H^+ or OH^- concentration.

pH = Logarithm of reciprocal of hydrogen ion concentration

$$= \log \frac{1}{H^+}$$

For example:

$$pH = \log \frac{1}{1 \times 10^{-7}\ mol} = \log \frac{10^7}{1} = 7$$

Acids and Bases

Acid. Substance that dissociates into hydrogen ions (H^+)

Base. Substance that dissociates into hydroxyl ions (OH^-)

Salt. Substance that dissociates into neither H^+ nor OH^- in solution

A major problem for the body is that many acids are produced during metabolism. One of the most important of these is carbonic acid, formed when carbon dioxide (CO_2) dissolves in body fluids:

$$\underset{\text{carbon dioxide}}{CO_2} + \underset{\text{water}}{H_2O} \rightleftarrows \underset{\text{carbonic acid}}{H_2CO_3} \rightleftarrows \underset{\text{hydrogen ion}}{H^+} + \underset{\text{bicarbonate ion}}{HCO_3^-}$$

This is one of the most important chemical reactions you will encounter in your study of physiology. Other acids that lower the pH of body fluids are phosphoric, sulfuric, hydrochloric, lactic, keto, and fatty acids.

Buffer Systems

Even though the body produces tremendous quantities of acid each day, blood pH usually remains within the range 7.35 to 7.45. This is made possible by the action of various buffer systems in the body fluids.

Buffer. Substance that prevents (resists) a drastic pH change when acids or bases are added to a solution.

Buffer mechanism. Mechanism that replaces strong acids or bases with weak acids or bases that produce fewer H^+ or OH^- in solution.

The major buffer systems found in most animals are the protein, phosphate, and bicarbonate systems. **Proteins** are the most abundant buffers, playing a critical role inside body cells as well as in the blood (e.g., albumins, hemoglobin). **Phosphates** are less abundant but perform important buffering in the intracellular fluid and kidney tubules.

Bicarbonates are most important in buffering of the extracellular fluid (interstitial fluid and plasma). The bicarbonate system is a unique buffer because its components (HCO_3^- and CO_2) can be regulated by the renal and respiratory systems. This makes it a very powerful and flexible buffer, one that deserves closer scrutiny.

$$\text{Bicarbonate buffer} = \underset{\text{(carbonic acid)}}{\overset{\text{Weak acid}}{H_2CO_3}} + \underset{\text{(sodium bicarbonate)}}{\overset{\text{Conjugate base}}{NaHCO_3}}$$

	Acid							Neutral	Basic (Alkaline)						
H+ conc.	10^{0}	10^{-1}	10^{-2}	10^{-3}	10^{-4}	10^{-5}	10^{-6}	10^{-7}	10^{-8}	10^{-9}	10^{-10}	10^{-11}	10^{-12}	10^{-13}	10^{-14}
pH	0	1	2	3	4	5	6	7	8	9	10	11	12	13	14
OH− conc.	10^{-14}	10^{-13}	10^{-12}	10^{-11}	10^{-10}	10^{-9}	10^{-8}	10^{-7}	10^{-6}	10^{-5}	10^{-4}	10^{-3}	10^{-2}	10^{-1}	10^{0}

FIGURE 1.3. pH scale.

Adding a strong acid or base to this system produces the following reactions:

HCl hydrochloric acid (strong acid)	+	$NaHCO_3$	→	H_2CO_3 (weak acid)	+	NaCl (salt)
NaOH sodium hydroxide (strong base)	+	H_2CO_3	→	$NaHCO_3$ (weak base)	+	H_2O

Thus, the bicarbonate buffer chemicals replace strong acids and bases with weak acids and bases that dissociate only weakly and therefore produce little change in pH. The following experiment demonstrates how buffers moderate the pH effects of acids and bases.

Experimental Procedure

1. Allow the pH meter to warm up with the selector switch on the "standby" position. Set the temperature selector to the room temperature.
2. Immerse the pH electrode in 150 ml of distilled water contained in a 250-ml beaker. Turn the selector switch to "pH" and record the pH of the water. Return the switch to standby.
3. Add 1 drop of concentrated hydrochloric acid (HCl) to the distilled water, mix with a stirring rod, and record the new pH. Repeat these procedures for 4 additional drops of HCl, recording the new pH after each drop is added. Be careful not to add the acid directly on the electrode, as this could damage the electrode.
4. Add 1 drop of concentrated sodium hydroxide (NaOH) to the beaker of distilled water, mix, and record the new pH. Repeat this with 4 additional drops of base, recording the new pH after each drop is added.
5. Remove the pH electrode from the solution, rinse with distilled water, and dry with a lint-free paper.
6. Immerse the electrode in 150 ml of phosphate buffer solution in a 250-ml beaker and record the pH. This buffer is a mixture of dibasic sodium phosphate (Na_2HPO_4) and sodium acid phosphate (NaH_2PO_4).
7. Repeat the addition of concentrated acid and base, as in steps 3 and 4, on this buffer solution. Record the new pH after each drop is added and mixed.

LABORATORY REPORT

Name ______________________

Date ____________ Section ____________

Score/Grade ______________________

1. Fundamental Physiological Principles

Units of Measurement

1. Provide the correct conversion units for the following measurements:

10 k run = 6.2 mi

55 mph = 88.55 km/hr

100 Å thickness of cell membrane = 10 nm

5 L cardiac output = 5.25 qt

600 ml urinary bladder volume = 20.3 oz

8 Å diameter membrane pore = 800 pm

2000 lb Honda car = 909.1 kg

15 g hemoglobin per 100 ml blood = 150 mg/ml

120 mm Hg blood pressure = 2.32 lb/in.^{2}

100 yd dash = 91.7 m

7 ft athlete = 2.14 m

7 micron RBC diameter = 7 µm

1 ml urine per minute = .38 gal/day

100 ml plasma = 0.1 L

70 kg man = 154 lb

25 °C room temperature = 77 °F

98.6 °F body temperature = 37 °C

2. A power lifter lifts 500 lb 6 ft off the ground. How much work has he performed?

3000 ft-lb work = 415 kg-m work

How many calories of energy did he use to produce this work? 972 cal

If he performs this feat 11 times in 1 minute, what is his power output? 1 HP = 746 W

Concentration of Solutions

Atomic weights: Na = 23 K = 39
Ca = 40 Cl = 35.5

Molecular weights: Glucose = 180 NaCl = 58.5
KCl = 74 $CaCl_2$ = 110

1. How many grams of glucose would you need to make 500 ml of an 8% solution? 40 g
2. If 6 g of NaCl is dissolved in 1 L of solution, what percent concentration is prepared? 0.6 %
3. How many grams of KCl would you need to make 250 ml of a 0.5 M solution? 9.25 g
4. A 9% solution of glucose would be what molar concentration? 0.5 M
5. A 300 mM solution of $CaCl_2$ would be what percent? 3.3 %

6. Sodium ions are found in the extracellular fluid (ECF) in a concentration of 150 mM. How many grams per liter is this? 3.45 ________ g/L
   How many milligrams per milliliter would this be? 3.45 ________ mg/ml
7. An 11.7% solution of NaCl would be what osmolar concentration? 4 ________ Osm
8. A 0.9% solution of NaCl is considered isotonic to mammalian cells. What molar concentration is this? .154 ________ M
9. What percent concentration of KCl would be isotonic to body cells? 1.11 ________ %
10. A 33% solution of $CaCl_2$ would be what osmolar concentration? 9 ________ Osm
11. You want to make 500 ml of an isotonic glucose solution to infuse into a patient. How many grams of glucose do you need? 27 ________ g
12. Potassium ion concentration in the interstitial fluid is 5 mEq/L. What milligrams percent concentration is this? 19.5 ________ mg %

Acid-Base Balance

1. Record the pH obtained for each experimental situation:

DISTILLED H_2O	DROPS OF CONCENTRATED HCl 1	2	3	4	5	DROPS OF CONCENTRATED NaOH 1	2	3	4	5

PHOSPHATE BUFFER	DROPS OF CONCENTRATED HCl 1	2	3	4	5	DROPS OF CONCENTRATED NaOH 1	2	3	4	5

2. Write the chemical reactions occurring as the HCl and NaOH are buffered by these phosphates:

HCl + ____________ → ____________ + ____________

NaOH + ____________ → ____________ + ____________

2 Movement Through Membranes

Caution! *Parts of this lab involve experiments with blood. If human blood is used, disposable latex gloves must be worn and the* Precautions for Handling Blood *must be followed. (See Appendix A.)*

Throughout the body we find many types of membranes, such as the capillary membrane, alveolar membrane, cell membrane, and nuclear membrane. These membranes serve as barriers between different compartments in the body to confine certain processes to specific locations. The transport of various molecules and ions through these membranes is of critical importance for the maintenance of a constant internal environment in the body (**homeostasis**). In the following experiments we will examine some of the physiological principles that govern movement through membranes in general and the cell membrane in particular.

DIFFUSION

Diffusion is the random movement of molecules due to their internal kinetic energy. This continuous movement allows molecules and ions to be distributed uniformly within a closed space (e.g., the plasma or interstitial space). A **net diffusion** of particles results when there is a difference in concentration between two regions of a system; that is, a **concentration gradient** is established. A net diffusion of particles will then take place from the region of higher concentration to the region of lower concentration, and this diffusion will continue until the system reaches equilibrium (when the concentration everywhere in the system is equal). The various factors influencing the rate of diffusion are conveniently grouped together in what is known as **Fick's law of diffusion:**

$$Q = DA\,(C_1 - C_2), \text{ in which}$$

$$Q = \text{Diffusion rate}$$
$$C_1 - C_2 = \text{Concentration gradient}$$
$$A = \text{Cross-sectional area}$$
$$D = \text{Diffusion coefficient}$$

The diffusion coefficient is a composite constant that takes into account the temperature, the molecular weight of the substance diffusing, and the nature of the substance through which the first substance is diffusing.

Experimental Procedure

Place several crystals of methylene blue in three beakers containing, respectively, water that is cold (5 °C), room temperature (25 °C), and hot (50 °C). In the Laboratory Report, record the time required for the dye to become evenly dispersed throughout the beakers.

OSMOSIS

The phenomenon of osmosis occurs whenever a higher concentration of solute is separated from a lower concentration of solute by a membrane that is either semipermeable or selectively permeable. In this situation, water molecules begin moving through the membrane into the region of higher solute concentration. This net movement of water is defined as **osmosis,** and the force of the water movement across the membrane is called the **osmotic pressure.** The amount of osmotic pressure developed depends on the *number* of particles present on either side of the membrane; the more particles there are, the higher the osmotic pressure. Osmotic pressure may be calculated using the following formula:

$$\pi = iRT\,(C_1 - C_2), \text{ in which}$$

T = Absolute temperature Kelvin

π = Osmotic pressure in atmospheres

i = The number of ions dissociated from each molecule in solution

R = Gas constant 0.082 L-atm/deg-M

$C_1 - C_2$ = Concentration gradient in moles per liter

One mole of an electrolyte can produce more osmotic pressure than can one mole of a nonelectrolyte, because the electrolyte will dissociate into two or more ions in solution, each of which acts as an osmotically active particle. To compare the osmotic ability of various solutions, we use the **osmolar** unit of concentration, which was developed for this purpose.

Experimental Procedure

1. Osmotic Pressure

To demonstrate the phenomenon of osmosis and osmotic pressure, set up two osmometers. Soak a piece of dialysis tubing, about 15 cm in length,

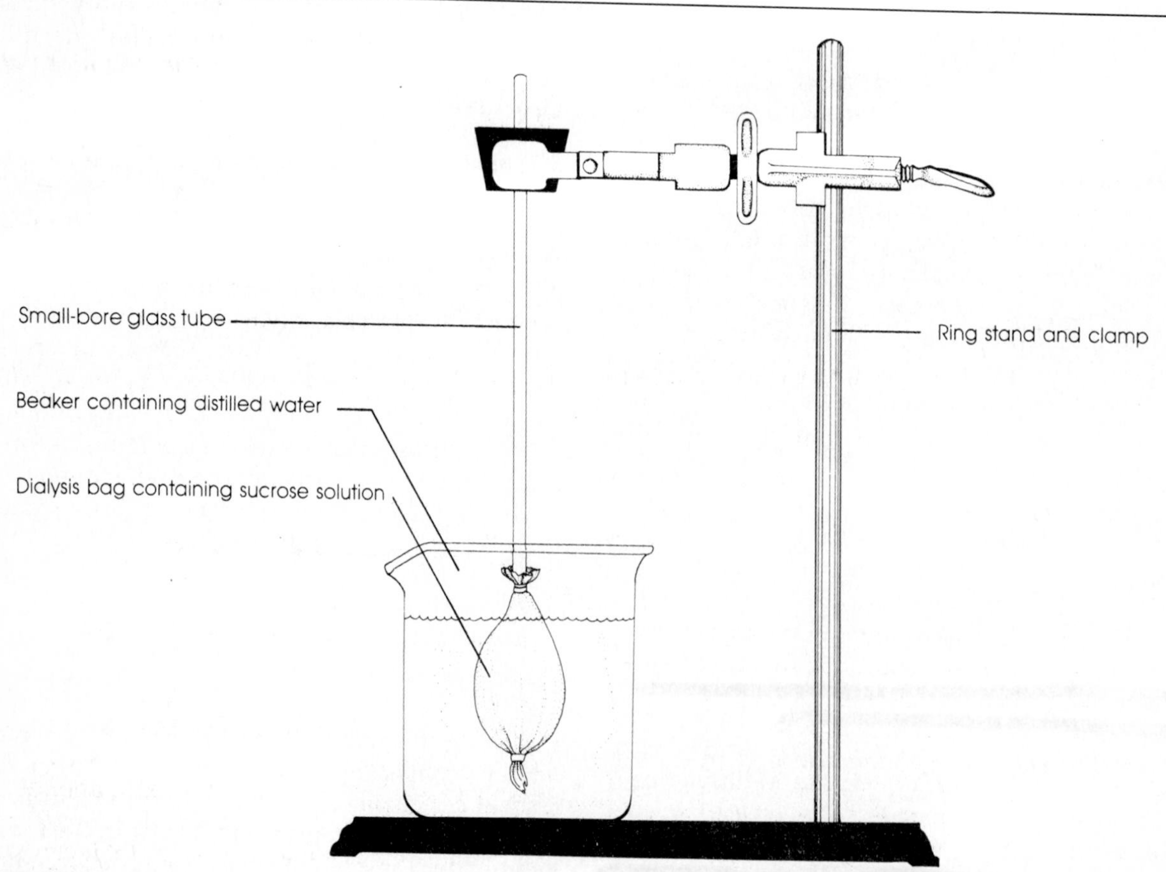

FIGURE 2.1. Osmometer setup.

in distilled water for a few minutes until it is pliable. Tie a knot in one end of the tubing and fill a dialysis bag with a 30% sucrose solution (containing a small amount of dye such as congo red). Tie the bag to one end of a 3- to 4-ft-long piece of glass tubing having a small bore. This is best done by wrapping heavy cord around the bag and tube several times and tying with a knot. Attach the osmometer to a burette clamp and lower the dialysis bag into a beaker of distilled water (Figure 2.1). If sucrose leaks from the bag, remove the bag from the beaker and tie it more tightly. Mark the height of the fluid in the glass tubing with a marking pencil and record the time. Repeat this procedure with the other osmometer using a 60% sucrose solution.

a. In the Laboratory Report, record the number of millimeters the fluid rises for the two solutions every 10 minutes for at least 1 hour. Which has the fastest rate of fluid movement and largest total movement?

b. Plot the millimeters of fluid movement for each 10-minute period against the time. How does the rate of fluid movement vary for each of the 10-minute periods of time? Why does this change occur? How does this rate of movement compare for the two solutions?

c. Calculate the initial osmotic pressure (in atmospheres) developed by each solution.

2. Osmosis in Plant Cells

The cell membranes of plants are similar in their structure to the cell membranes of animals and therefore can also be used to demonstrate various phenomena of membrane transport.

Using a cork borer of approximately 8- to 10-mm diameter, cut five pieces out of a potato, each measuring about 50 mm long. Determine the volume of each piece by immersing it in a known volume of water in a 10-ml graduate cylinder and noting the volume rise of water in the cylinder (Figure 2.2). Place one potato piece in each of the following five solutions: distilled water, 0.4% NaCl (sodium chloride), 0.9% NaCl, 5% NaCl, and 10% NaCl. After 2 hours, measure the volume of each piece of potato again and express the change as a percent of the original volume.

$$\% \text{ change} = \frac{\text{Change in volume (ml)}}{\text{Original volume (ml)}} \times 100$$

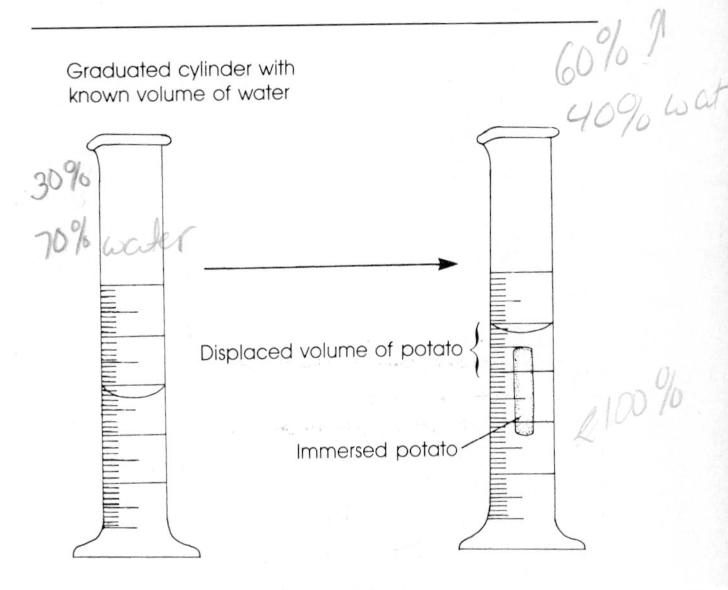

FIGURE 2.2. Osmosis by potato.

TONICITY

Within a living cell, all of the molecules, ions, and particles combine to produce a certain total osmotic pressure. Any solution that contains an equal number of osmotically active particles will produce the same osmotic pressure as produced by the cellular constituents and is said to be **isotonic** to the cell. Isotonic means "same tension" and implies that the solution will produce no change in the cell size due to net water movement in or out of the cell. A 0.3-M (300 mM) solution of a **nonelectrolyte,** such as glucose, is isotonic to mammalian cells, as is a 0.9% solution of NaCl (an **electrolyte**). A **hypertonic** solution is one that exerts a greater osmotic pressure than do the cell contents, whereas a **hypotonic** solution produces a lower osmotic pressure than that within the cell.

Experimental Procedure

1. Macroscopic Observations

Arrange in order a series of labeled test tubes containing 2 ml each of the following solutions: soap solution, distilled water, and NaCl solutions of 0.2%, 0.4%, 0.6%, 0.9%, 2%, and 5%.

Add 2 drops of citrated mammalian blood to each of the test tubes and mix by gently inverting the test tube several times. Record the lysis time

(time when the membranes of all red blood cells have been ruptured). For the end point of lysis, use the time when you can first see the blue lines on a piece of ruled notebook paper held behind the test tube. When the red blood cells are intact the solution has an opaque, milky appearance; when the cells are lysed the solution becomes transparent.

2. Microscopic Observations

Place a small drop of each of the following solutions on separate, clean microscope slides: 5% NaCl, 0.9% NaCl, 0.4% NaCl, and distilled water. Add a few red blood cells to each drop by dipping a clean toothpick into a drop of blood and then washing it in the drop of solution on the microscope slide. Gently stir. Add a coverslip and examine as *rapidly* as possible. Compare the cell size and shape in each solution and explain what has happened. Which of the solutions are isotonic, hypertonic, or hypotonic? Figure 2.3 shows possible observations through the microscope.

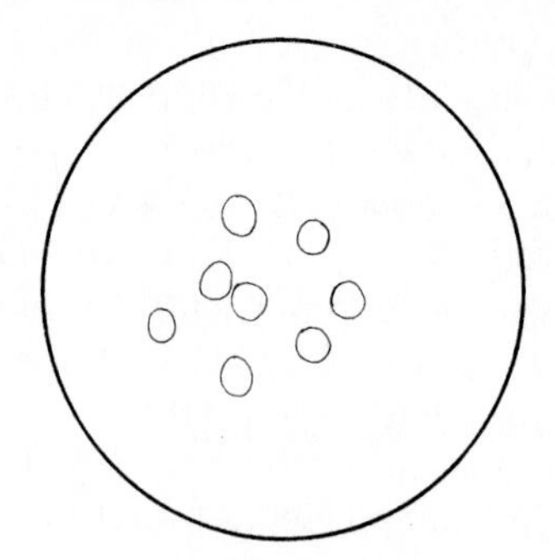

Normal RBCs

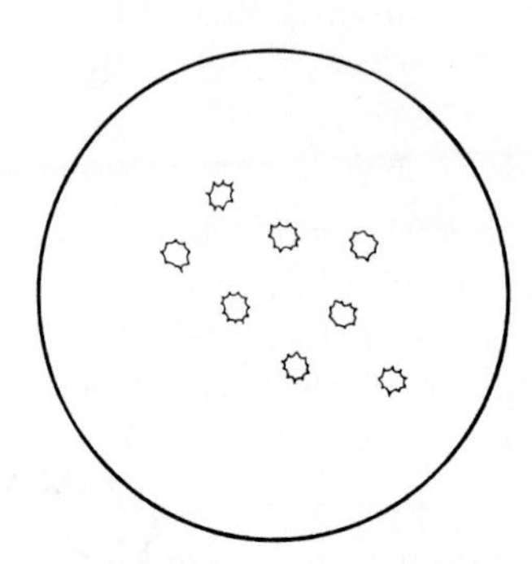

Crenated (shrunk) RBCs

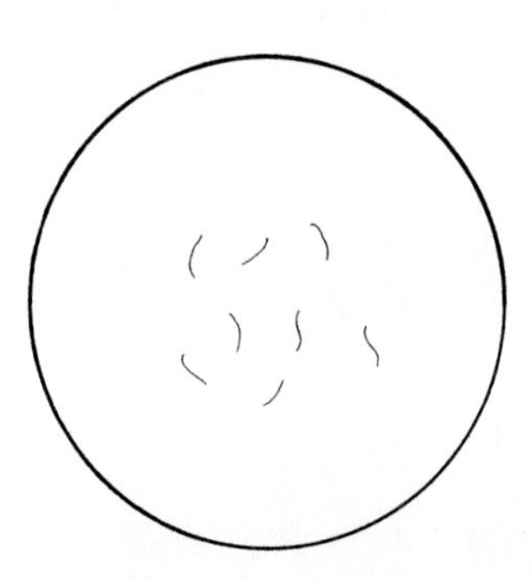

Fragments of lysed RBCs

FIGURE 2.3. Red blood cells (RBCs) in solutions of various concentrations of sodium chloride.

CELL PERMEABILITY

For a substance to enter the cell, it must pass directly through the cell membrane (which is largely lipid) or through the aqueous "pores" in the membrane. The effect of molecular size and lipid solubility on the movement of substances through cell membranes is demonstrated in the following experiments.

Experimental Procedure

1. Effect of Molecular Size on Cell Permeability

Set up test tubes containing 2 to 3 ml of each of the following solutions (all 0.5 M): urea (MW 60), glycerine (MW 92), glucose (MW 180), and sucrose (MW 342).

Add two drops of citrated mammalian blood to each of the four solutions and mix gently. Determine the time required for hemolysis of the cells (clearing of the initial murky solution). Observe each solution carefully for the first 3 minutes after adding the blood. If hemolysis does not occur within 3 minutes, go on to the next tube. If you do not observe hemolysis in a given tube in 30 minutes, you can assume it will not occur. Because time requirements for different substances vary, start the experiment early in the laboratory period.

Determine from the data whether any correlation exists between hemolysis time and the molecular weights of the substances. Plot hemolysis time against molecular weight.

2. Effect of Lipid Solubility on Cell Permeability

Alcohols have the ability to rupture the cell membranes of beets and thereby allow the red pigment **anthrocyanin** to be released from the cells. Cut small slices of fresh beet and *rinse them thoroughly* in 0.9% NaCl to remove any anthrocyanin that may be present on the beet surface. Place one slice of beet into test tubes containing each of the series of alcohols in the following table. Record the time required for the first release of the red anthrocyanin from the beet cells. This will be designated as the **penetration time.** Calculate the **penetration coefficient** for each alcohol by dividing the time in minutes for release of the pigment from the beets by the concentration of each alcohol. Plot the penetration coefficient against the partition coefficient for each alcohol.

CONCENTRATION OF ALCOHOL	PARTITION COEFFICIENT
22 M methyl alcohol	0.01
8.5 M ethyl alcohol	0.03
3 M propyl alcohol	0.13
1.1 M isobutyl alcohol	0.18
1.1 M n-butyl alcohol	0.58
0.38 M amyl alcohol (or iso-amyl)	2.0

hemolyzed when blood cells blow up & solution becomes clear enough to read through.

LABORATORY REPORT

Name ______________________

Date __________ Section __________

Score/Grade ______________________

2. Movement Through Membranes

Diffusion

1. Time in minutes for methylene blue to become evenly dispersed throughout the beakers.

5 °C	25 °C	50 °C
3	2	1
75mins	20mins	5min

2. What causes the more rapid dispersion? greater kinetic activity caused by increased temperature

3. What is the principal driving force for net diffusion? concentration gradient
the movement from ↑ to ↓

Osmosis

1. Osmotic Pressure – pressure need to prevent movement of water

a. Data

	FLUID MOVEMENT (MM)	TIME (MIN) 10	20	30	40	50	60
30% sucrose	Total movement	1.8	2.6	3.3	3.9		
	10-min movement	1.8	0.8	0.7	0.6		
60% sucrose	Total movement	3.6	5.2	6.6	7.8		
	10-min movement	3.6	1.6	1.4	1.2		

b. Plot

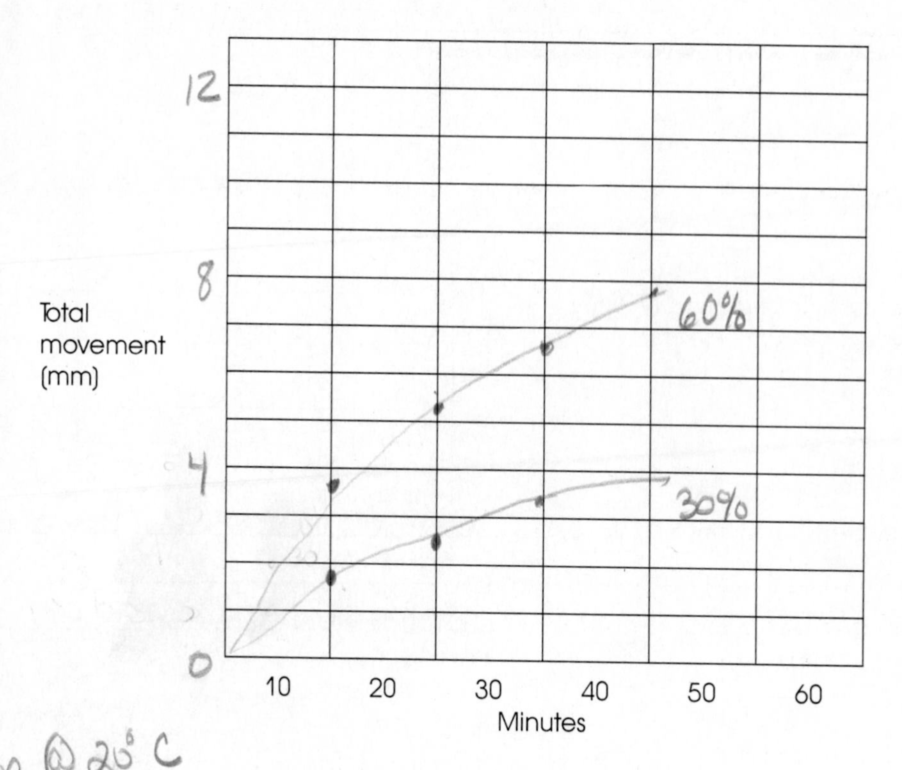

temp. @ 20° C

c. Osmotic pressure developed initially by each solution

30% sucrose = 21.1 atm $\pi = iRT(C_1 - C_2) = (1).082(293)(.88-0)$

60% sucrose = 42.2 atm " " " (1.76-0)

How is diffusion related to osmosis? osmosis is net diffusion of water

Where in the body do we find osmosis operating? across all cell membranes

What causes the fluid movement to decrease with time? decreas in concentration

2. Osmosis in Plant Cells

POTATO SOLUTIONS	INITIAL VOLUME (CC)	FINAL VOLUME (CC)	% CHANGE IN VOLUME
Distilled Water			
0.4% NaCl			
0.9% NaCl			
5% NaCl			
10% NaCl			

Name

Date Section

Score/Grade

If the final cell size were examined, how would it compare for the potato cells in each solution? Explain.

Tonicity

2cm of sol.
+2 drops
into solution
time added
until time
clear (read thru)

SOLUTION	LYSIS TIME
Soap solution	4 sec.
Distilled water	8 sec.
0.2% NaCl	25 sec.
0.4% NaCl	60 sec.
0.6% NaCl	73 sec.
0.9% NaCl	180+ sec.
2% NaCl	180+ sec
5% NaCl	180+ sec

} more than 3 mins.

1. Explain the differences in cell size you observed under the microscope for cells in the 5% NaCl and distilled water.

5% crenated cell due to water exiting from the cell flowing from ↑ concentration to ↓ concentration in 5% NaCl

dH_2O water entered the cell causing it to enlarge then rupture

2. Compare the mechanisms causing lysis of cells in these solutions:

0.2% NaCl

Soap solution Soap dissolved the bilipid cell membrane
→chemical hemolysis

3. In which of the solutions would crenated cells be found? 2% 5%
4. Which solutions would be rated as:

   Hypertonic 2% NaCl 5% NaCl

   Hypotonic .2% NaCl, dH₂O, 0.4% NaCl, 0.6% NaCl
5. Which solution would you rather receive in an intravenous injection? Why? 0.9% because it is isotonic

Cell Permeability

1. Effect of Molecular Size

SOLUTION	LYSIS TIME
Urea 60 MW	10 sec.
Glycerine 92 MW	
Glucose 180 MW	3 min.
Sucrose 342 MW	

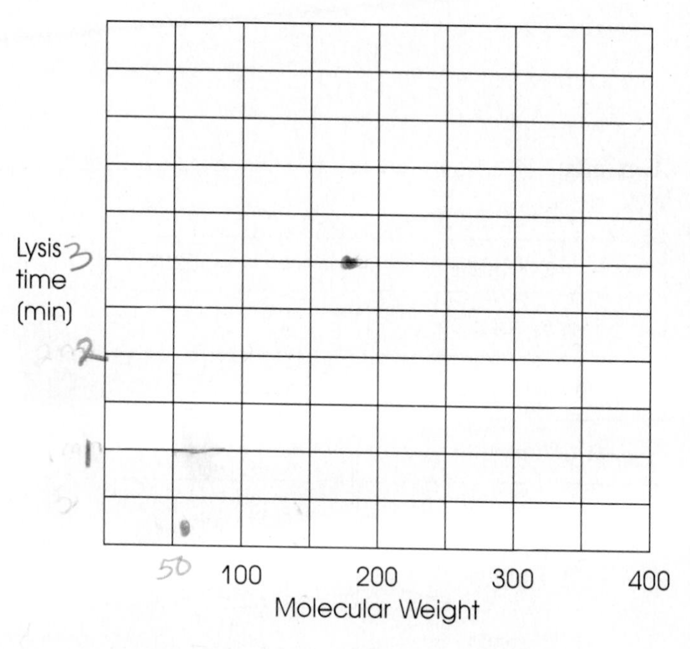

Why are some hypertonic solutions able to cause hemolysis?

Solutions composed of small molecules will diffuse down their concentration gradient thus causing an increase in water concentration which then follows the other small molecule into the cell resulting in hemolysis

Why does molecular weight affect the movement of molecules through membranes?

Molecular weight is indicator for molecular size; Low Molecular weight means molecule may be small enough to pass through membrane pores. Molecules larger than approx. 100 M.W. unable to penetrate membranes

Name ______________________

Date __________ Section __________

Score/Grade ______________________

2. Effect of Lipid Solubility

Observe a tint of pink

.5/22 = time/conc

CONCENTRATION AND TYPE OF ALCOHOL	PARTITION COEFFICIENT	PENETRATION TIME (min)	PENETRATION COEFFICIENT
22 M methyl alcohol	0.01	4.4	0.2
8.5 M ethyl alcohol	0.03	2.1	0.25
3 M propyl alcohol	0.13	1.2	0.4
1.1 M isobutyl alcohol	0.18	0.5	0.5
1.1 M n-butyl alcohol	0.58	1.3	1.2
0.38 M amyl alcohol (or iso-amyl)	2.0	1.52	4.0

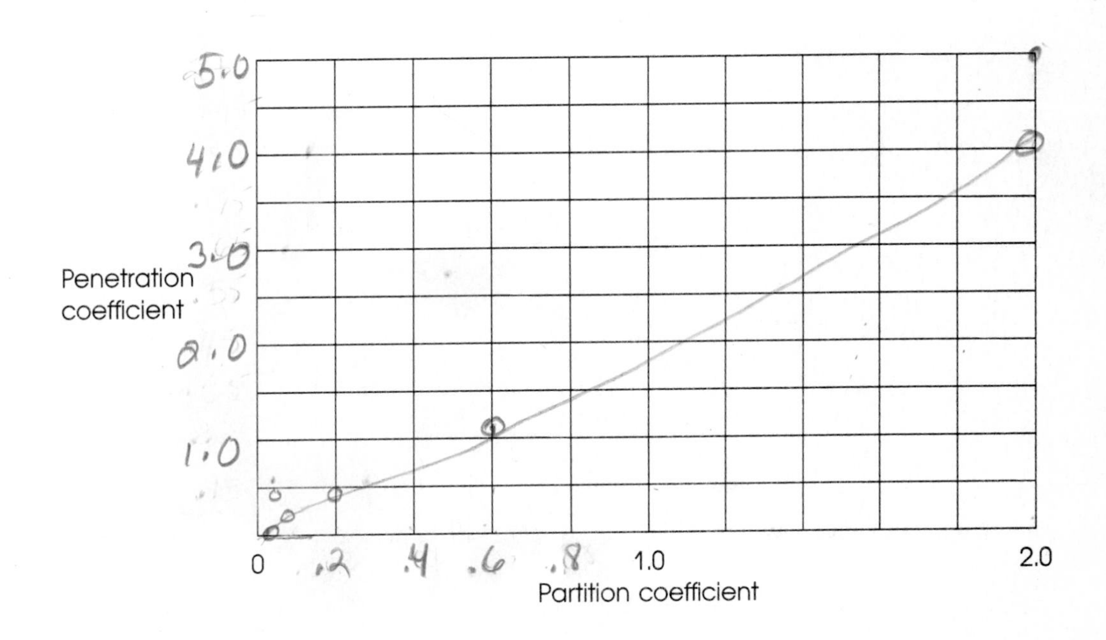

What is the primary factor governing penetration of alcohols (and similar compounds) through biological membranes? lipid stability which is reflected in partition coefficient

Why are different concentrations of alcohols used in this experiment? to shorten penetration times so experiments may be completed in laboratory time

3 Renal Physiology

KIDNEY REGULATION OF OSMOLARITY

One of the kidney's main functions is to regulate the osmolarity of the body fluids at around 300 milliosmoles (mOsm)/L. Regulation of osmolarity will be demonstrated in the following experiment by presenting the kidney with an excess water or salt load and recording its response as reflected in the concentration and volume of urine produced.

Experimental Procedure

1. Limit your fluid intake on the day of the experiment. Empty your bladder 1 or 2 hours before the laboratory begins and record the exact time. Do not save this urine sample.
2. On entering the laboratory, take a urine collection bottle to the restroom and void into the bottle, emptying the bladder completely. Record the exact time. This will be designated the "control" urine.
3. Return to the laboratory and immediately drink the solution assigned to you as quickly as possible. The class will be divided into three groups as follows:

   Group 1 drinks 800 ml of water

   Group 2 drinks 800 ml of water plus 7 g of NaCl (coated salt tablets)

   Group 3 drinks 80 ml of water plus 7 g of NaCl (coated salt tablets)

   Caution! *Students with health problems or who are on medication should check with their physician before ingesting salt.*
4. Every 30 minutes after drinking the solution, empty your bladder into a clean collection bottle. If you are unable to void, retain the urine in the bladder until the next 30-minute collection time.
5. Analyze the urine from each collection for:
   a. Volume. Measure the total volume with a graduated cylinder and express it as milliliters per minute excreted.
   b. Specific gravity. Pour some of the urine sample into a urinometer cylinder and measure the specific gravity as described in the following section on specific gravity.
   c. Chloride concentration. Place 10 drops of urine into a test tube (use a standard medicine dropper—20 drops/ml).

      Add 1 drop of 20% potassium chromate.

      Add 2.9% silver nitrate solution drop by drop, shaking the mixture continuously while the silver nitrate is being added. Count the number of drops of silver nitrate required to change the bright yellow solution to a brown color.

      Each drop of 2.9% silver nitrate added to produce the brown color represents 1 g/L (1 mg/ml) of NaCl in the urine. The silver nitrate should be made fresh daily.

      Calculate the total grams of NaCl in the urine collected at each 30-minute period and the NaCl concentration in milligrams per milliliter.

6. Record the data from all students on the board and the data sheet, calculate average values for each experimental group, and graph the results in the Laboratory Report. Make *three separate plots* of milliliters of urine per minute, specific gravity, and chloride concentration (mg/ml), each plotted against time. Use different colors on each plot to represent the average results of groups 1, 2, and 3. Be able to explain how the results illustrate the kidney's processing of the water or salt loads.

URINALYSIS

The kidneys are the chief regulators of the internal environment of the body. They do this by regulating the concentration of ions, water, and pH in the various body fluids. In addition, they provide for the elimination of the waste products of metabolism. The million nephrons in the kidneys contain two main structures, the **glomerulus** and the **renal tubule.** As blood passes through the kidneys, it is first filtered through the glomerulus (120 ml/min), and the filtrate then passes into the renal tubule.

The tubular filtrate is similar to blood plasma in composition, except that large molecules (having molecular weights of more than 70,000) are excluded (e.g., plasma proteins). As this filtrate passes along the proximal and distal tubules, most of the water is reabsorbed, and many essential substances are actively or passively reabsorbed into the bloodstream. Toxic by-products of metabolism and substances in excess are retained in the filtrate or are secreted into the filtrate and finally excreted in the urine (1 ml formed per minute). Thus, the final composition of the urine is quite different from that of the glomerular filtrate and reflects the integrity of kidney function and changes in blood composition.

An analysis of urine can yield valuable information about the health of the kidney and of the body in general. Various diseases are characterized by abnormal metabolism, which causes abnormal by-products of metabolism to appear in the urine. For example, phenylpyruvic acid appears in the urine in phenylketonuria (PKU), a disease resulting in mental retardation. In diabetes mellitus, deficient production of insulin by the pancreas results in the appearance of glucose in the urine (glycosuria). The volume of urine produced and its specific gravity give information on the state of hydration or dehydration of the body.

In this experiment you will analyze your own urine for some of its clinically important constituents. A sample of urine containing abnormal quantities of these constituents will also be examined to allow comparison with your urine.

Experimental Procedure

Take a urinalysis bottle to the restroom and collect a 15- to 25-ml sample of your urine.

1. Specific Gravity

Fill a urinometer cylinder to about 1 in. from the top with the collected urine. Holding the urinometer float by its stem, slowly insert it into the cylinder. Do not wet the stem above the water line or an inaccurate reading will result. Give the float a slight swirl and read the specific gravity from the graduated marks on the stem as it comes to rest. Do not accept a reading if the float is against the side of the cylinder. Remove the float, rinse it in tap water, and dry it. Measure the temperature of the urine specimen immediately. Return the urine specimen to the urinalysis bottle and rinse and dry both the urinometer cylinder and the thermometer.

The urinometer is calibrated to give a correct reading only if the urine is at 15 °C. If your urine is at a different temperature, you will need to correct the specific gravity by adding 0.001 for each 3 °C above, or by subtracting 0.001 for each 3 °C below the calibration temperature (15 °C). Record your results in the Laboratory Report.

The normal range of urine specific gravity is 1.0015 to 1.035. Readings above or below these limits may indicate a pathological condition. For example, a low reading is found in chronic nephritis.

2. Labstix Test

Recent advances in urinalysis techniques have made it possible to perform in a few seconds tests that previously took hours. The Labstix test is a combined test of urinary pH, protein, glucose, ketones, and occult blood. Abnormally low pH, along with a high level of glucose and ketones, indicates diabetes mellitus. Alkaline urine is found in many conditions, an example being cystitis, in which urine decomposes in the bladder with the production of ammonia. Urinary pH usually is slightly acid (around pH 6), but the pH may be lowered by a diet rich in proteins or citrus fruits, so pH is not very informative in itself. Protein and occult blood in the urine are much more definite, indicating nephritis, a disease in which the glomeruli are damaged and plasma proteins and erythrocytes leak into the kidney tubules.

Obtain a Labstix reagent strip and bottle with the color standards. Examine the strip carefully before making the test so you will know which portions to read first. When ready, dip the reagent portions into the well-mixed urine specimen, wetting all five reagents completely. Wipe the excess urine off on the lip of the urinalysis bottle. In *exactly 10 seconds* read the glucose test portion against the appropriate color standard. Exactly 5 seconds later read the ketone portion, and in exactly 15 more seconds read the occult blood portion (at the end of the strip). The pH and protein portions may be read after this at your leisure, because time is not so critical with these two.

If the urine glucose or pH is found to be beyond the normal range, make a more accurate analysis for glucose by using the **Clinitest** tablets, and for pH by using the **pHydrion** paper.

Obtain a sample of the abnormal urine and run the Labstix test on it so that you can compare your urine results with some non-normal results.

LABORATORY REPORT

Name ________________

Date ________ Section ________

Score/Grade ________________

3. Renal Physiology

Kidney Regulation of Osmolarity

1. Plot the average values for each group. Use the following colors in the plot:

   Group 1 = blue   Group 2 = green   Group 3 = red

Urine volume (ml/min excreted)

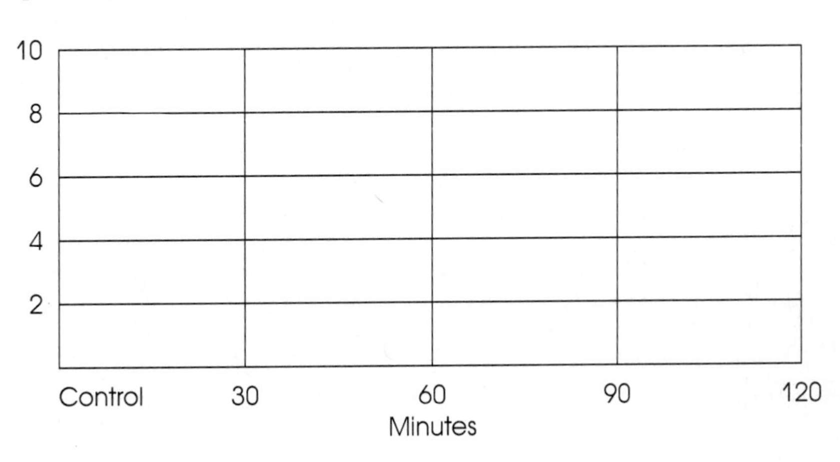

Specific gravity

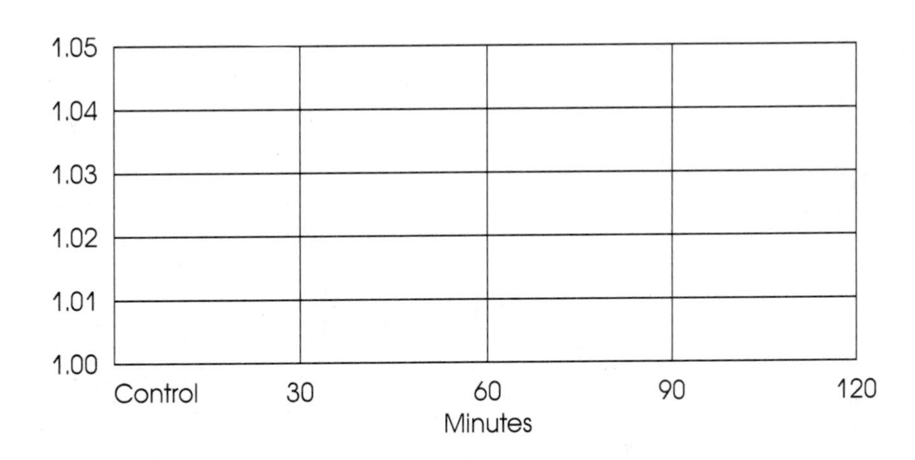

Sodium chloride (mg/ml)

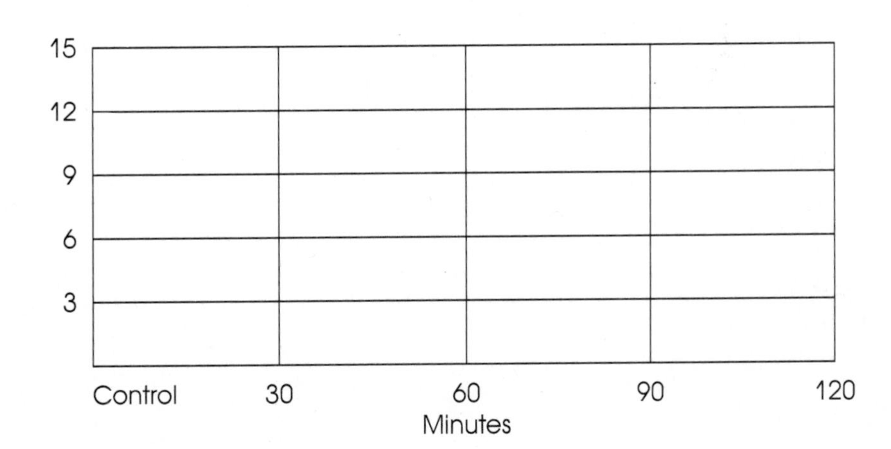

		VOLUME EXCRETED					URINE VALUES									
		(ml/min)					Specific Gravity					NaCl (mg/ml)				
SOLUTION INGESTED	STUDENT	C*	30 min	60	90	120	C	30 min	60	90	120	C	30 min	60	90	120
Group 1 800 ml of water																
	Average															
Group 2 800 ml of water + 7 g of NaCl																
	Average															
Group 3 80 ml of water + 7 g of NaCl																
	Average															

*C = control urine.

Name ______________________

Date __________ Section __________

Score/Grade ______________________

Urinalysis

1. Record the values obtained for your urine and for the unknown urine.

   Specific Gravity: Measured SG __________ Urine temp. __________

   Corrected SG __________

   Labstix test:

	GLUCOSE	KETONES	BLOOD	PROTEIN	pH	SG
Student's urine						
Unknown urine						

2. Examine the data obtained in the experiment on kidney regulation of osmolarity. Are the results consistent with what you would expect when a subject imbibes fluids that are hypotonic, isotonic, and hypertonic? Explain.

3. The loss of water during sweating on a hot day causes the blood volume to decrease and the osmolarity of body fluids to increase. Outline the mechanisms operating to restore homeostasis via the release of antidiuretic hormone (ADH) in this situation.

4. Two symptoms present in a person with diabetes mellitus are hyperglycemia (elevated blood glucose) and diuresis (increased urine production). What causes this increase in urine output?

5. Explain how the Renin-Angiotensin-Aldosterone mechanism restores homeostasis when there is a decrease in blood volume and blood pressure.

6. Transport of materials across membranes occurs by several processes listed below. Where do we see these processes operating in the kidney? Give one example of each transport process.

   Active Transport:

   Osmosis:

   Filtration:

   Diffusion:

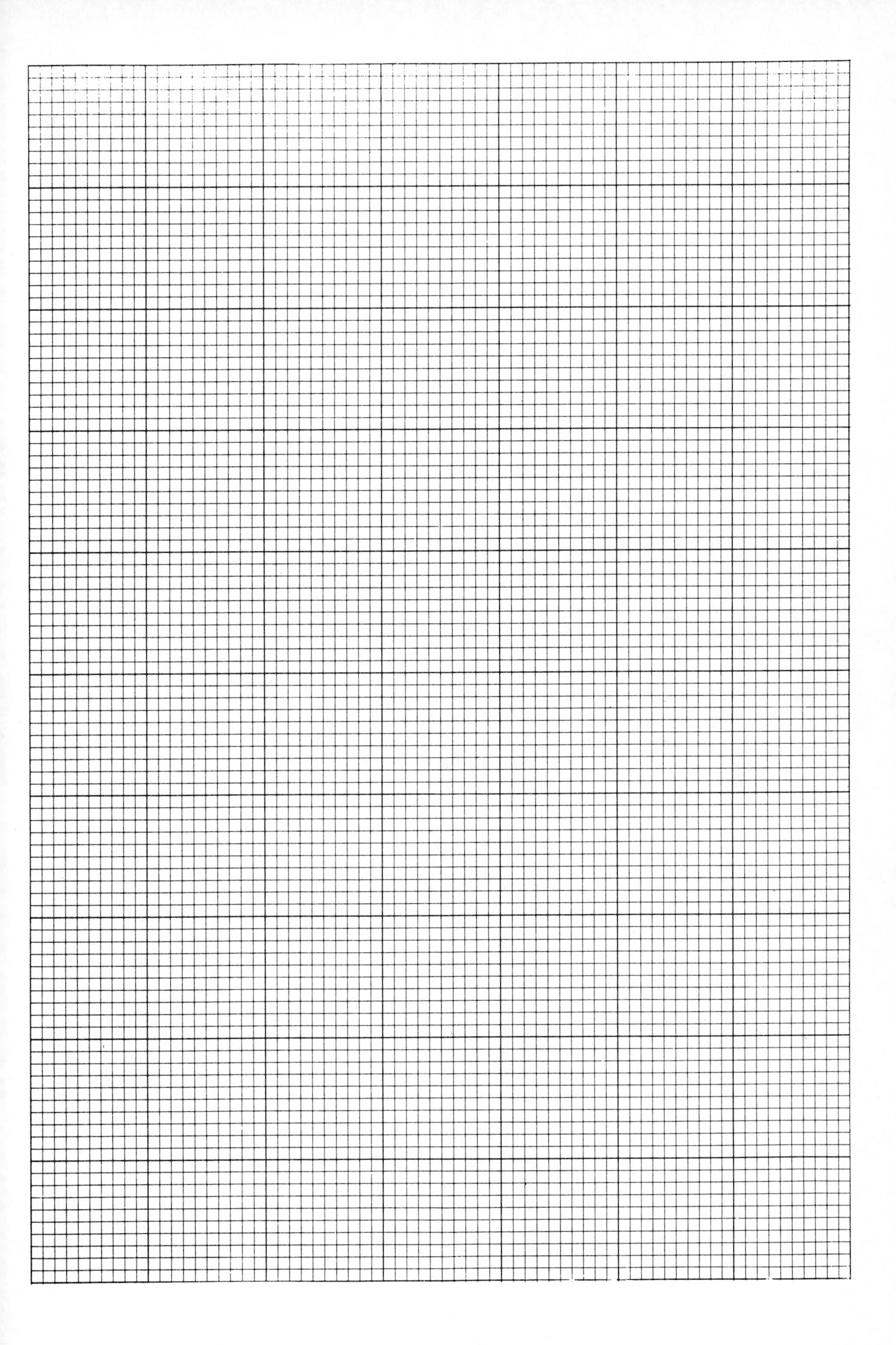

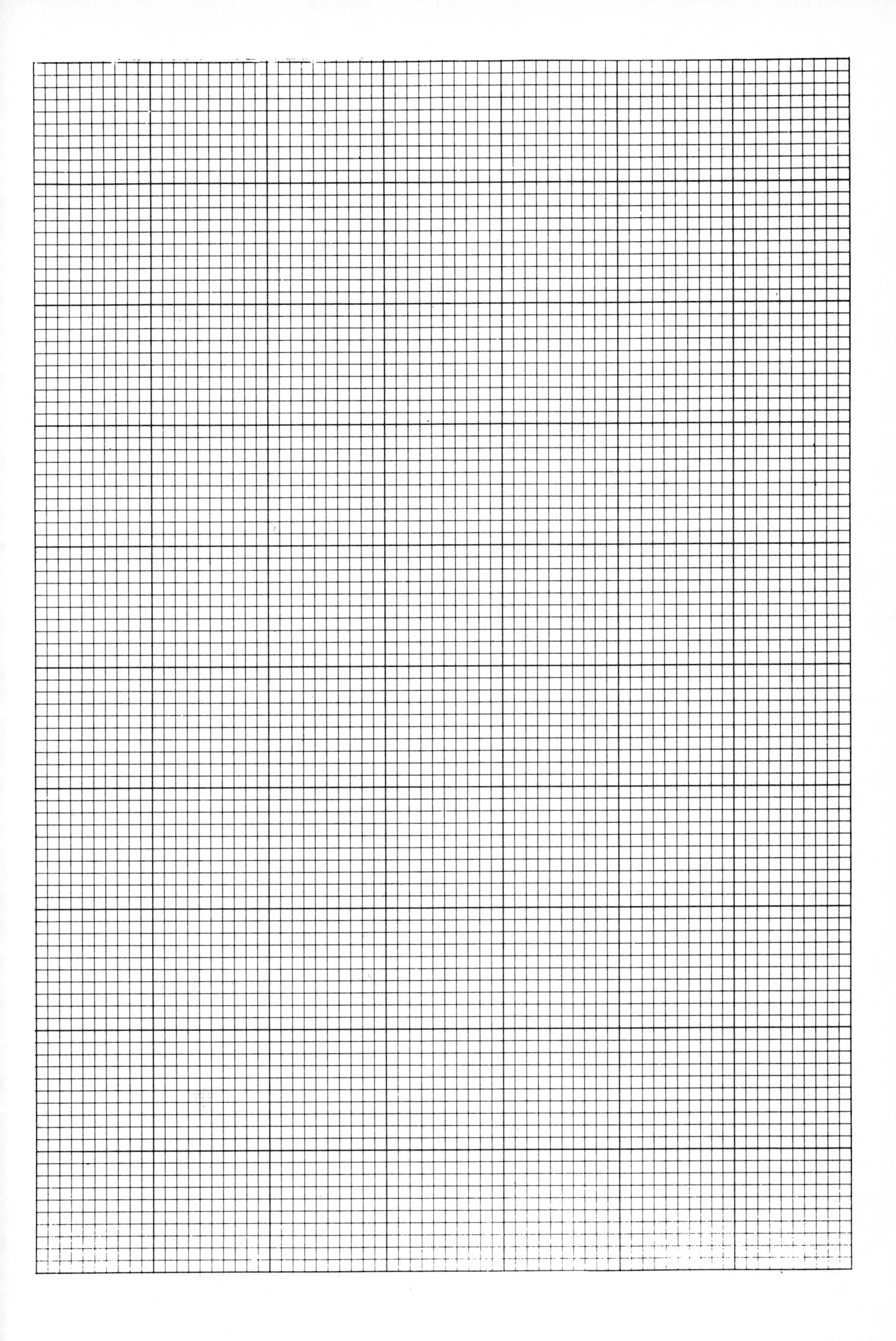

4 Neuroanatomy

ORGANIZATION OF THE NERVOUS SYSTEM

The nervous system is classically organized into the central and peripheral nervous systems as outlined here:

Central nervous system (CNS)

- Brain
- Spinal cord

Peripheral nervous system (PNS)

- Sensory (afferent)
- Motor (efferent)
    - Somatic nervous system
    - Autonomic nervous system (ANS)
        - Sympathetic system
        - Parasympathetic system

Neurons that lie completely within the brain or spinal cord are in the CNS, whereas neurons that lie partially or wholly outside the CNS are considered peripheral neurons. Cell bodies and synapses are often grouped together as functional integrative centers called **nuclei** in the CNS or **ganglia** in the PNS. Sensory neurons carry nerve impulses "directed toward" (**afferent**) the CNS, whereas motor neurons conduct impulses "away from" (**efferent**) the CNS to control the various organs of the body. **Somatic** neurons innervate skeletal muscles to produce voluntary body movements. **Autonomic** neurons provide the involuntary control of the internal organs that regulate homeostasis of the internal environment. The individual peripheral neurons that relay information to and from the integrating centers of the CNS are grouped into bundles called the **cranial nerves** and the **spinal nerves.**

SPINAL NERVES AND SPINAL CORD

The 31 pairs of spinal nerves are all "mixed" nerves (i.e., they contain both sensory and motor neurons) and are grouped from the top of the body down, as shown in Figure 4.1.

If the spinal cord is cut in cross section anywhere along its length, a similar anatomical pattern is seen, as diagrammed in Figure 4.2.

Note that near the spinal cord, the spinal nerve branches into the **dorsal root,** which contains only sensory neurons, and the **ventral root,** containing the motor neurons. In rare instances, the sensory neuron synapses directly with a motor neuron, thus producing a **monosynaptic** reflex arc. More often, however, sensory neurons synapse with **association neurons (interneurons),** which then synapse with the motor neurons. This produces a more complex, **multisynaptic** (polysynaptic) spinal reflex. The **gray matter** of the spinal cord (and the brain) is a region containing mainly nerve cell bodies and synapses; the **white matter** is made up of bundles of myelinated axons, which give a white color to the area.

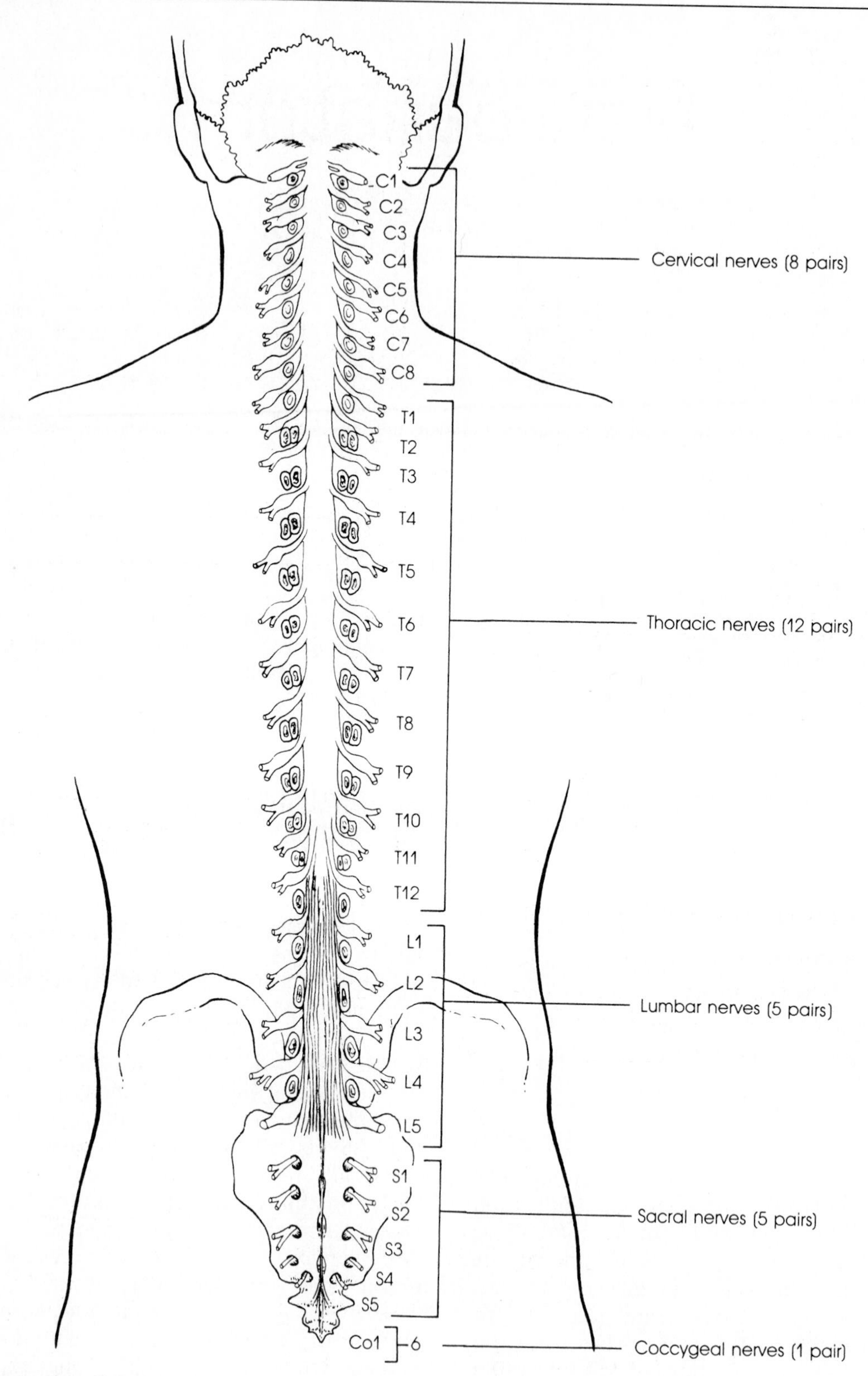

FIGURE 4.1. Spinal nerves.

Sensory neurons that enter the spinal cord synapse not only with motor neurons to produce spinal reflexes but also with association neurons, which extend up the cord to provide the brain with sensory information. Figure 4.3 is a simple schematic of the synapses involved in one set of ascending neurons. Note that three neurons are involved and that nerve impulses cross over

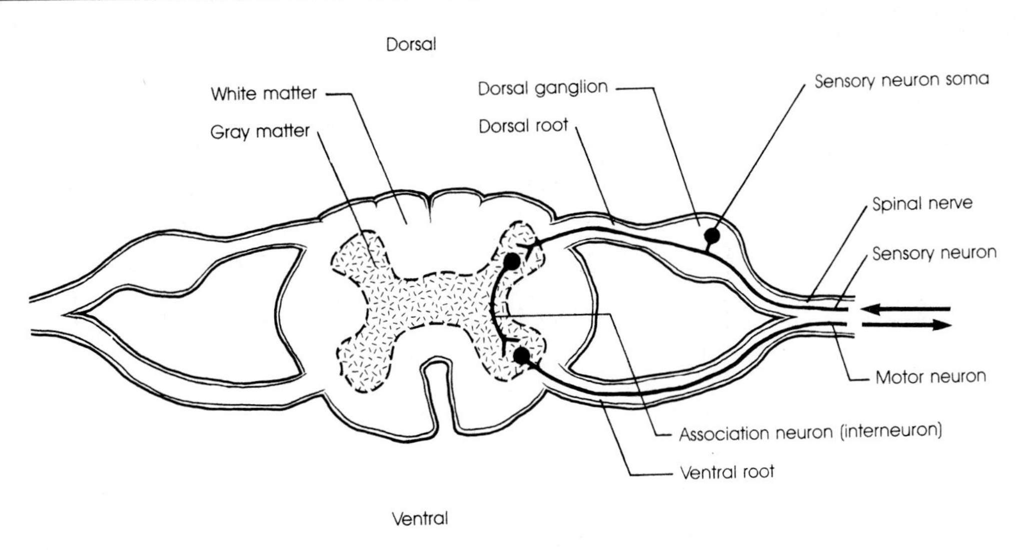

FIGURE 4.2. Spinal cord cross section.

from one side of the body to the other side before reaching the brain.

A simplified view of one set of descending neurons is also shown in the diagram. Again, the motor pathways cross over before synapsing with the spinal motor neurons.

The ascending and descending axons in the white matter of the spinal cord are grouped into bundles called **tracts** or **fasciculi.** Most of these tracts are named in a rather simple manner: The prefix indicates where the neuron originates and the suffix tells where it terminates. Some of the major spinal cord tracts are shown in Figure 4.4.

Use the spinal cord model and diagrams from your text to become familiar with the anatomy of the spinal cord and the spinal nerves.

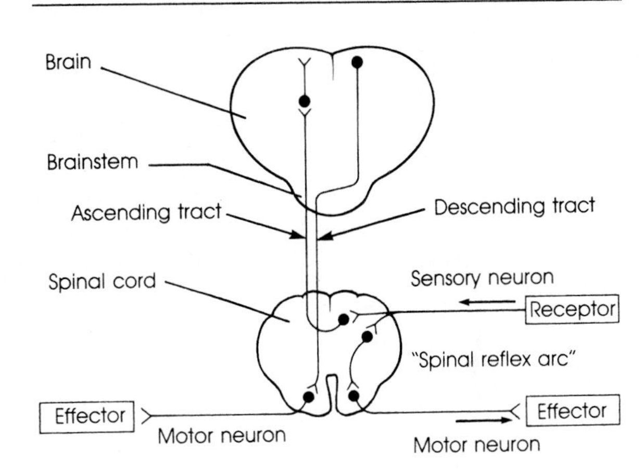

FIGURE 4.3. Generalized scheme of ascending and descending nerve tracts.

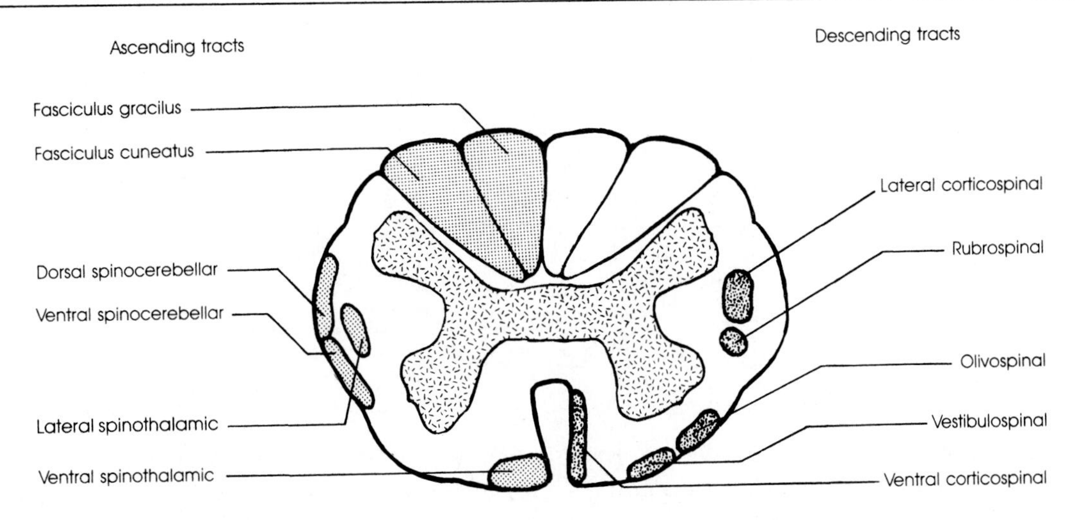

FIGURE 4.4. Major spinal cord tracts.

CRANIAL NERVES

The 12 pairs of cranial nerves originate from various structures on the ventral surface of the brain and are numbered anterior to posterior as follows:

1. Olfactory
2. Optic
3. Oculomotor
4. Trochlear
5. Trigeminal
6. Abducens
7. Facial
8. Vestibulocochlear
9. Glossopharyngeal
10. Vagus
11. Accessory
12. Hypoglossal

Examine the sheep brain and try to identify as many of these nerves as possible. (See Figure 4.5.) Some of the nerves may have been lost or damaged when the brain was removed from the skull. Several of the cranial nerves are mixed nerves; some are strictly sensory or motor in their functions. Using your text and other resources, identify the functions of each cranial nerve and its classification (mixed, sensory, motor). To help you remember the cranial nerves in their proper order, you might use this old mnemonic rhyme: "On Old Olympus' Towering Top A Fat Vicious Giant Vaults And Hops." The first letter of each word is the first letter of a cranial nerve.

EXTERNAL STRUCTURES AND LANDMARKS OF THE BRAIN

During embryonic development, the brain changes from a simple neural tube to a complex mass with several prominent enlargements. Examine the brain and identify these structures: spinal cord, medulla oblongata, pons, cerebellum, cerebrum, pituitary, and olfactory bulbs. In higher animals, such as sheep, monkeys, and man, the cerebrum has grown so large that its surface has developed **fissures** (deep folds) and **sulci** (small folds), which increase its surface area and produce some key landmarks and lobes of the cerebrum. The deepest groove is the **longitudinal fissure,** which separates the left and right cerebral hemispheres. The **lateral fissure of Sylvius** separates the **frontal lobe** from the **temporal lobe.** The **central sulcus,** which separates the frontal lobe from the **parietal lobe,** is less obvious in the sheep brain than in the human brain. The **occipital lobe** at the rear area of the cortex is not separated from the parietal and temporal lobes by any major sulci or fissures. Study the external features of the sheep brain until you are familiar with the key structures and landmarks.

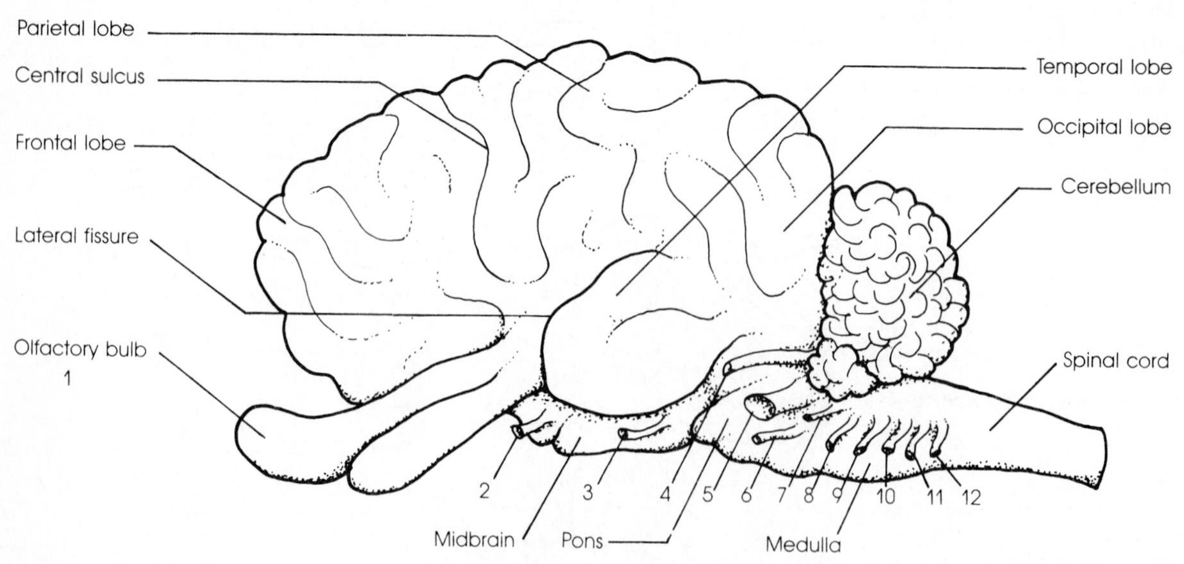

FIGURE 4.5. Sheep brain (lateral view). Numbers correspond to 12 pairs of cranial nerves.

SECTIONING OF THE BRAIN

Experimental Procedure

1. Midsagittal Section

Using a sharp scalpel, section the sheep brain into right and left halves by cutting through the longitudinal fissure, the corpus callosum, and the entire brainstem (Figure 4.6). To make a clean cut, pull the knife toward you with a continuous movement, anterior to posterior. The midsagittal section allows one to view the major areas of the brain:

Forebrain:	Telencephalon—cerebral hemispheres
	Diencephalon—thalamus, hypothalamus
Midbrain:	Mesencephalon
Hindbrain:	Metencephalon—pons, cerebellum
	Myelencephalon—medulla

Note the relationship of the third and fourth ventricles to these brain structures. The ventricles of the brain contain the cerebrospinal fluid (CSF), which transports nutrients to the neurons of the CNS. The first and second lateral ventricles in each cerebral hemisphere are not visible in this section but will be seen in the cross sections.

2. Cross Sections

Make several cross sections of the left or right hemisphere from the front of the brain to the back. Do not be concerned with detailed structures, but try to develop in your mind a three-dimensional view of the major structures you have examined previously.

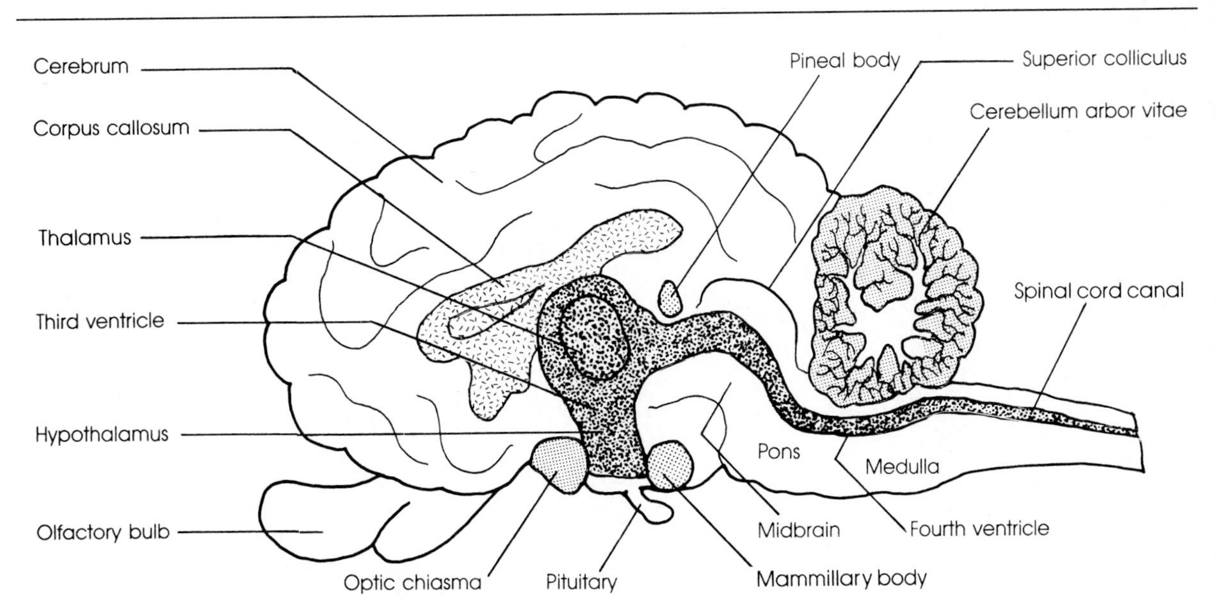

FIGURE 4.6. Sheep brain (midsagittal section).

LABORATORY REPORT

Name ______________

Date ______ Section ______

Score/Grade ______________

4. Neuroanatomy

Spinal Nerves and Spinal Cord

1. Give the origin and termination of each of the following spinal nerve tracts.

TRACT	ORIGIN	TERMINATION
Spinothalamic		
Rubrospinal		
Corticospinal		
Spinocerebellar		
Fasciculus cuneatus		
Olivospinal		

Cranial Nerves

1. Give the name, type and function of each cranial nerve.

NO.	NAME	TYPE	FUNCTION
1			
2			
3			
4			
5			
6			
7			
8			
9			
10			
11			
12			

Cerebral Cortex

1. Briefly describe the location in the cerebrum of the following areas.

   Primary motor area ______

   Primary sensory area ______

   Primary hearing area ______

   Broca's speech area ______

   Primary visual area ______

   Premotor area ______

Sectioning of the Brain

1. What is the overall function of the following structures?

   Cerebellum ______

   Corpus callosum ______

   Thalamus ______

   Hypothalamus ______

   Pituitary (hypophysis) ______

   Optic chiasma ______

   Pineal body ______

   Medulla ______

   Hippocampus ______

   Basal ganglia ______

2. What would happen to the area supplied by a spinal nerve if the dorsal root were cut?

   ______

   If the ventral root were cut? ______

3. If someone wants to compliment you on how smart you are, they might say "You sure have a lot of gray matter." Is this statement truly a compliment? Explain why or why not.

5 Reflex Functions

HUMAN REFLEXES

The essentials of a reflex mechanism are a receptor organ, an effector organ, and some type of communications network connecting the two. Reflex action is initiated by an input stimulus and results in an output response. Reflex activity ranges from the simple axon reflex to the complex reflexes in which the cerebrum participates.

Many reflexes might be regarded as being programmed; that is, the appropriate response to the stimulus has been built into the nervous system. The spinal reflexes that require transmission from the periphery to the spinal cord and then back to the appropriate effector organ are examples of this kind of programming. For instance, if one experiences a painful stimulus such as burning a finger on a hot object, the spinal reflex immediately causes withdrawal of the finger from the offending object. No action is required by the central nervous system, and the reflex functions equally as well in an animal whose spinal cord has been divided above the location of the cell bodies of the participating nerves.

Other reflexes, such as eye reflexes and labyrinthine reflexes, require action of centers in the brain. In these instances the appropriate response may need to be determined after several different inputs have been evaluated; hence, integrative function of the central nervous system is required.

In this experiment you will investigate several types of human reflexes to demonstrate their integrative function at several levels of integration in the body.

Experimental Procedure

Eye Reflexes

1. Pupillary Reflex

Observe the size of your partner's pupils in a given intensity of light. Flash a light into one eye and observe the pupillary responses. Do both eyes change simultaneously? What is the receptor in this reflex? The effector?

Observe the diameter of the pupils in a given light. Without changing either the light intensity or the focus, place your hand over one eye. Observe the pupil of the uncovered eye. What happens to its diameter? Explain the change in the uncovered pupil. This is called the consensual reflex.

2. Accommodation Reflex

Observe the size of your partner's pupils when the eyes are focused on a distant object (more than 20 ft away). Watch carefully while the focus is shifted to a near object. Do not change the light intensity. How does the pupil size change? What is the advantage of this change?

3. Corneal (Blink) Reflex

Move your hand suddenly toward your partner's eyes. The automatic closing of the eyelids is a protective reflex to prevent eye injury.

Spinal Reflexes

1. Patellar Reflex

Have your partner sit in a chair or on a laboratory stool with legs crossed. Gently tap the patellar tendon of the crossed leg with a reflex mallet and note the response. Compare responses of the right and left knees. Diagram the reflex arc in the Laboratory Report.

Repeat, but instruct your partner to perform **Jendrassik's maneuver,** that is, to clasp his or her hands in front and, with fingers locked, try vigorously to pull the hands apart at the same time that the tendon is tapped. How do you explain the responses obtained?

2. Achilles Reflex

Have your partner kneel on a chair or stool with the feet hanging free and relaxed over the edge. Bend his or her foot downward to increase tension on the gastrocnemius muscle. Tap the Achilles tendon lightly with a reflex hammer or the side of your hand. The contraction of the gastrocnemius causes plantar extension of the foot. Have your partner grasp the back of the chair and repeat. Results?

3. Biceps and Triceps Reflexes

Place your forearm on the laboratory table, so that the elbow is bent to approximately 90 degrees; push vigorously against the top. Palpate the biceps muscle. Try to contract the biceps. Record your results.

Flex your arm to 90 degrees and attempt to lift the table. Palpate the triceps muscle. Try to contract the triceps. Record your results and explain them.

Ciliospinal Reflex

Pinch the skin on one side of the nape of your neck and note the dilation of the pupil of the eye on the ipsilateral side. This is a reflex response mediated over the sympathetic nervous system in response to a painful stimulus.

Plantar Reflex and Babinski's Reflex

Scratch or stroke sharply the sole of your foot near the inner side, using a blunt probe. The normal adult reflex response is a plantar (downward) flexion of all toes. If the toes fan out with the big toe flexed dorsally (upward), the response is referred to as the positive Babinski reflex; this reflex is often associated with damage to the pyramidal tract fibers. Babinski's reflex is the normal response of a child in its first year because the nerves are still undergoing myelination at this time.

Labyrinthine Reflexes

The purpose of this demonstration is to show the role of the labyrinths in orientation of the body during movement and how labyrinthine reflexes correlate to muscular movements and eye movements reflexly to maintain equilibrium.[1]

Angular acceleration during rotation is detected by the cristae ampullaris receptors in the semicircular canals. These receptors contain hair cell filaments embedded in a gelatinous mass (cupula) that extends into the endolymph in the semicircular canals. When the filaments are bent, the hair cells are depolarized and send nerve impulses to the brain.

During rotation of the horizontal canal to the right, the inertia of the endolymph causes it to lag behind the canal movement. The relative movement of the endolymph to the left displaces the cupula, thus bending the hair cell filaments and altering the nerve activity to the brain (Figure 5.1a). This provides the brain with sensory information that the head is being accelerated. After 15 to 20 seconds of rotation, when a constant velocity is reached, the endolymph movement catches up with the canal movement. The cupula then assumes its normal upright position and the sensation of acceleration is lost.

After rotation is stopped, the canal is stationary but the endolymph continues moving to the right, due to its inertia. This postrotatory endolymph movement displaces the cupula in the opposite direction (Figure 5.1b); this modifies the flow of nerve impulses to the brain, which produces a sensation of spinning to the left.

Through reflex neural connections, the cristae ampullaris receptors are linked to centers in the brain that control eye movements and muscle tone. These relationships are readily demonstrated in the following experiments.

[1]People who are subject to motion sickness are good subjects for the labyrinthine reflex tests because they show a strong response. Because of the discomfort they might experience, however, these persons should be subjected to only one test in which they are rotated.

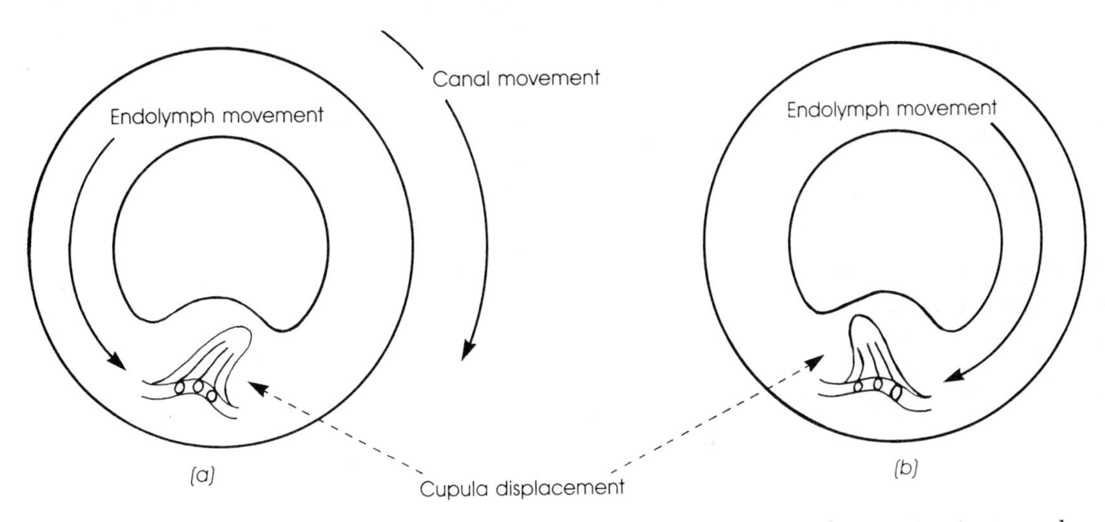

FIGURE 5.1. Horizontal semicircular canal (a) during rotation to the right, (b) after rotation is stopped (postrotatory).

1. Nystagmus

One of the functions of the semicircular canal mechanism is to aid visual fixation on moving targets. If the canals are stimulated under experimental conditions, reflex responses result in movement of the eyes, called **nystagmus.** Nystagmus has two components, a fast and a slow phase. The *direction of nystagmus is designated* as that of the fast phase. If a person is angularly accelerated in one direction, the eyes will move very slowly in one direction as though to maintain fixation on a moving target, and then very rapidly swing back in the other direction. This action is *rotatory nystagmus* and is caused by the acceleration of fluid in the semicircular canals, which stimulates the crista in the ampulla and produces the sensation of turning. It is very difficult to observe rotatory nystagmus, but the same phenomenon can be obtained by suddenly stopping a spinning person and observing *postrotatory* nystagmus; in observing postrotatory nystagmus, however, you must remember that all movements have been reversed and the direction of nystagmus is reversed.

1. Seat the subject in an office swivel armchair from which the casters have been removed. Several members of the group should stand close around to catch the subject in case he or she topples from the chair.
2. With the subject's feet off the floor and the head bent forward at an angle of 30 degrees, rotate the subject to the right at the rate of once every 2 seconds for ten turns. Stop the rotation suddenly and observe the motion of the eyes the moment the movement of the chair stops. Note the direction of nystagmus.
3. Repeat the rotation with the head bent to the right shoulder at an angle of about 90 degrees. What is the direction of nystagmus?

2. Past Pointing

With his eyes open, have the subject extend his arm to touch the finger of the operator; then with the eyes closed, have him repeat the process. The normal person is able to perform this demonstration without difficulty.

Now rotate the subject to the right as before. Stop the rotation abruptly and let the subject reach out, as previously, to touch the operator's finger. Let him immediately close his eyes and try to touch the finger. In which direction did the subject err in relation to the direction of rotation? How do you explain the direction of pointing?

3. Equilibrium

With the subject's head bent forward at an angle of 30 degrees, rotate him to the right as before. Immediately on cessation of rotation, have the subject attempt to rise and walk. Observe in which direction he appears to fall. Ask the subject to describe the sensation he experiences as he attempts to walk after discontinuation of rotation.

Repeat the procedure, but with the head bent to the right shoulder at an angle of about 90 degrees. Record your observations and explain.

Proprioception and Spatial Orientation

Have the subject hold her arms stretched forward with index fingers pointed toward each other and then try to bring the fingertips together. Test this reaction first with the subject's eyes open and then with the eyes closed. Record the results.

Have the subject stand with feet together and arms outstretched. Observe body sway and the corrective motions required to maintain balance, and then test the subject first with her eyes open and then with the eyes closed. Then have the subject stand first on one foot and then on the other. Again perform the test with the eyes closed and open. These tests evaluate the static equilibrium of the person and point out the contribution made by the eyes to static equilibrium.

Have the subject look at the ceiling and stand on one foot. Again have her stand on one foot, with her head in the same position as before, but this time with the eyes closed.

REACTION TIMES

The time required for a person to react to a stimulus depends on several factors: the responsiveness of the receptor, nerve conduction velocity, synaptic delay, number of neurons and synapses involved, nerve distance to be traveled, the efficiency of neuromuscular transmission, and the speed of muscle contraction. The type of response is also of importance. In an automatic reflex, relatively few synapses are traversed and the response time is short. In a response that requires thought, decision making, and choice, more neural pathways must be traversed and the response time is longer. In this experiment you will measure and compare the times required to complete various reaction responses.

The reaction time apparatus consists of the operator's initiate console, the subject's response panel, and a stop clock. The operator's console and the clock should be placed as far as possible from the subject's panel so the subject cannot see the operator's console. When the operator depresses the initiate button, the stop clock and the stimulus (colored lamp or buzzer) are activated simultaneously. When the subject depresses the proper response button, the clock is stopped and the reaction time can be read on the clock in hundredths or thousandths of a second. The operator can select the buzzer sound or any one of four colors (red, white, blue, or green) by turning the knob on the operator's console.

In each of the following tests, obtain a minimum of three measurements, calculate the average reaction time for each test, and record your results in the Laboratory Report. If the subject reponds prematurely, disregard that particular time.

Experimental Procedure

1. Reaction Time to Sight

a. Test the subject's reaction time to a single color stimulus using one response button.

b. Determine the reaction time to a choice of two colors using two response buttons (with a specified response button for each color).

c. Determine the reaction time to a choice of three colors using three response buttons.

d. Determine the reaction time to a choice of four colors using four response buttons.

2. Reaction Time to Sound

Test the subject's response time to the buzzer sound produced when the initiate button is depressed.

3. Reaction Time to Word Association

When you depress the initiate button you will simultaneously give a stimulus word—for example, "bread." The subject will reply with a word she associates with the stimulus word (e.g., "butter") and will simultaneously depress a previously designated response button. Record the stimulus and response words used in each test and the response time required. You will see quite a variation in time depending on the stimulus words used—and on the subject.

LABORATORY REPORT

Name ______________________

Date ____________ Section ____________

Score/Grade ______________________

5. Reflex Functions

Human Reflexes

Spinal Reflexes

1. Patellar Reflex (Stretch Reflex) 313 Fox text

Diagram the reflex arc operating in the patellar reflex.

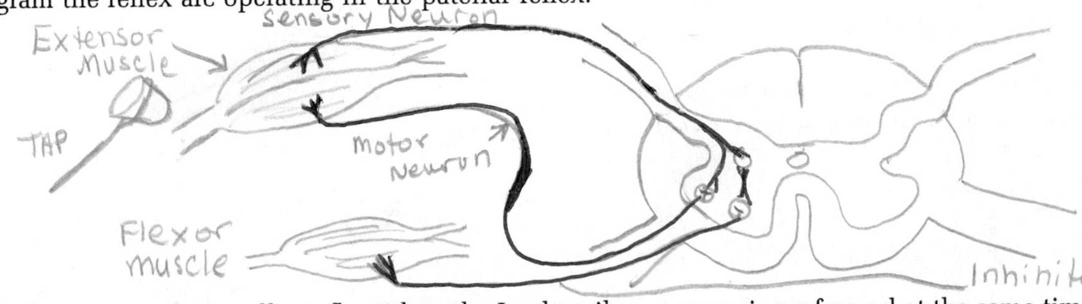

What happens to the patellar reflex when the Jendrassik maneuver is performed at the same time the patella is struck? Explain the mechanism of the response. Pulling on hands provides increased sensory input resulting in facillitation of synapses resulting in faster response when patellar reflex is activated

2. Biceps and Triceps Reflexes

What part does reciprocal innervation play in the biceps and triceps reflexes?
When the prime mover is contracting the antagonist can't contract

Diagram the reflex arc in the flexion reflex and the crossed extensor reflex.

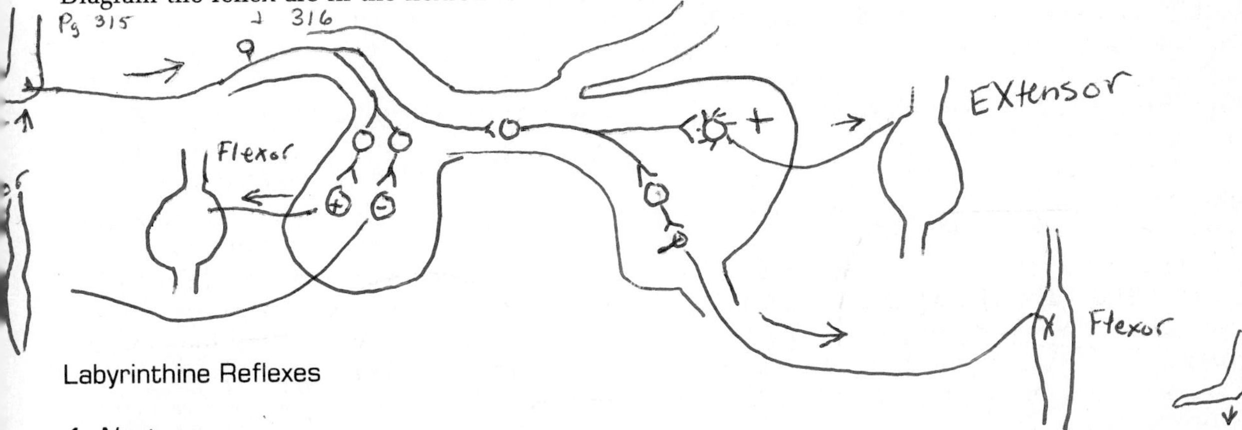

Labyrinthine Reflexes

1. Nystagmus

What is the direction of nystagmus when the subject is stopped after rotation to the right with the head bent forward 30 degrees? to the left

With the head bent to the right shoulder at a 90-degree angle? down

What practical function does the nystagmus reflex play in a person's physiological function?

allows vision to be focused on distant objects
Preventing blurred vision + nausea

2. Past Pointing

After rotation to the right and stopping, did the subject point to the left or right of the operator's finger? to right

How do you explain this direction of pointing based on teleology (that is, some practical purpose for the observed reflex)? on stopping, perception would be in rotation to the left. When moving forward an object the hand is placed forward of anticipated point of contact
Carousal rotation - person reaching for the ring reaches well before they arrive at ring location

3. Equilibrium

In which direction did the subject move or fall when he stood up after being rotated to the right and then stopped with the head bent forward 30 degrees? ~~forward~~ right

With the head bent to the right 90 degrees? ~~to left~~ backward

A person lays her head on her left shoulder and then is rotated to the left for 20 seconds and stopped abruptly. What will be the postrotatory direction of nystagmus? ~~to right~~ downward

If the subject is instructed to stand up right after rotation is stopped, what will probably happen to her? sit back down, fell backward

Explain the latter response. Describe the receptor response, the sensation perceived by the subject, and the final effect on muscle tone.
During rotation superior semicircular canal perceives backward. Somersault - on stopping - perception is falling forward muscles activated to prevent "false" forward fall causes backward fall

Diagram and label the labyrinthine receptors responsible for detecting static postural changes and angular acceleration.

Static Receptor

otoliths
hair cells
nerve
Macula in Utricle

Angular Acceleration Receptor

capula
hair cells
Nerve
Crista ampullaris ampella of Semicircular canal

Name ______________________

Date __________ Section __________

Score/Grade ______________________

Reaction Times

TRIAL	RESPONSE TIME					
	1	2	3	4	5	AVERAGE
Reaction time to sight, one color	20	22	21	20	13	19.2
Reaction time to sight, choice of two colors						
Reaction time to sight, choice of three colors						
Reaction time to sight, choice of four colors	13					
Reaction time to sound	13	14	18	18	23	17.2

.21

.24

Reaction time to word association	STIMULUS WORD	RESPONSE WORD	RESPONSE TIME
	car	truck	67
	Purce	Thief	50
	Paper	Pen	56
	test	quiz	85

.5

66.5

1. How do you explain the differences obtained for the various reflex and reaction times?

Sight and sound only require automatic reflexes word association requires thought & decision making therefore using more neural pathways delaying the response time

In general the greater the number of synapses traversed (longer path) the longer the reaction time,

ie Patellar reflex (monosynatic) reflex time very short

6 Membrane Action Potentials

RESTING AND ACTION POTENTIALS

Various body activities such as secretion, muscle contraction, mental activity, cardiac function, and sensory perception involve electrical changes across membranes. These electrical changes are produced by movement of ions, which produces the **resting** and the **action potentials.**

Resting Potential

This is a polarization of the resting membrane due to the distribution of positively and negatively charged ions across the membrane. The outside surface of the cell membrane has a larger net concentration of positive ions and is normally positive (+). The inside surface of the membrane has a greater net concentration of negative ions and is therefore negative (−) with respect to the outside of the membrane. The average mammalian cell has a resting potential of −85 millivolts (mV) across its membrane.

Action Potential

When a membrane is stimulated to alter its permeability to various ions, a series of local changes called the **action potential** occurs. A rapid influx of positive sodium ions into the cell takes place, causing the inside of the membrane to become more positive (with respect to the outside) in that particular portion of the membrane. This change of polarity is termed **depolarization.** Depolarization does not remain localized but spreads rapidly (is propagated) down the membrane in both directions. This propagation of the action potential is called the **impulse** (nerve impulse in the case of a nerve membrane). Within a millisecond or so after this initial movement of sodium, the permeability of the membrane to potassium ions increases and potassium rushes out of the cell. This potassium movement reverses the potential again and restores the original resting potential (inside [−] with respect to outside). This restoration of the original resting potential is referred to as a **repolarization** of the membrane. The movement of potassium out of the cell provides for a temporary repolarization of the membrane so that it can quickly respond to a second stimulation and produce a second action potential. Thus, a membrane is capable of transmitting many thousands of impulses (action potentials) before it becomes fatigued. For a more permanent type of repolarization, the sodium and potassium ions must be returned to their original locations and concentrations in and out of the cell. This return is accomplished, at a time of rest, by the sodium-potassium pump, which moves sodium out and potassium back into the cell.

STIMULATION OF TISSUES

In many experiments it is necessary to stimulate living tissue in order to record a particular re-

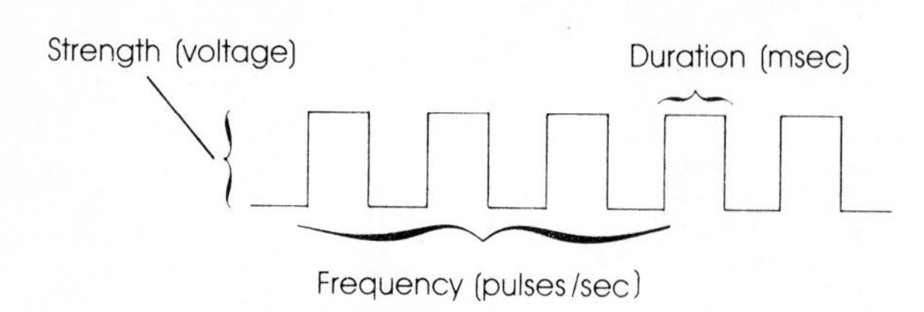

FIGURE 6.1. "Square wave" stimulation pulse.

sponse. Although tissue may be excited (depolarized) by various types of stimuli (e.g., chemical, thermal, mechanical), the method of choice is electrical stimulation. It is advantageous because it is more closely related to the electrical phenomena occurring in tissues (resting and action potentials) and because it is easy to apply and regulate. The electronic stimulator provides quantitative control over the **voltage, frequency,** and **duration** of the stimulation pulse. The pulse wave most commonly generated is the "square wave" pulse (Figure 6.1).

The stimulator allows one to apply either (+) or (−) polarity to the tissue and either a monophasic ( ) or diphasic ( ) type of pulse wave. For most physiology experiments, a (−) polarity and monophasic type of stimulation is employed, with a duration of 1 msec. More elaborate stimulators provide a means for delivering a pair of pulses to the tissue and have a **delay knob** that allows one to control the time interval between stimulus pairs. Two types of stimulators commonly used in physiology are illustrated in Figure 6.2.

OSCILLOSCOPE

The changes in electrical potential that occur in the propagation of an action potential are so small that they must be measured in units of millivolts (mV)—thousandths of a volt—or even microvolts (μV)—millionths of a volt. In addition, the speed with which these changes take place is so fast that it must be measured in units of milliseconds—thousandths of a second. To measure such small and rapid changes, highly sensitive instruments, such as the oscilloscope, must be used. The oscilloscope is an instrument that plots variations in electrical potential against time. Its component parts and the accompanying circuitry for the recording of the nerve compound action potential are shown in Figure 6.3. The oscilloscope is similar to a television set in that its essential component is the cathode ray tube. Within the evacuated tube, a narrow beam of electrons is ejected from a rear-heated cathode and strikes the phosphor coating of the oscilloscope face, where it creates a glowing spot. The electron beam passes between two pairs of plates arranged at right angles to each other.

An electrical potential applied to one pair of plates moves the electron beam up or down. The input signal will be applied to these plates, and thus the vertical trace indicates the magnitude of the signal being received from the biological source being recorded. An internal electrical circuit, called the **sweep circuit** or time base circuit, regularly applies a varying potential to the second pair of plates. This varying potential

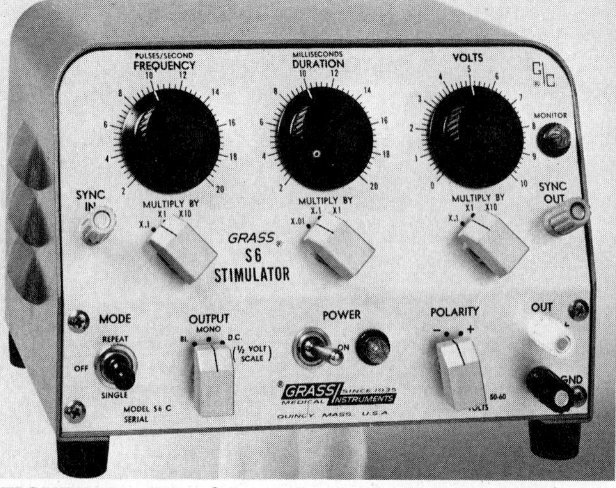

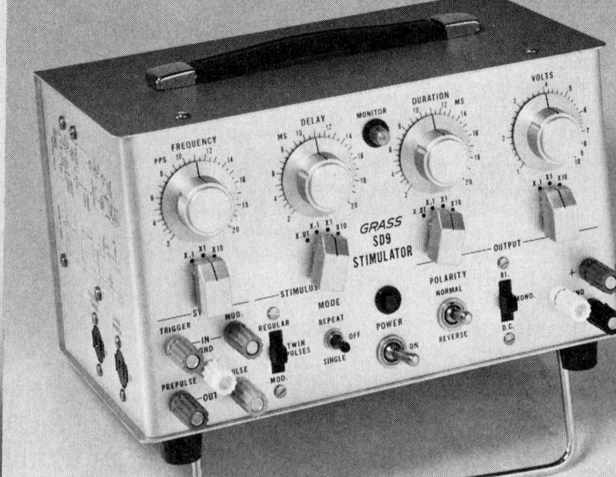

FIGURE 6.2. Stimulators. (Courtesy of Grass Instruments, Quincy, Mass.)

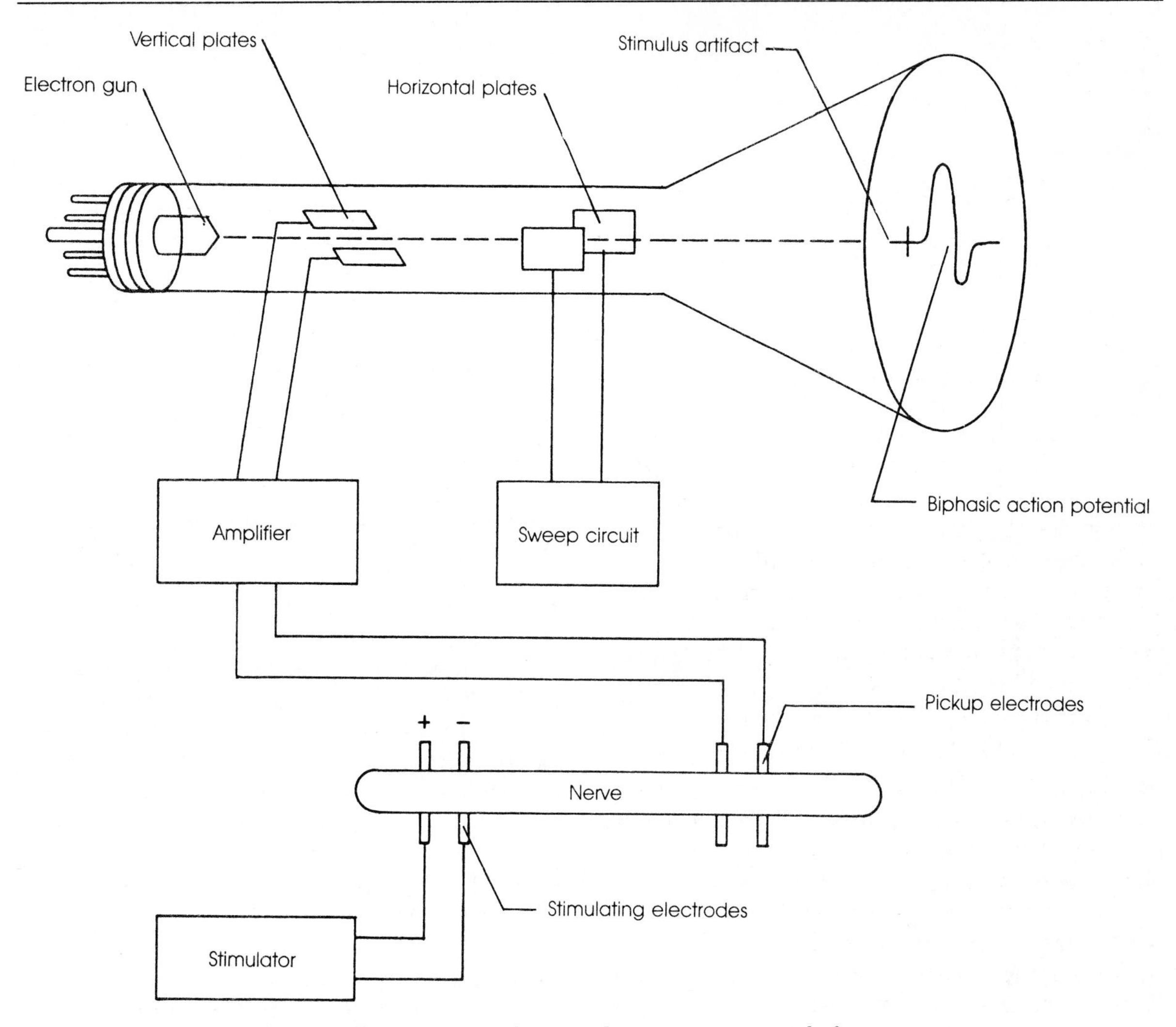

FIGURE 6.3. Diagram of an oscilloscope setup for recording action potentials from nerves.

causes the beam to move from left to right and can be regulated to move the beam at a precise speed.

SCIATIC NERVE COMPOUND ACTION POTENTIAL

Experimental Procedure

Connect the stimulator, oscilloscope, and nerve chamber as shown in Figure 6.4.

Double pith a large frog and dissect out (using glass probes) the sciatic nerve, which extends from the spinal cord to its insertion into the gastrocnemius muscle.

Caution! *Use extreme care not to damage the nerve by touching it with dry fingers or metal probes, or by excessive stretching.*

Your success in this experiment requires a nerve that has not been damaged or depolarized excessively. Place the nerve in a beaker of aerated frog Ringer's solution and allow it to equilibrate for 30 to 60 minutes.

1. Oscilloscope and Stimulator Controls

Familiarize yourself with the oscilloscope and stimulator controls. Your instructor will demonstrate the major controls.

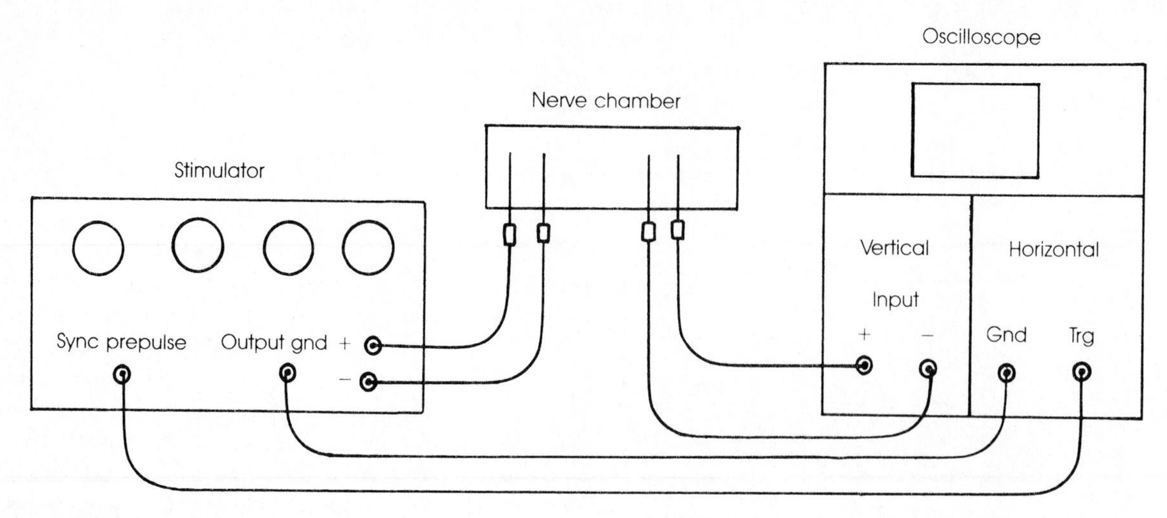

FIGURE 6.4. Connected stimulator, oscilloscope, and nerve chamber.

2. Calibration of Oscilloscope

The vertical and the horizontal sensitivities of the oscilloscope may both be calibrated by using an internal voltage provided by the oscilloscope or by using the stimulator square wave pulse. To use the internal calibration, disconnect the lead wires from the nerve chamber to the oscilloscope and connect a lead from the calibration outlet on the scope to either the (+) or the (−) of the vertical channel. This connection will put a 1-V square wave pulse into the scope for use in calibration of the trace.

To use the stimulator pulse for calibration, connect the stimulator (+) and the output (−) terminals directly to the vertical channel input terminals of the oscilloscope (bypassing the nerve chamber). Using the stimulator controls, you can vary the amplitude and duration of the square wave and adjust your vertical sensitivity and horizontal sweep speed to the desired calibration.

3. Stimulus Artifact

In the stimulus artifact experiment you will use an artificial or "shirttail" nerve to demonstrate how stimulus artifact voltage looks on the oscilloscope. This voltage sometimes resembles the compound action potential generated by the nerve and thereby confuses the observer.

Connect the equipment as shown in Figure 6.4. Lay a thread that has been thoroughly soaked in Ringer's solution across the stimulating and recording electrodes in the nerve chamber. Place a wet towel at the bottom of the nerve chamber to keep the air in the chamber saturated. Stimulate the shirttail nerve with a pulse of 1-msec duration, increasing the voltage until the stimulus artifact appears on the oscilloscope screen. Vary the vertical sensitivity and sweep speed of the scope so that you become familiar with the shape of this artifact and can recognize it at different scope settings.

4. Threshold, Submaximal, and Maximal Stimuli

Remove the shirttail nerve and carefully lay the sciatic nerve across the electrodes in the nerve chamber. Lift the nerve only by the threads at each end. Be careful not to stretch it. Apply Ringer's solution to the nerve to ensure good conduction between nerve and electrodes.

Set the oscilloscope vertical sensitivity at 2 mV per division and the sweep speed at 1 msec per division. Stimulate the nerve with pulses of 0.1-msec duration at a frequency of 15 to 30/sec. Gradually increase the stimulus voltage until the compound action potential appears. If the potential is not stationary on the oscilloscope, alter the sweep speed until a standing wave is obtained. Distinguish between the stimulus artifact and the action potential. Determine the threshold voltage for the most sensitive nerve fibers and record it in the Laboratory Report. Now increase the stimulating voltage gradually, observing the changes in the action potential as you do. You will eventually reach a point at which the action potential does not increase further (at maximal stimulus). Does the sciatic nerve follow the all-or-none law of neurons? What is the maximal height of the action potential in millivolts? What is its duration in milliseconds?

5. Strength-Duration Curve

Determine the threshold voltage of the nerve at a variety of pulse durations (e.g., 0.02, 0.03, 0.04 msec). Use as the threshold response either the first appearance of the nerve action potential or half the maximal height of the action potential (when approximately half of the neurons have reached threshold). Tabulate the data and plot a strength-duration curve in the Laboratory Report. From this plot, determine the rheobase voltage and chronaxie time for the nerve membrane. Rheobase voltage is the lowest voltage capable of reaching threshold and producing an action potential, no matter how long the duration of the stimulus. Chronaxie is the duration of a stimulus of twice rheobase voltage.

How are strength and duration of stimulus related in obtaining a threshold response? What is the significance of the chronaxie time?

6. Conduction Velocity

Stimulate the nerve with a maximal voltage to obtain a full compound action potential. Adjust the delay knob on the stimulator so the stimulus artifact can be seen clearly. If your stimulator does not have a delay knob, use the origin of the oscilloscope trace as the beginning of the stimulus artifact. Determine the speed of conduction of the action potential over the nerve as follows:

a. Measure the distance (T) in centimeters from the origin of the stimulus artifact to the peak of the potential spike.

b. Measure the distance (E) in centimeters between the cathode (−) stimulating electrode and the proximal recording electrode.

c. Record the sweep speed (S) in milliseconds per centimeter.

d. Calculate the conduction velocity using the following formula. Express your results in meters per second.

$$\text{Impulse velocity (m/sec)} = \frac{E(\text{cm}) \times 1000 \text{ msec/sec}}{S(\text{msec/cm}) \times T(\text{cm}) \times 100 \text{ cm/m}}$$

How does the conduction velocity you obtained compare with published values for the sciatic nerve of around 25 to 35 m/sec?

7. Refractory Period

Stimulate the nerve using maximal voltage of 1-msec duration at a frequency of 10 pulses per second. Change the stimulator setting to "twin pulses" instead of "regular." In this mode the nerve is stimulated by twin pulses, the second pulse following the first pulse by the time (msec) indicated on the delay knob of the stimulator. When the delay time is of sufficient length, you will be able to see two full-size compound action potentials on the scope. Now begin slowly reducing the stimulus delay time and you will see the second potential gradually move toward the first potential. As it nears the first potential, its height becomes smaller (why?) until the entire second potential disappears. Measure the distance on the scope from the point of disappearance of the second potential to the beginning of the first potential, and from this, determine the refractory period (msec) of the nerve. What mechanism is responsible for the refractory period? What is the significance of this period?

8. Summation of Subliminal Stimuli

Using a short pulse duration (e.g., 0.01 to 0.1 msec), determine the threshold voltage for your nerve. Stimulate the nerve with twin pulses using a voltage that is slightly below threshold (subliminal). Decrease the delay time between the pulses until threshold is reached and an action potential is produced. Summation of subliminal stimuli is sometimes difficult to obtain, and you may have to try several combinations of duration and voltage until you are successful.

9. Temperature Effects

Progressively cool or warm the nerve by dropping cool (10 °C) or warm (35 °C) Ringer's on it. What effect does temperature have on the threshold, size, or shape of the action potential? How does temperature affect the conduction velocity and refractory period?

10. Direction of Propagation

Reverse the stimulating and recording electrodes or turn the nerve around. Stimulate the nerve.

Is the impulse conducted in the opposite direction? How can you reconcile this result with the concept of unidirectional propagation between neurons?

11. Monophasic Action Potential and Nerve Fiber Types

Place a drop of isotonic potassium chloride (KCl) solution on, or crush (using forceps), the portion of the nerve lying over the distal recording electrode. Gradually increase the stimulus voltage from threshold to maximal and note the order of appearance of the various peaks, or elevations, in the action potential. What is the threshold for each of these peaks? What do the peaks represent? Why are these elevations seen only when a monophasic potential is produced? What is meant by a compound action potential?

12. Nerve Conduction Blockade

The propagation of nerve impulses can be blocked by many agents, such as the local anesthetics procaine and lidocaine (Xylocaine) and the general anesthetics ether and chloroform. Even alcohol can depress nerve conduction. Of the different nerve types, the fibers that have relatively small diameters are the most sensitive to anesthetic action and larger fibers are the most resistant. Thus, in a mixed nerve, sensory neurons are the first to be anesthetized and motor neurons are the last to be blocked. Nerve conduction can also be blocked by local application of pressure greater than 130 mm Hg. With pressure blockade, however, the sensitivity picture is reversed—motor fibers are blocked before sensory fibers are affected.

a. Drug Blockade

Dampen a small piece of cotton with ether, squeeze out the excess fluid, and place the cotton in the nerve chamber. Avoid touching the nerve with the cotton or your fingers. Cover the chamber. Stimulate the nerve continuously and observe the action potential for several minutes. What happens? Is there a sequence of changes? Remove the cotton before complete nerve block occurs and apply fresh frog Ringer's solution to the nerve. Is recovery complete?

Similar effects can be produced by placing a small piece of cotton soaked in 50% ethanol or 1% procaine on the nerve between the stimulating and recording electrodes. Stimulate the nerve and observe the changes in the action potential over a period of time. Remove the cotton and rinse with Ringer's solution to bring about recovery of the nerve.

b. Pressure Blockade

Cover the jaws of a Gaskell clamp with rubber tubing and moisten the tubing with Ringer's solution. Place the clamp over the nerve between the stimulating and recording electrodes. Stimulate the nerve continuously and gradually apply pressure by closing the jaws of the clamp. Apply pressure until conductivity is completely blocked. Release the pressure and test to determine if nerve conduction returns.

LABORATORY REPORT

6. Membrane Action Potentials

Name ______________________

Date __________ Section __________

Score/Grade ______________________

Sciatic Nerve Compound Action Potential

1. Threshold, Submaximal, and Maximal Stimuli

Stimulator:

Threshold voltage = ____________ Maximal voltage = ____________

Oscilloscope:

Maximal voltage of compound potential = ____________ Duration = ____________

What is the relationship between the all-or-none response of an action potential spike and the formation of the compound action potential?

2. Strength-Duration Curve

Duration (msec)																
Threshold voltage																

Strength (V)

Duration (msec)

Rheobase voltage = ____________ Chronaxie time = ____________

What is the significance of the chronaxie time?

3. Conduction Velocity

$$\text{Impulse velocity (m/sec)} = \frac{\underset{(E)}{\_\_\_\_} \times 1000 \text{ msec/sec}}{\underset{(S)}{\_\_\_\_} \times \underset{(T)}{\_\_\_\_} \times 100 \text{ cm/m}} = \_\_\_\_\_\_\_\_ \text{ m/sec}$$

How does the conduction velocity obtained compare with the conduction velocities of individual neurons?

4. Refractory Period

Refractory period obtained = ____________ msec

What is the significance of the refractory period?

5. Summation of Subliminal Stimuli

How is it possible to produce an action potential using a subthreshold level of stimulation?

6. Temperature Effects

	COOL 10 °C	ROOM TEMPERATURE 25 °C	WARM 35 °C
Conduction velocity			
Refractory period			

What mechanism is responsible for these temperature changes?

7. Direction of Propagation

How can you reconcile your result with the concept of unidirectional propagation between neurons?

Name ____________________

Date __________ Section __________

Score/Grade ____________________

8. Monophasic Action Potential and Nerve Fiber Types

What do the various peaks of the compound action potential represent?

Why are these peaks at different distances from the stimulus artifact?

Why are these peaks seen only when a monophasic potential is produced?

9. Nerve Conduction Blockade

Diagram the sequence of changes in the action potential when the nerve is exposed to ether.

Explain the mechanism of action of ether and ethanol.

How is this effect of ethanol evidenced in a drunken individual?

How does pressure produce nerve blockade? Can you think of any practical applications of such a pressure blockade?

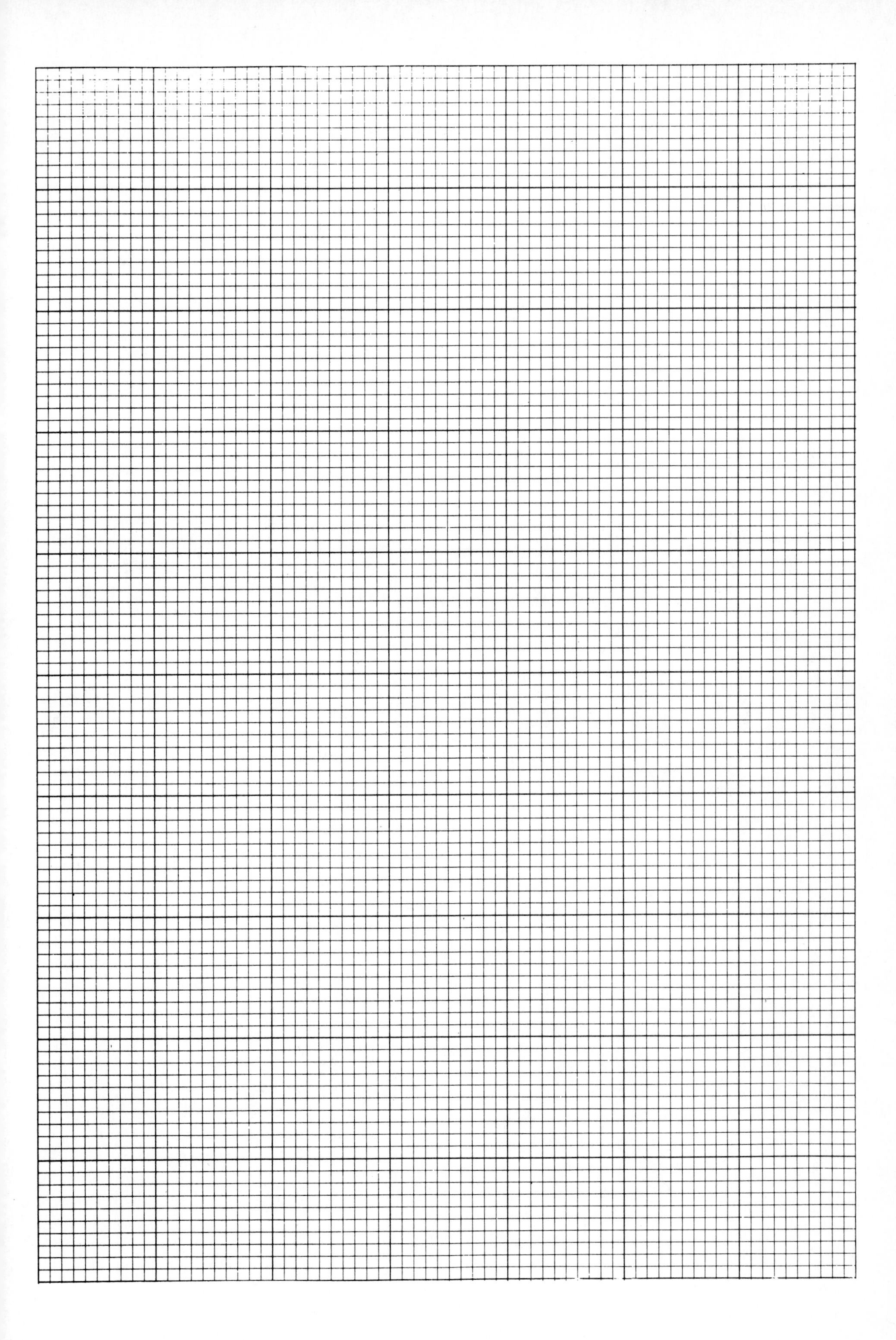

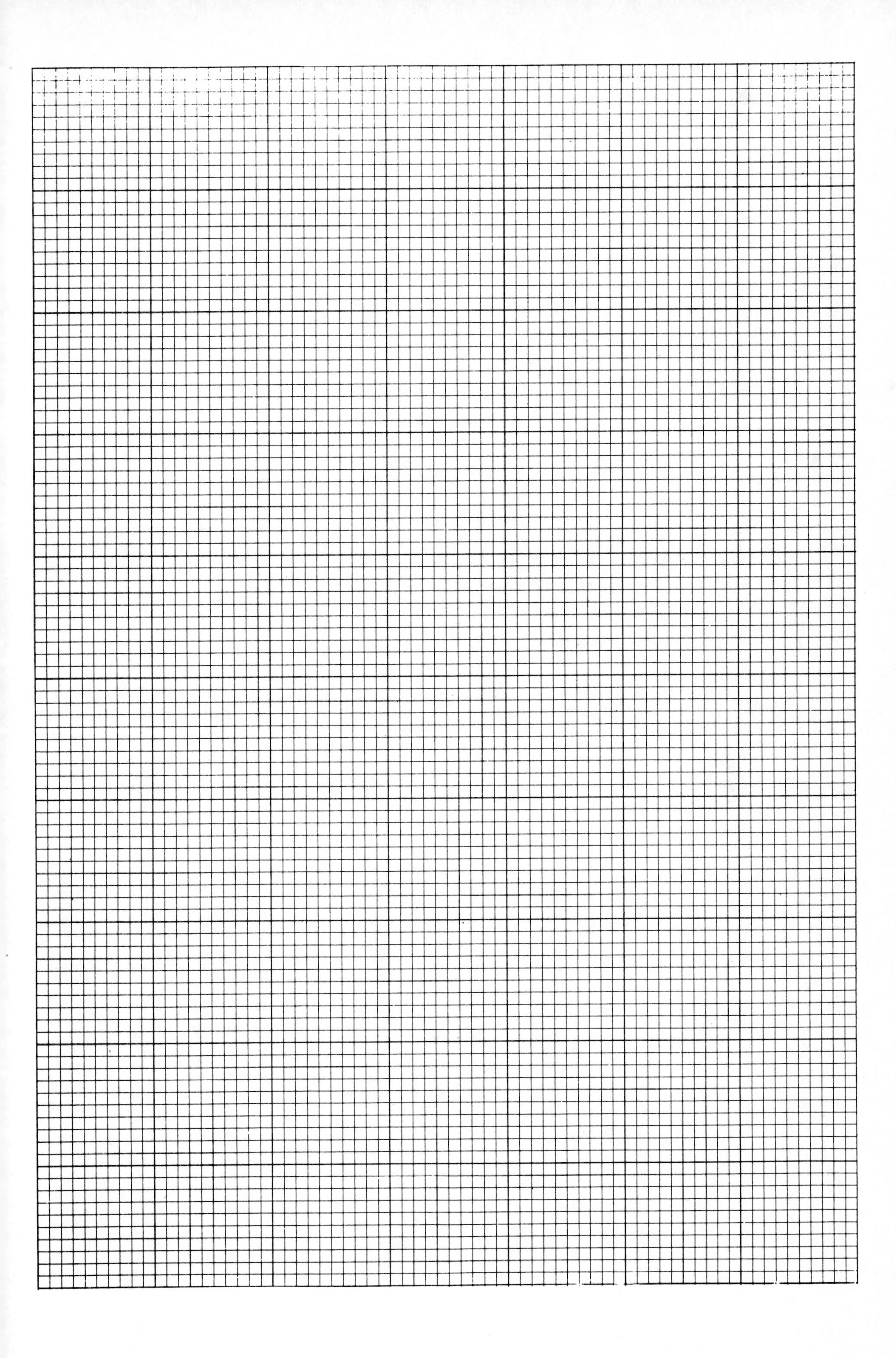

171-fox
197-200-fox

7 Sensory Physiology I: Cutaneous, Hearing

SENSORY RECEPTORS

Our knowledge of changes in our environment depends on the sensory nervous system and its receptors. Classically, we speak of five senses: sight, hearing, touch, taste, and smell. They are often listed in this order of importance. In this laboratory exercise you will study the mechanism of action of several **exteroceptors**—receptors that receive stimuli from outside the body. Exteroceptors are usually located on the surface of the body, in contrast to **interoceptors,** which are located deep within muscles, tendons, and other structures and which detect changes within the body.

CUTANEOUS RECEPTORS

Experimental Procedure

1. Tactile Distribution: Two-Point Sensibility

Receptors for touch vary in their density of distribution over various areas of the body surface. Areas of the body having many touch receptors, such as the fingers, have a finer sense of "feel" or tactile discrimination.

Have the subject seated with the eyes closed. Use a compass that has blunted tips, a caliper, or an aesthesiometer to apply tactile stimuli to the subject's skin. Start with the points close together and then increase their distance apart until the subject feels *two distinct points.* Be sure that the two points are applied simultaneously each time. Apply one and two points randomly to test whether the two-point threshold has been reached or whether the subject's imagination is operating overtime. Record in the Laboratory Report the point distance in millimeters for the following body areas:

Back of neck

Fingertip

Forearm (palmar surface)

Tip of nose

Palm of hand

Tongue

Upper arm

2. Tactile Localization

Have the subject close her eyes. Touch her skin with a pointed pencil to leave an indentation. Remove the pencil. Then have the subject try to touch this exact spot using another pencil. Measure her error of location in millimeters. Repeat, using the same stimulation point. Does her localization improve the second time? Record the localization error distance (two trials) for the following body regions:

Palm

Fingertip

Forearm

Lips

3. Adaptation of Touch Receptors

Close your eyes and have your partner place a small coin on the inside of your forearm. Determine (in seconds) how long the initial pressure sensation persists. Repeat the experiment at a different forearm location, and when the sensation disappears have your partner add two more coins of the same size. Does the pressure sensation return, and if so, how long does it last with the added coins? What receptors are functioning here, and why is the sense of pressure soon lost?

Another illustration of sensory adaptation is provided by the touch receptors around the root of each hair. Using a pencil point, move one hair as slowly as possible until it springs away from the pencil. Is the touch sensation greater when the hair is slowly bent or when it springs back? These hair receptors have one of the fastest adaptation times to stimulation of all receptors.

What is meant by sensory adaptation? What is its function? How do receptors vary in their ability to adapt?

4. Weber's Law

Blindfold the subject and have her place her hand, palm up, on a table. Place a 2-in. square piece of cardboard on the distal phalanges of her index and middle fingers. On the cardboard place a 10-g weight. After the subject feels the weight, remove the cardboard and weight, add additional weights of 1 to 5 g, and replace the cardboard and weights on the fingers. Repeat this procedure until the subject reports a perceptible sensation of increased weight compared with the initial weight of 10 g. Record the increment (added weight) required to produce the sensation of added weight. This is called the **just noticeable difference (JND)** or **intensity difference (ΔI).** The ratio of intensity difference to the initial weight intensity is called **Weber's fraction (ΔI/I).**

Note: The cardboard and weights should be lifted from the fingers while additional weights are being added, and then all placed back on the fingers.

Repeat the experiment, starting with an initial weight of 50 g, then 100 g, and finally 200 g. Record the weight increments required for the subject to perceive a JND in weight for each of these initial weights. In each trial, what is the ratio of the JND to the initial weight (Weber's fraction)? How does this ratio compare between the various initial weights? Note that Weber's law holds only for weights in the medium range.

5. Temperature Receptors: Adaptation and Negative After-Image

Prepare three 1000-ml beakers half full of ice water (0–5 °C), water at room temperature (25 °C), and water at 45 °C, respectively. Place your left hand in the ice water and your right in the warm water. What happens to your sensation of cold or warmth in each hand after 2 minutes? Which hand seems to adapt fastest?

Now rapidly place both hands in the water at room temperature. What are your sensations in each hand? This experiment illustrates that the sensations of heat and cold are not absolute but depend on how rapidly the skin gains or loses heat and on the magnitude and direction of the temperature gradient.

6. Referred Pain

Referred pain is the strange phenomenon of perception of pain in one area of the body when another area is actually receiving the painful stimulus. We say the pain is "referred" to the other, more remote area.

Place your elbow in ice water and over a period of time note any changes in location of sensation perceived. Does it change location? If so, where is the referred pain felt? The ulnar nerve, which supplies the ring finger, little finger, and inner side of the hand, passes over the elbow joint. The ulnar nerve serves as the mediator for this referred pain sensation. You may have experienced other examples of referred pain, such as pain in the forehead after swallowing ice cream.

HEARING

Experimental Procedure

1. Watch Tick Test for Auditory Acuity

This should be performed in a quiet room. Have the subject close one ear with cotton and close his eyes. Hold a watch in line with his open meatus. Gradually move the watch away from his ear until he just loses the ability to hear the ticking. Measure the distance between watch and ear and record. Move the watch farther away and then begin moving it nearer the subject's ear until he first hears the ticking. Is this distance the same as when the watch is moving away? Test the other ear in like manner. Is the acuity the same for both ears? This is really not a fair test for an elderly person, because high tones are lost first in old age.

2. Localization of Sound

With the subject seated and blindfolded, bring a watch within hearing range from several different angles around his head. (The clicking of two coins together can be used in place of the watch in this test.) Ask the subject to point to the direction from which he hears the sound. Is his judgment better in the median plane or at the side of the head? In the median plane, is he more accurate with sound above the head or in front of it?

3. Auditory Adaptation

Place a stethoscope in the subject's ears. Place a vibrating tuning fork near the bell of the stethoscope so that the sound seems equally loud to both ears. Remove the tuning fork and wait a minute or two. Pinch the tube to one ear to occlude the tube and place the vibrating tuning fork in its former position near the bell. When the sound becomes nearly inaudible to the open ear, open the pinched tube. Is this ear also adapted to the sound? Explain.

4. Tuning Fork Tests

These two simple hearing tests are used to distinguish between conduction and nerve deafness. In conduction deafness, transmission of sound waves through the middle ear to the oval window is impaired. In nerve deafness, transmission of nerve impulses from the cochlea to the auditory cortex is impaired. A 512-Hertz (Hz) tuning fork is preferred for these tests because most people have difficulty telling whether the vibrations are felt or heard when forks of lower frequency are used. The fork should be struck on a soft surface such as the heel of the hand, because striking a hard surface produces overtones.

a. Weber Test

This test should be performed in a room with a normal noise level (not quiet). Strike the tuning fork and place the tip of the handle against the middle of the subject's forehead. If both ears are normal, the tone will be equally loud in both ears and the sound will be localized as coming from midline position. An individual with conduction deafness will hear the sound as louder in the deaf ear than in the normal ear because in the normal ear the sound will be partially masked by environmental noise to which the defective ear is less sensitive. Also, in the normal ear the sound is damped or softened by the tensor tympani and stapedius muscles, which prevent the full amplitude of vibration of the auditory ossicles. This attenuation reflex is less effective or is absent in conduction deafness.

If an individual has a defect in the auditory nerve or cochlear apparatus (nerve deafness), the sound will be heard better in the normal ear than in the deaf ear, because neural activity is essential for hearing.

Conduction deafness can be simulated by plugging one ear with cotton.

b. Rinne Test

This test compares air conduction of sound with that of bone conduction. It should be performed in a quiet room. Locate the mastoid process of the temporal bone behind the ear. Strike the tuning fork and place the handle against the mastoid at the level of the upper portion of the ear canal. As soon as the sound is no longer audible through the bone, hold the vibrating prongs of the tuning fork about one inch from the ear and the subject should again be able to hear the sound. A person who has normal hearing will hear the sound several seconds longer by air conduction than by bone conduction because the threshold for air conduction is lower. A per-

TABLE 7.1. Guide for Interpreting Weber and Rinne Tests.

CONDITION	FINDING	
	WEBER TEST	RINNE TEST
No hearing loss	No lateralization	Sound perceived longer by air conduction
Conduction deafness	Lateralization to the deaf ear	Sound perceived as long or longer by bone conduction
Nerve deafness	Lateralization to the normal ear	Sound perceived longer by air conduction

son who has conduction deafness will hear the sound as long or longer by bone conduction. The person with nerve deafness will hear longer by air conduction, but usually requires a louder sound to hear at all.

A concise guide for interpreting the Weber and Rinne tests is given in Table 7.1.

5. Audiometry

a. Audiogram

A pure-tone audiometer is an instrument for measuring hearing acuity. It consists of an earphone connected to an electronic oscillator capable of producing pure tones in the range of 250 to 8000 Hz. This is a narrower range than the hearing range for normal humans of 20 to 20,000 Hz. Dogs and very young children often can hear frequencies as high as 40,000 Hz.

In this test, the subject listens to several tones in the range of 250 to 8000 Hz, usually at octave or half-octave intervals. His threshold of perceiving each tone is determined and recorded as decibels (dB) of loudness needed to just hear each tone. If the loudness of a tone must be increased to 20 dB above the normal tone level, the subject is said to have a **hearing loss** of 20 dB for that particular tone. The amount of hearing loss is plotted on the **audiogram** below the normal hearing line at zero.

Test the auditory acuity of a subject, starting with high frequencies (8000 Hz). Determine the number of decibels required for him to hear each tone. Decrease the frequency in octave intervals (4000—2000—1000—500 Hz, etc.). Plot his audiogram of hearing loss on the chart provided in the Laboratory Report. The audiometric zero (normal) is based on extensive tests of young persons having normal hearing. A hearing threshold of 15 to 25 dB above the reference level in the range of 500 to 2000 Hz is considered to be a slight hearing impairment. How would an audiogram for old-age deafness look?

b. Evaluation of Hearing Impairment (Shorter Method)

In 1959, the American Academy of Ophthalmology and Otolaryngology issued a statement describing a simpler method for evaluation of hearing impairment. This method recognizes that hearing impairment should be evaluated in terms of ability to hear everyday speech under everyday conditions, and that *everyday speech* does not encompass the entire audiometric range but primarily the frequencies from 500 to 2000 Hz. Therefore, for a practical evaluation of everyday speech impairment, the academy recommends the simple average of the hearing levels at the three frequencies 500, 1000, and 2000 Hz using a pure-tone audiometer for testing.

Audiometric zero, which is usually considered the average normal hearing threshold level, is not actually the point at which impairment begins. Therefore, the academy considers that *for every decibel for which the hearing level exceeds 15 dB, there is a 1.5% impairment for everyday speech.* Thus, at 82 dB there would be a 100% or total hearing impairment. This system applies to the *American Standard (ASA)* of 1951. When using the *International Standard (ISO)* of 1964, 26 dB is the starting point; that is, *for every decibel that the hearing loss exceeds 26 dB, there is a 1.5% hearing impairment,* and total impairment is reached at 93 dB. No allowance is made for presbycusis (old-age deafness) in this estimation of hearing impairment. For estimation of the biaural hearing impairment, the percentage of hearing impairment in the better ear is multiplied by 5, the resulting figure is added to the percentage impairment in the poor ear, and the sum is divided by 6. The following example illustrates the calculations used to determine hearing impairment (using the *ISO*).

Decibel threshold levels

Hz	Right ear	Left ear
500	30	40
1000	45	50
2000	40	55
Total	115	145
Average (÷3)	$38\frac{1}{3}$	$48\frac{1}{3}$

Calculation of monaural hearing impairment:

Right ear	**Left ear**
$38\frac{1}{3}$ dB	$48\frac{1}{3}$ dB
−26 dB	−26 dB
$12\frac{1}{3} = 12$ dB	$22\frac{1}{3} = 22$ dB
12 dB × 1.5% per dB	22 dB × 1.5% per dB
= 18% impairment in right ear	= 33% impairment in left ear

Calculation of biaural hearing impairment:

$$= \frac{\left(\begin{array}{c}\text{\% impairment}\\ \text{in better ear}\end{array} \times 5\right) + \left(\begin{array}{c}\text{\% impairment}\\ \text{in worse ear}\end{array}\right)}{6}$$

$$= \frac{(18\% \times 5) + (33\%)}{6}$$

$$= \frac{90\% + 33\%}{6}$$

$$= 20.5\%$$

Hearing impairment may also be calculated using slide rule calculators available with the different types of audiometers.

Have your partner test your hearing level for the three frequencies in both ears and calculate your hearing impairment, if any.

LABORATORY REPORT

Name ______________________
Date __________ Section __________
Score/Grade ______________________

7. Sensory Physiology I: Cutaneous, Hearing

Cutaneous Receptors

1. Tactile Distribution: Two-Point Sensibility

Record the distance in millimeters.

Back of neck 12 mm 35 mm

Fingertip 4 mm 2 mm

Forearm 9 mm 30 mm

Tip of nose 6 mm 6 mm

Palm of hand 14 mm 12 mm

Tongue 1 mm

Upper arm 24 mm 55 mm

2. Tactile Localization

Record the error distance in millimeters for two trials.

	TRIAL 1	TRIAL 2
Palm	28 mm	15 mm
Fingertip	6 mm	4 mm
Forearm	32 mm	32 mm
Lips		

7 8
3 2
10 8
2 2

3. Adaptation of Touch Receptors

How is the brain informed that a receptor is adapting to a stimulus?

when a receptor adapts to a stimulus, it sends fewer action potentials per second over sensory nerves to the brain

What is meant by a phasic receptor? A tonic receptor? Give two examples of each type.

Phasic receptors adapt rapidly (touch & hair)
tonic receptors adapt slowly (pain + muscle spindle organs)

4. Weber's Law

INITIAL WEIGHT INTENSITY (I)	JUST NOTICEABLE DIFFERENCE (JND) OR INTENSITY DIFFERENCE (ΔI)	WEBER'S FRACTION (ΔI/I)
10 g	3g	.3
50 g	17g	.34
100 g	33g	.33
200 g	64g	.32

What is the importance of Weber's law in the physiology of sensory perception?

most extero receptors obey Webers law, the frequency of action potentials is proportional to log of stimulus intensity. It allows a greater range of receptor response. decibel is based on log scale

5. Temperature Receptors: Adaptation and Negative After-Image

Which hand seems to adapt fastest? warm hand

Describe the sensation in the right and the left hand when they are placed in room-temperature water after adapting to hot and cold water. Explain.

warm hand feels cool, receptors were warm but begin to lose heat to surrounding water hence feels cool

Cool hand fells warm, receptors were cold but gain heat from surrounding water, hence feel warm

6. Referred Pain

What physiological mechanism is responsible for referred pain?

neurons from peripheral areas may converge on the same second order neurons as those from visceral areas. Thus the brain interprets the sensation from a different region

Give two examples of referred pain, giving the actual location of the pain and the location where the pain is perceived.

actual	percieved
Heart	Left shoulder
Kidney	back
elbow	fingers

Name ______________

Date __________ Section __________

Score/Grade ______________

Hearing

1. Watch Tick Test

Record distance from ear when first unable to hear sound.

Left ear 13 cm Right ear 33 cm

30 25

2. Localization of Sound

Is localization better in the median plane or at the side? median

Is localization better directly in front of the head or above the head? in front

3. Auditory Adaptation

Does the ear on the side of the pinched tube also adapt to the sound? No

Why or why not? during the time when pinched, the ear was not receiving stimuli

4. Tuning Fork Tests

a. Weber Test

Is the sound localized in the midline, left, or right? midline

b. Rinne Test

Is the sound perceived longer by air or bone conduction? air

What do you conclude concerning your possible deafness? Explain your conclusion.

Normal – No lateralization sound perceived longer by air conduction

Lateralization would indicate hearing defect in one ear

If bone conduction were longer it would indicate conduction defect

5. Audiometry

a. Audiogram

Construct your own audiogram by plotting your threshold for hearing in each ear on the following graph. Use a different color for plotting the left and right ear thresholds. Using another color, make a theoretical plot for the hearing of a person who has presbycusis.

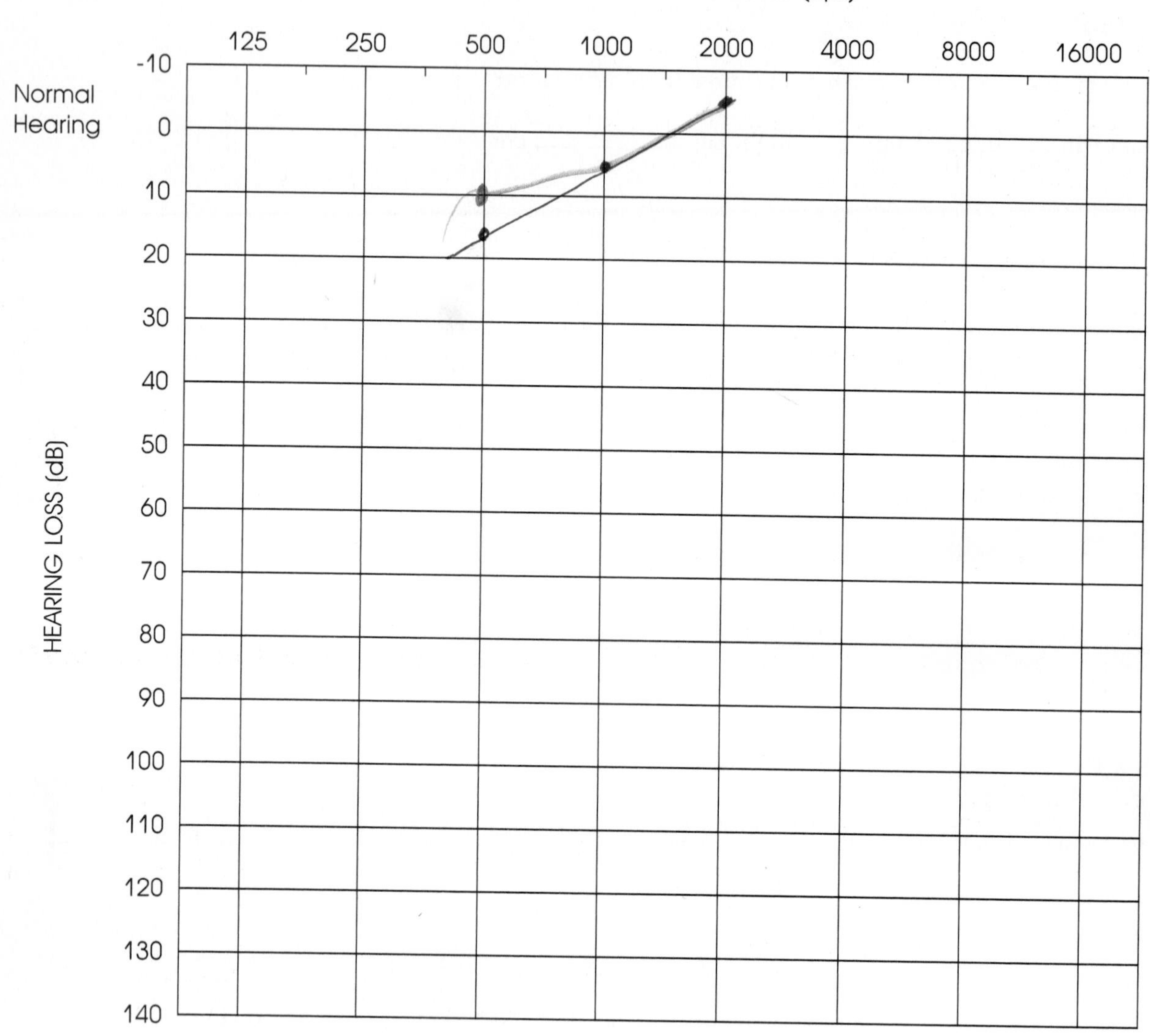

b. Evaluation of Hearing Impairment (Shorter Method)

Lynn

CYCLES PER SECOND (HERTZ)	DECIBEL THRESHOLD LEVELS	
	RIGHT EAR	LEFT EAR
500	10	15
1000	5	10
2000	-5	-5
	3.3	6.6

Name

Date   Section

Score/Grade

Hearing Impairment: Right ear = _______ % Left ear = _______ %

Biaural = _______ %

Why does partial deafness occur when one has a cold?

Blockage of eustacian tube
Prevents free movement of air within the middle ear
behind the tympanic membrane

What is presbycusis? What causes it?

old age hearing - loss of hair cells & decreased
elasticity of Basular membrane
diminished hearing in high frequencies

How do most hearing aids increase auditory acuity? For which type of deafness are they best suited?

thru bone conduction
best suited for conduction deafness

8 Sensory Physiology II: Vision

FUNCTIONS OF THE EYE

Vision provides our most important link with the external world. The eye is often compared with a camera in its functions, but this is a gross oversimplification. Indeed, the fantastic complexity of the eye is said to have given Charles Darwin the "cold shudders" because he could not imagine how the eye had developed through the random processes of evolution. In this lab you will examine the anatomy and physiology of the eye, which make it such a unique receptor for the focusing and processing of light energy.

Experimental Procedure

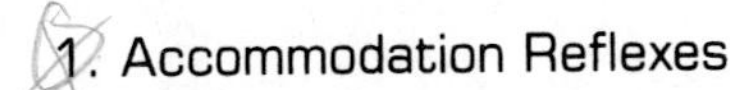

1. Accommodation Reflexes

Accommodation reflexes are often studied under the topic of reflexes because they represent a programmed response to a stimulus.

Observe the pupil of your partner's eye under normal lighting. Shine a light into the eye and notice the constriction of the pupil. The purpose of this pupillary reflex is quite obvious. Now have your partner focus on an object across the room and observe the size of the pupil. Then have him or her focus on an object 6 in. from the eye. What happens to pupil size? Measure the size of the pupil again and record your results in the Laboratory Report. What mechanism produces the changes in pupil size? How sharp is the image of the far object when the near object is being viewed, and vice versa?

2. Near Point of Accommodation

To produce a sharp image on the retina, the lens of the eye must be able to change its focusing power for viewing objects at different distances. When viewing near objects, the lens becomes more spherical than when viewing a distant object. As we age, the lens becomes less elastic and therefore less able to form the spherical shape needed for near vision. To compensate for this loss of accommodation power with age (presbyopia), persons over 45 begin wearing bifocal or trifocal glasses to give them better near vision. Determination of the near point for the eye gives us a measure of the elasticity of the lens and its accommodation power.

The *near point* is the closest distance at which one can see an object in sharp focus. Print a small letter 5 mm high on a white card. With one hand hold a meter stick, directed forward at the bridge of your nose. Hold the card at a distance on the meter stick and and close one eye. Move the card toward your eye until the letter becomes blurred; then move it away until the letter has a clear image. Measure the distance from the card to your eye. Repeat the test for the other eye and record your results.

Compare your near point measurements with the following normal values:

AGE (YR)	NEAR POINT (CM)
10	9
20	10
30	13
40	18
50	53
60	83
70	100

3. Binocular Vision and Space Perception

Hold a cube 3 in. in front of your nose and focus on it. Close one eye, open it, and close the other. Do the views seen with the right and left eyes appear different? Why? Repeat this procedure while looking at a picture in a stereoscope.

Try to thread a needle, first with both eyes open and then with one eye closed. These simple experiments illustrate the advantage of binocular vision in providing depth of field and space perception.

4. Blind Spot

The blind spot is an area on the retina where the optic nerve and blood vessels enter and leave the retina, and hence where there are no rods or cones for visual reception.

Close your left eye and focus your right eye on the cross at the left. Hold the paper about 15 in. from the eye and slowly bring it closer. Soon the dot will disappear. At this distance the dot image is being focused on the blind spot area of the right eye where there are no rods or cones to perceive it.

5. Visual Acuity and Astigmatism

Visual acuity is the power to discriminate details. You have all tested your visual acuity many times using the **Snellen** test letters, but you may not have understood exactly what the score means. The basis for the Snellen test is that letters of a certain size should be seen clearly at a specific distance by eyes that have normal acuity. For example, line 1 should be read easily at 200 ft, and line 8 at 20 ft (Figure 8.1).

A person's visual acuity is stated as V = d/D, in which "d" is the distance at which the person can read the letters and "D" is the distance at which a normal eye can read the letters. Stand 20 feet from the full sized Snellen chart on the wall (not the reduced chart in the manual). Cover one eye and attempt to read line 8. If you can read it your eye is normal and is rated as 20/20. If you can read line 9 at 20 feet your visual acuity is above average and rated at 20/15. If you can't see line 8 but can see line 7 your visual acuity is below average and rated as 20/25. Repeat the test for the other eye and record your results.

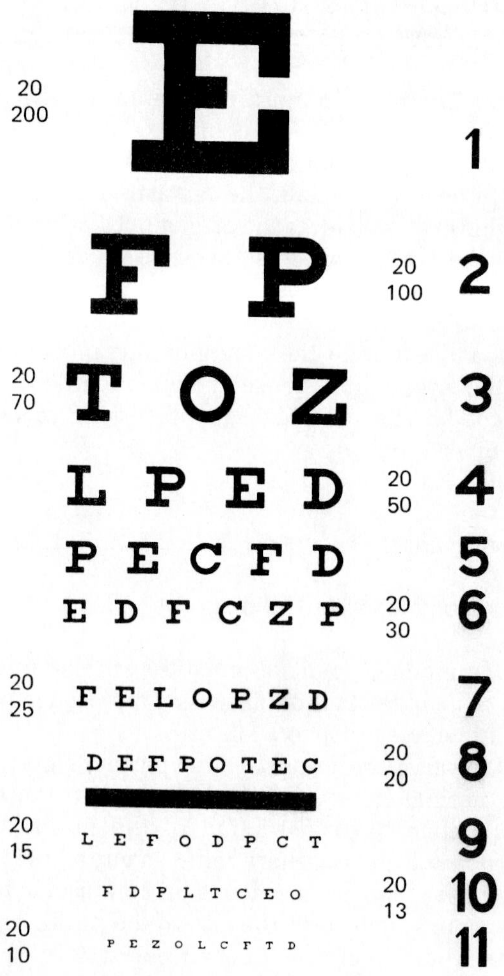

FIGURE 8.1. Snellen eye chart.

Astigmatism is a condition where there is an uneven curvature of the surface of the lens or cornea. This causes a greater bending of light rays as they pass through one axis of the lens or cornea than when light passes through another axis. This causes the image viewed to be blurred in one axis and sharp in the other axes.

Remove any corrective lenses, cover the left eye and look at the center of the astigmatism test chart (Figure 8.2). If some of the lines appear blurred or lighter, this indicates that astigmatism is present. If all lines are equally sharp and black, no astigmatism exists. Repeat the test with the right eye. If you wear corrective lenses, repeat the tests with them on.

6. Negative After-Images: Complementary Colors

Look at a very bright scene (such as a burning candle) for around 30 seconds; then shift your gaze quickly to a blank white surface. What sort of image do you perceive? A negative "after-image" should be seen in which the light areas of the original scene appear dark and the dark areas light. This perception is explained in the following manner: While you are viewing the bright scene, cones receiving light from the bright areas become **light adapted** or "bleached" of their visual pigment, and when you shift your gaze to the white paper, these cones cannot initiate impulses. Thus, this area of the scene now appears dark. The unadapted cones are still sensitive and are stimulated by the white background.

Color vision is initiated by the activation of cone cells in the retina that are sensitive to red, green, and blue wavelengths of light. Other colors are perceived by differential stimulation of these three types of cone cells. The sensation of white results when all the visible light rays are combined and the three types of cones are equally stimulated. White is also perceived when two particular colors are mixed. These are the so-called complementary colors. The relationship between the red, green, and blue colors and the complementary colors that produce a white sensation are clearly portrayed in the physiologist's color wheel shown in Figure 8.3.

Stare at a bright red color for a minute or so and then shift your gaze to a blank white sheet. What color is the after-image? Repeat this procedure using a green color and then a yellow color. Record the color of the after-image seen in each case. Why are these particular colors produced?

7. Tests for Color Blindness

Color blindness is an abnormality that is transferred genetically, resulting from the lack of a particular gene in the X chromosome. The most

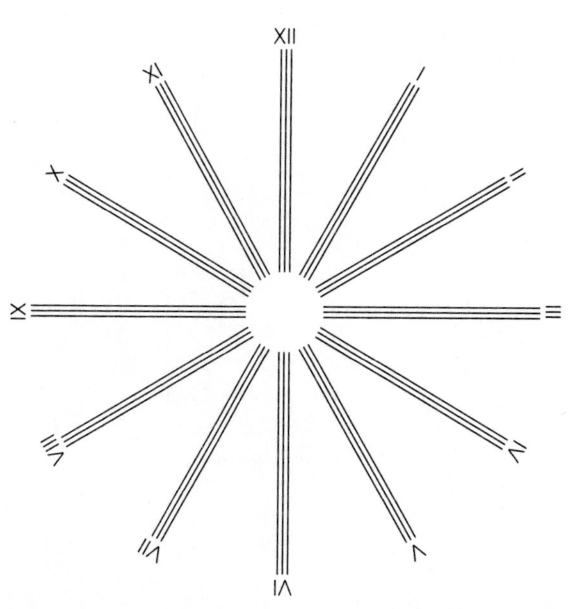

FIGURE 8.2. Astigmatism test chart. (From Tortora, *Human Physiology Lab Manual,* MacMillan, p. 157.

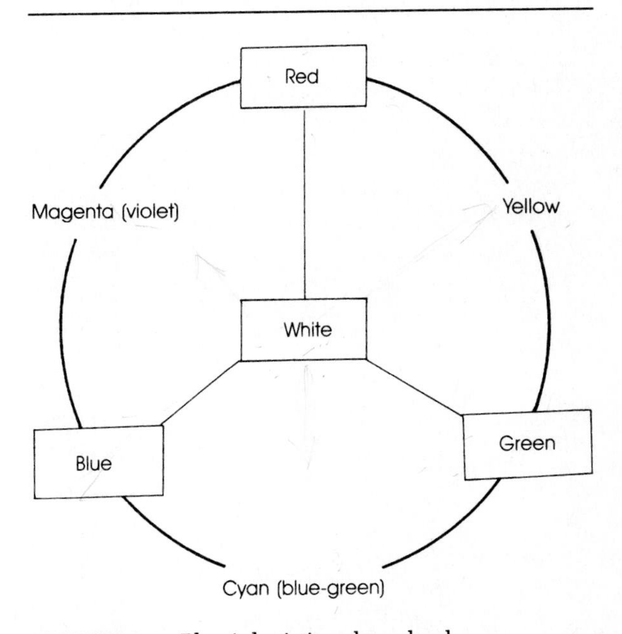

FIGURE 8.3. Physiologist's color wheel.

common type is red-green color blindness, in which the person lacks either the red or the green cones in the retina. If the red cones are lacking, red wavelengths of light stimulate primarily green cones and the person perceives red as green. If the green cones are missing, he or she will perceive only red colors in these wavelengths of light.

a. Ishihara Test

These test charts are probably the most widely used in the testing of color blindness. Each chart contains different-colored dots arranged so that the person with normal color vision reads one number and the color-blind person perceives a different number.

Hold the charts about 30 in. away from you and in a good reading light (avoid intense light, because the colors will fade). Read each plate. How many color-blind persons are detected in the class?

b. Holmgren's Test

This test consists of several skeins of colored yarn that must be matched accurately with three sample colors, A, B, and C. The odd-numbered skeins are the same color as the A, B, and C colors, but the even-numbered skeins are "confusion" skeins and should not be mistaken for the true color-matched skeins. A color-blind person will match some of the "confusion" skeins with the A, B, or C colors because they appear to be the same.

8. The Optics of Vision

The primary task of most of the structures in the eye (except the retina) is simply to aid in the focusing of light rays on the retina. The function of various refractive parts of the eye can be demonstrated by the Ingersoll or Cenco eye models. These models consist of a water-filled black tank with chambers representing the aqueous and vitreous humors. In the anterior end, a curved piece of glass performs the function of the cornea. Behind and in front of this "cornea" are slots for holding various "eye" lenses (and irises) and corrective lenses. At the posterior end of the model is a movable metal screen that serves as the retina, on which the image is focused. It can be placed in three positions, to simulate normal, farsighted, and nearsighted vision. In most models, a white spot in the middle of the retina represents the fovea and a dark spot off to one side represents the blind spot. See Figure 8.4.

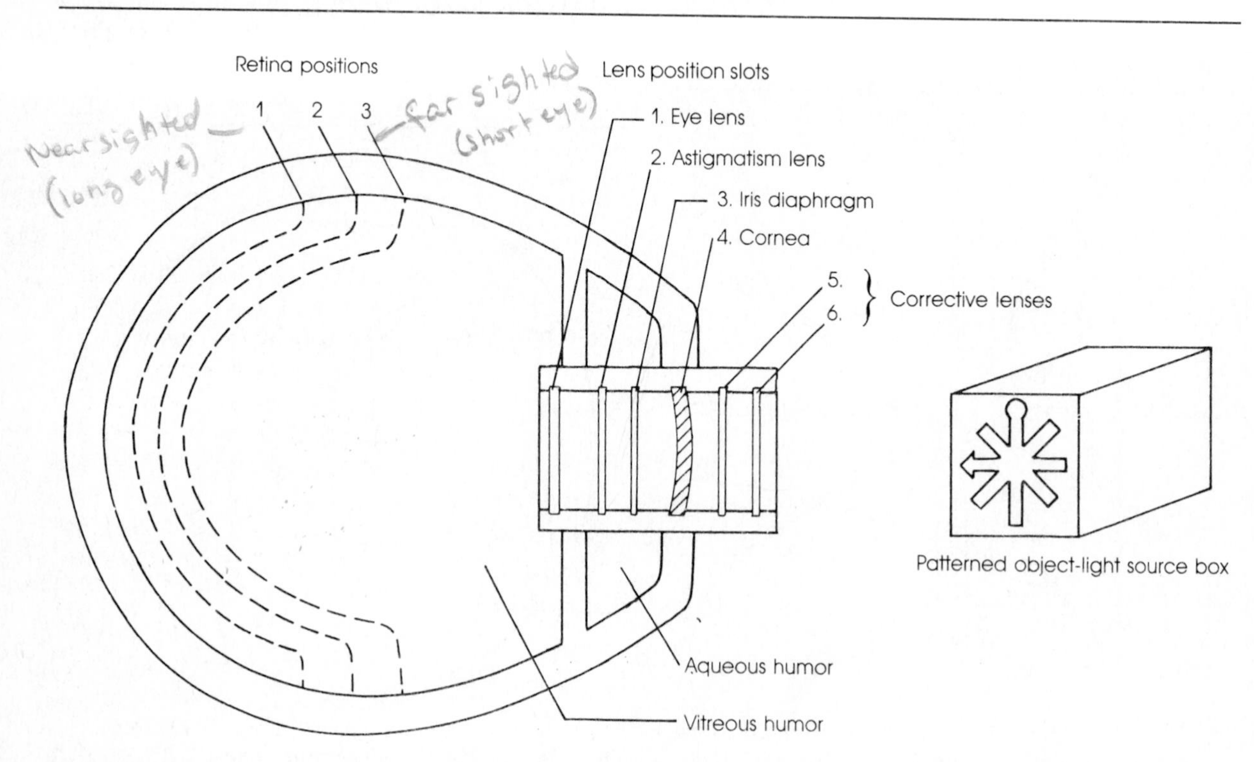

FIGURE 8.4. Top view of Ingersoil eye model.

A complete set of lenses includes the following:

1. Double convex, +7.00 diopters, spherical convergent.
2. Double convex, +20.00 diopters, spherical convergent.
3. Double convex, +2.00 diopters, spherical convergent.
4. Double concave, −1.75 diopters, spherical divergent.
5. Concave cylindrical, −5.50 diopters, divergent.
6. Convex cylindrical, +1.75 diopters, convergent.
7. Iris diaphragm disk, 13-mm diameter aperture.

The refractive power of a lens is measured in **diopters:**

$$\text{Diopters} = \frac{1}{\text{Focal length in meters}}$$

Thus, a lens that can focus light at 17 mm (0.017 m) has a power of 1/0.017 m = 59 diopters. This is about the average refractive power of the normal adult eyeball (the length of the eyeball is about 17 mm). The larger the diopter rating, the stronger the refractive power of the lens.

a. Normal Distance Vision

Place the eye model on a table facing a bright object (a window) 4 to 5 m away. Fill the tank to within 2 cm of the top with clear water. Add a few drops of eosin to make it possible to trace the path of the light rays from the lens to the retina.

Place the retina in the normal position (slot 2) and the +7.00 lens in slot 1. Describe the appearance on the retina of the image being viewed. Is it in focus? Is it upright or inverted? How is the image size related to the object size? What name do we give to normal eye vision?

b. Normal Near Vision and Accommodation

Face the eye model toward the patterned object–light source box, placed at a distance of exactly 33 cm from the cornea (lens position slot 4). Is the image in focus on the retina? Replace the +7.00 lens with the +20.00 lens in position 1. Compare the shapes of these two lenses. The image should now be in focus; if it is not, move the light source a little. This comparison illustrates how the eye lens thickens to accommodate for viewing near objects. What internal changes take place in the eye during such accommodation? Leave the +20.00 lens in position 1 for the following experiments c through h.

c. Effect of Pupil Size

Move the light source slightly until the image on the retina is a little out of focus. Insert the iris diaphragm disk (13-mm diameter hole) in position 3 behind the cornea. How does this affect the retina image? What is the function of the iris? What is spherical aberration? What is chromatic aberration?

d. Hypermetropia (Farsightedness)

Place the object light source 33 cm from the cornea as before and move the retina to position 3. Note the character of the retinal image and its size. Correct this defect by moving the object nearer to or farther from the cornea. Which works? Have you ever seen anyone use this technique for correcting farsightedness? Now return the object to the 33-cm position and try to correct the image using a corrective lens in front of the cornea. Try a +2.00 or −1.75 lens. Which corrects it—a converting or a diverging lens?

e. Myopia (Nearsightedness)

Remove the corrective lens and place the retina in position 1. Note the character and size of the image. Try to focus the image by moving the object. Must it be moved nearer to or farther from the eye? Return the object to the 33-cm position and try to correct the defect using a +2.00 to −1.75 lens in front of the cornea. Which provides correction in this case? Where is the light being focused in myopia and hypermetropia—behind or in front of the retina? How do these lenses provide correction?

f. Astigmatism

Remove the correcting lens and leave the object lamp at 33 cm from the cornea. Place the retina in the normal (2) position. Now insert the cylindrical concave −5.50 lens in position 2 behind the cornea. Note that now the image is blurred along certain lines of the object. Such astigmatism in the human eye is produced by an irregular curvature of the cornea. The −5.50 lens allows you to simulate such a refractive defect.

Now rotate the cylindrical lens and note how different lines of the object become blurred while others remain sharp. Correct for this astigmatism by placing the convex cylindrical +1.75 lens in front of the cornea and rotating it to focus the image. Compare the relative directions of the cylindrical axis in each of these lenses when the correction is made. A similar corrective lens is used in your own eyeglasses to correct your astigmatism.

g. Compound Defects

Produce a compound defect in the model eye of astigmatism plus either myopia or hypermetropia. Try to correct this compound defect by using the proper combination of lenses. Such compound defects are common in human vision, and the corrective lenses are combined in one eyeglass lens or contact lens.

h. Action of a Magnifier

Place the retina in the normal position, the object at 33 cm, and the +20.00 lens in position 1. Use the +7.00 lens as a magnifying lens in front of the cornea. Try to focus the image on the retina by moving the object lamp either closer to or farther from the cornea. How near or far must it be moved? How has the size of the image changed on the retina? How does the new image size compare with the original size? How is this new image size related to the distance of the object from the cornea?

i. Lens Removal

In cataract disease, the lens of the eye becomes opaque and vision diminishes to near blindness. Cataracts are corrected by removing the lens and substituting a corrective eyeglass lens to take the place of the eye lens. Simulate this correction by removing the +20.00 lens. How does this affect the image? Now place the +7.00 lens in front of the cornea and describe the image appearance. Bring the object closer and observe the image. These adjustments give you some idea of the corrections and relative degree of restored vision for a person who has cataract disease.

j. Examination of Eyeglass Lenses

Look through the +2.00 lens and move it from side to side. Note whether the object viewed moves in the same direction as the lens is moved or in the opposite direction. Repeat this, using the −1.75 lens. Examine your own eyeglasses, or those of your partner, and determine whether they are concave or convex and what defect they are meant to correct. Do the two eyeglasses have the same kind of correction? Rotate the eyeglasses. Is there any correction for astigmatism? Compare it with the rotation of the +1.75 convex cylindrical lens.

At the end of the experiment, remove all lenses from the model, clean them with lens paper, and replace them in the lens case.

9. Visual Fields of the Eye: Perimetry

The field of vision is the entire area seen by the eye when fixed in one position. Charting of the visual field is useful in localizing brain lesions or determining blindness in various retinal areas. The instrument used to map the field of vision is called a **perimeter.** It consists of a black metal semicircle graduated in degrees from 0 degrees at the center to 90 degrees at the edges. The subject focuses on a white dot in the center of the perimeter during the test. The operator moves a pointer along the perimeter until it just enters the subject's field of vision; then the degrees from center are read and plotted on a perimetry chart. By repeating this in different planes and for different colors, a map of visual fields for rod and cone vision is obtained.

1. Have the subject seated at a table with her chin on the chin rest. Adjust the perimeter so the white dot is level with the eye and the ends of the perimeter semicircle are aligned with the eye. The untested eye is covered or closed. The perimeter should be in the horizontal or 0-degree position.

2. While the subject stares fixedly at the white dot in the center, the operator (seated behind the perimeter) slowly pushes the carrier with its white disk along the perimeter, from the outer edge toward the center. When the dot is just visible, the subject informs the operator, who then records the position on the perimeter. Usually, three tests are run at each perimeter angle and the average is recorded on the perimetry chart. Note that the read-

ings taken from the right side of the semicircle are plotted on the left side of the chart, and vice versa. Repeat the procedure for the opposite side of the perimeter.

3. Replace the white disk with a colored disk (red, blue, or green) and repeat the perimetry measurements. Do this for all three colors. *The subject must correctly identify the color,* not just the disk movement, before the degree of angle is recorded. Randomly mix the colors so that the subject cannot memorize any sequence.
4. Repeat the procedures using the same eye but with the perimeter rotated 30, 60, 90, 120, and 150 degrees from the original position. Do this with the white disk and with the three colored disks. Plot the visual fields for these on the same chart.

How do the fields compare in size? What might cause a decrease in the size of the visual field? If one of the members of the class knows he or she has an altered visual field, it would be interesting to plot the field and compare it with a normal field.

A perimeter chart of the right eye with the fields of vision for green, red, blue, and white is shown in Figure 8.5. Note that the field for green is the most limited, with red, blue, and white having a more expanded visual field. How do your fields compare with those shown for a normal eye?

ANATOMY OF THE EYE

Perhaps more than for the other organs of the body, the student needs a thorough knowledge of the anatomy of the eye to be able to understand how it works. Although some knowledge can be obtained by studying diagrams and models, there is no adequate substitute for actual dissection of an eye. Cow or sheep eyes will be used, because they are large and available. Fresh or frozen eyes are superior to those preserved in formalin, because the lens retains its clarity and elasticity.

External Features

Notice the stump of the **optic nerve** at the posterior of the eye. Because this nerve is directed toward the midline of the body in its natural position, you can use it as a guide in orient-

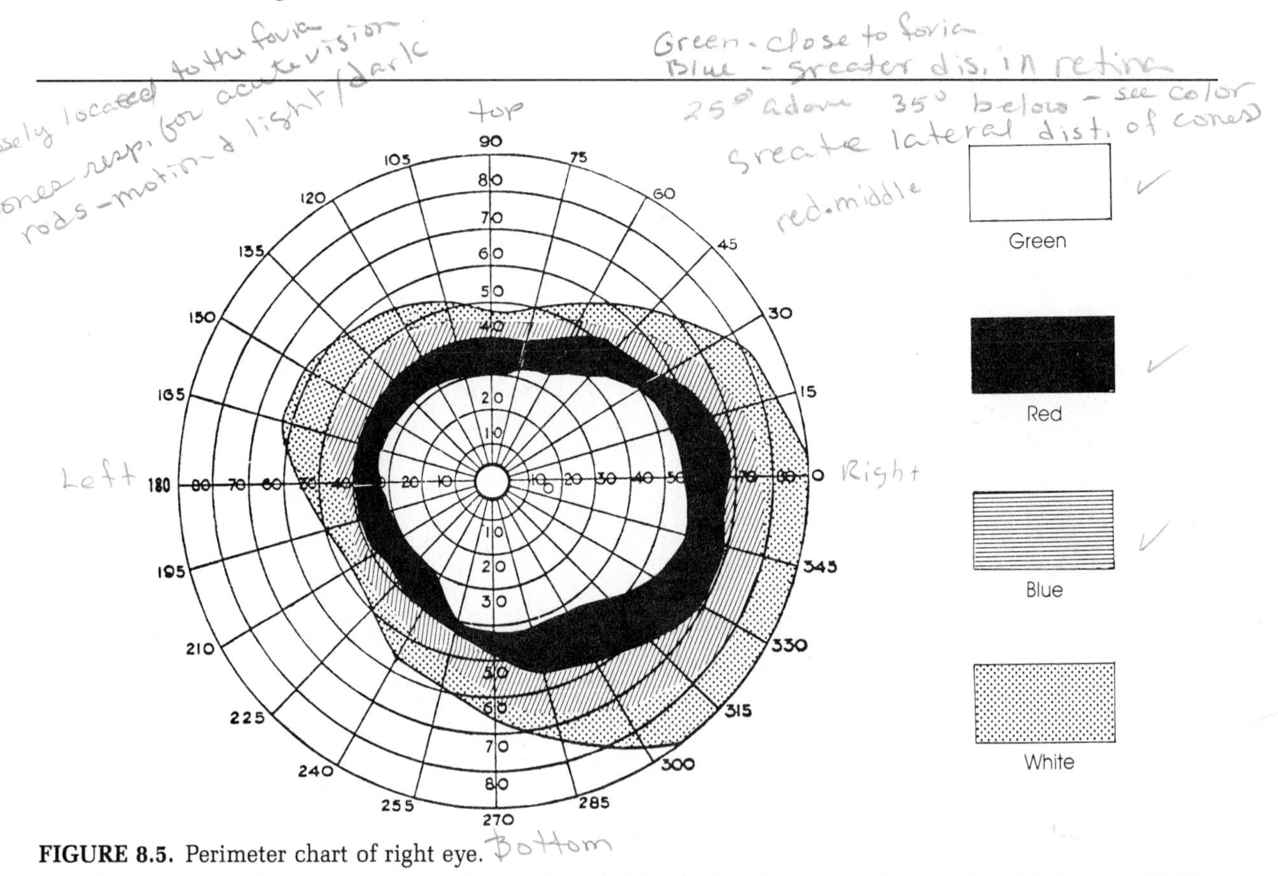

FIGURE 8.5. Perimeter chart of right eye.

ing the eye. The front surface of the eye has a clear window, the **cornea,** through which light enters. The cornea is covered by a thin transparent epithelium, the **conjunctiva,** which merges on all sides with the connective tissues that aid in holding the eye in place. These tissues are also called "conjunctiva," although in the strict sense, the term applies only to the epithelium. The tough white outer wall of the eye, composed of fibrous connective tissue, is the **sclera.** Attached at various points on the sclera are the **extrinsic** or **extraocular muscles,** which move the eye as a whole. There are three pairs of these muscles and their names describe their positions on the eye: the **superior** and **inferior recti** (singular, **rectus**) on the top and bottom, respectively; the **lateral** and **medial recti;** and the **superior** and **inferior obliques.** The last two muscles are attached to the eye near the superior and inferior recti, respectively, but the direction of their fibers is toward the medial wall of the eye socket. Look into the eye through the cornea and find the **iris,** a membranous curtain that divides the eye into anterior and posterior chambers, and the **pupil,** an opening in the iris that allows light to pass further into the eye.

Internal Features

Using scissors, carefully cut the eye into anterior and posterior halves. Refer to Figure 8.6 while examining the internal structures. The semigelatinous substance behind the lens is called the **vitreous humor.** Examine the posterior half of the eye first. Note that the wall of the eye is actually made of three main layers: the **sclera,** a vascular and darkly pigmented layer called the **choroid,** and the dark innermost layer called the **retina.** The retina actually has two layers: the **sensory layer,** which contains the modified neurons (rods and cones) that are the light receptors, and the **pigmented layer,** which always separates from the retina and remains attached to the choroid, giving it its dark color. Examine the surface of the retina closely and find where the fibers of the optic nerve and numerous blood vessels enter and fan out over its surface. This area, called the **optic disk,** contains nerve fibers and blood vessels instead of rods and cones, and thus a blind spot occurs. Lateral to the optic disk you may find a slight depression that has a yellowish color. This is the **macula lutea.** The central part of this area, called the **fovea centralis,** contains only cones and provides the best focus of the image.

Examine the anterior half of the eye from the inside. The most prominent structure to be seen is the **crystalline lens.** This structure is held in place by about 70 radially arranged **suspensory ligaments.** The suspensory ligaments are attached marginally to **ciliary bodies,** which are actually continuations of the choroid. The cil-

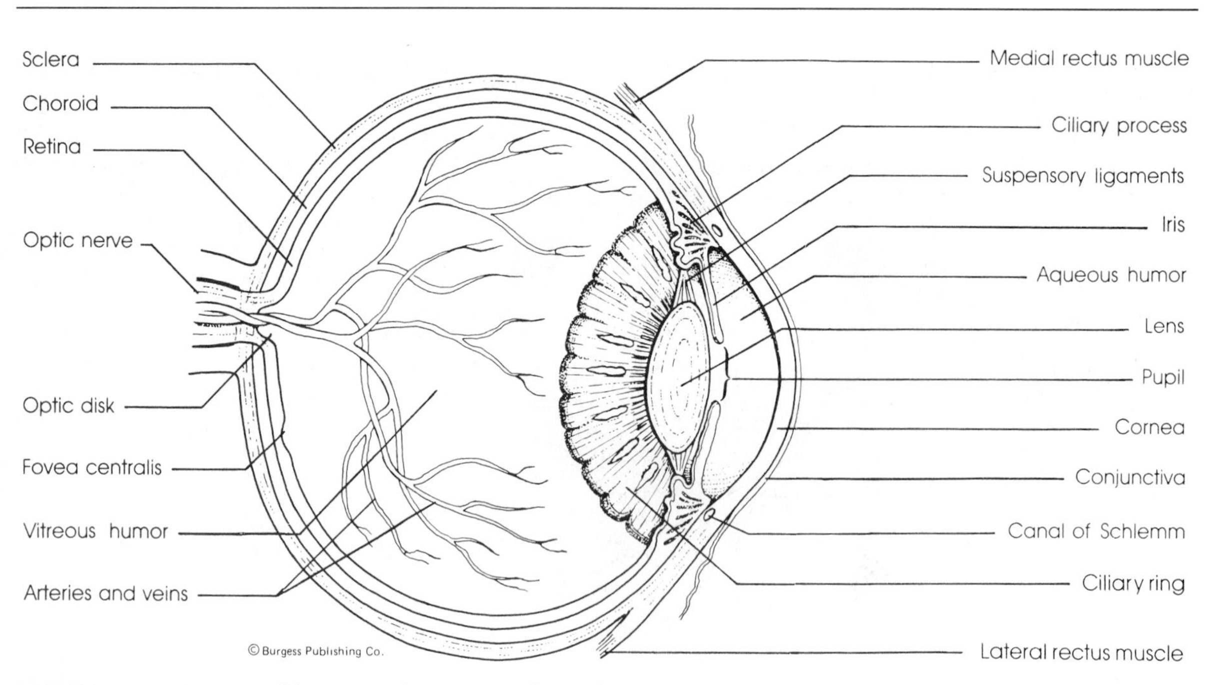

FIGURE 8.6. Anatomy of human right eye seen from above.

iary bodies contain **ciliary muscles,** which adjust the tension of the suspensory ligaments and thus change the shape of the very elastic lens. Carefully remove the lens and note the delicate **iris** just in front of it. Can you determine that there are **radial** and **circular muscle fibers** in the iris? The space between the iris and the cornea is the **anterior chamber** and is ordinarily filled with a fluidlike substance called the **aqueous humor.** Grasp the edges of the lens with forceps and test its elasticity. Cut the lens open to determine its consistency.

OPHTHALMOSCOPY

Examination of the interior (fundus) of the eye with an **ophthalmoscope** can provide important information about the anatomy of the inner eyeball. It can also reveal some general health problems, because the retina is one area of the body where blood vessels may be examined without surgery. The view of the retina seen with an ophthalmoscope is called the **fundus oculi** (Figure 8.7). The main structures viewed in the fundus are the following:

Optic disk. The spot on the retina where nerve fibers leave the eyeball and blood vessels enter and leave the eyeball. It contains no rod or cone cells and hence is a "blind spot" on the retina.

Macula lutea. The "yellow spot" lateral to the optic disk, composed only of cone cells for **photopic vision** (bright-light–color vision). The central pit of the macula, called the **fovea centralis,** contains a denser packing of cone cells, which provide greater visual acuity when light is focused on the fovea.

Peripheral retina. The retinal areas peripheral to the macula that contain fewer cones and more rod cells and provide our **scotopic vision** (dim-light–night vision) owing to the high sensitivity of the rods.

Blood vessels. Retinal arteries and veins. These can alert a clinician to potential health problems. For instance, in diabetes the peripheral vessels may be fewer and have small hemorrhage areas. In hypertension, vessels are of smaller diameter and the veins are constricted where they cross the arteries. The larger, darker vessels are the veins.

The Ophthalmoscope

Before you examine your partner's eyes you will need to become acquainted with the operation of the ophthalmoscope (Figure 8.8). Light is beamed into the eye by means of a mirror inside the top of the instrument. A slit in the mirror allows the observer to see the interior of the subject's eye. The light intensity is varied by depressing the red lock button and rotating the **rheostat control** clockwise. Sharp focus on retinal structures is obtained by rotating the **lens selection disk** until the retina comes into focus.

The diopter rating of the lens is shown in the window, black numbers for convex and red numbers for concave lenses. If the eyes of the subject and examiner are both normal (**emmetropic**), no lens is needed and an "O" appears in the window. The **aperture selection disk** on the front of the instrument changes the character of the light beam. A green spot is usually best, because it is less irritating to the eye and the blood vessels are seen more clearly.

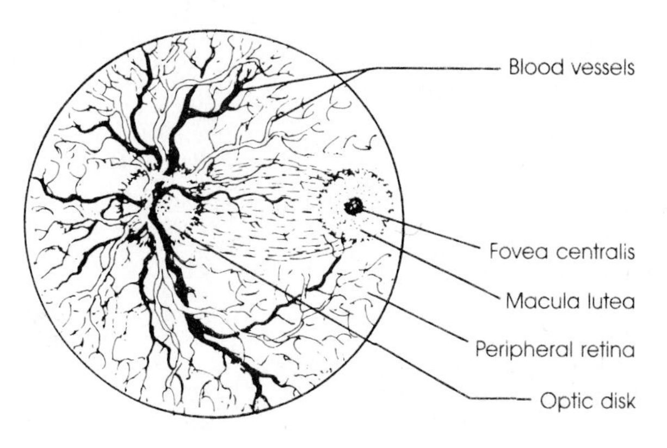

FIGURE 8.7. Fundus oculi.

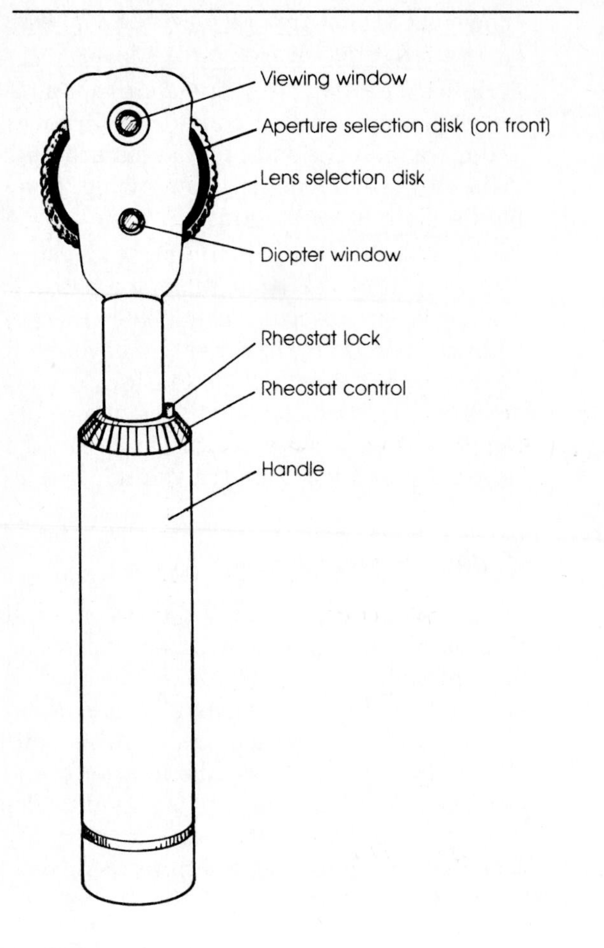

FIGURE 8.8. Ophthalmoscope.

The light beam should be directed toward the edge of the pupil, rather than through the center, to reduce the light reflection from the retina. It is important not to overexpose the retina with the strong light from the ophthalmoscope. **Limit the exposure time to 1 minute** and then give the subject several minutes for rest before resuming the examination.

Experimental Procedure

1. You and the subject sit facing each other in a darkened room. Instruct the subject to focus on a distant object so that the pupil will be maximally dilated.
2. Use your right eye to examine the subject's right eye and your left eye to examine the left eye. Hold the ophthalmoscope with the index finger on the lens selection disk and your eye as close to the viewing window as possible. Retract the subject's upper eyelid with the thumb of your other hand.
3. Start viewing at a distance of around 12 in. Focus the beam on the pupil and examine the vitreous body and lens.
4. Keeping the pupil in focus, move close to the subject's eye (within 2 in.) and direct the beam near the edge of the pupil.
5. Rotate the lens selection disk to obtain a sharp image. Examine the fundus oculi for the optic disk, macula lutea, and blood vessels. Look for any irregularities on the retinal surface. To view the peripheral areas of the retina ask the subject to look up, down, laterally, and medially. *Remember to limit your examination time to 1 minute.*
6. Examine the subject's other eye, then change roles so the subject can examine your eyes.

LABORATORY REPORT

Name ____________________

Date __________ Section __________

Score/Grade ____________________

8. Sensory Physiology II: Vision

Functions of the Eye

1. Accommodation Reflexes

Relative pupil size: Near vision smaller Distant vision larger

Of what benefit is the pupil change occurring in near vision?

By narrowing the pupil, peripheral light is prevented from stimulating the retinal receptors (rod) on periphery and light is concentrated near fovea (area where cones ar concentrated)

Describe the changes in the internal eye structures that permit accommodation for near vision.

Ciliary muscles contract causing reduction in tension on suspensory ligaments & allowing the elastic nature of the lens to cause it to become more spherical

2. Near Point of Accommodation

Near point distance: Left eye 40 cm Right eye 50 cm

Why does the near point become longer with increasing age?

the lens loses elasticity as aging occurs and it is unable to become as spherical as at a younger age.

Define the following:

Hypermetropia: Farsighted - distant objects are seen clearly, unable to focus on near

Myopia: Nearsighted - close objects seen clearly, unable to see or focus far

Presbyopia: old age vision. unable to focus on near objects as easy as when young

3. Binocular Vision and Space Perception

Hold your finger about 6 in. in front of your nose. Look at a distant object with both eyes open. The finger will double. Now cover the right eye.

Which image disappears? left

Why? During focusing on a distant object near objects are seen double due to binocular vision with the left image

What disadvantage would a one-eyed watchmaker have in his or her occupation?

depth perception is lost

4. Blind Spot

What causes the blind spot in the visual field? no receptors in optic disk, the site where optic nerve exits from retina

Where is the blind spot located in relation to the fovea? medial

5. Visual Acuity (no glasses or w/ contact lenses)

Rating for Left eye 29 Right eye 10

What does a rating of 20/20 mean? A person, with normal vision, tested is able to see (20/20) at 20 ft what the normal person can see at 20 ft.

6. Negative After-Images: Complementary Colors

After-image color for Red blue green Green purple Yellow blue

Explain the physiological mechanism that produces these after-image colors.
as you view a color (ex. Red), cones which are susceptable to red wavelengths are stimulated and pigment in them begins to depleted, the others are not affected. When stimulated by all wavelengths of light (white) the red cones have been adapted & do not respond while the unadapted blue green cones become active

7. Tests for Color Blindness

What is the trichromatic theory of color vision? Color vision is produced by 3 types of cones, each sensitive to a particular wave length. other color produced by relative frequency of impulses reaching the brain from each of the cone type,

Which wavelengths of light excite the cone cells the most?
" " 450 mm stimulate Blue cone
" " 525 mm " Green cone
" " 555 mm. " red cone

If a person lacks green cones in the retina, she or he has difficulty distinguishing between which colors? green & and other colors dependent on mixture of green
see color wheel

8. The Optics of Vision

If the lens in your eye had a refractive power of 25 diopters, what would be its focal length?

$25 = \frac{1000}{\text{Focal lgth}}$ 40 mm $1 \text{ diopter} = \frac{1 \text{ meter } (1{,}000)}{\text{Focal Length}}$

What vision defect would this cause? farsightedness

How could this defect be corrected? with a positive lens (Convex lens)

Name ______________

Date __________ Section __________

Score/Grade ______________

What are cataracts? lens lacks transparency

What is glaucoma and what is its potential danger to vision?

Excessive intraocular pressure (fluid in eyeball) may cause blood vessels in retina to be compressed + interruption of circulation to the eye, resulting in blindness

9. Visual Fields of the Eye: Perimetry

Place your perimetry chart in the following space.

	0	90	180	270
Object	80°	40°	50°	50°
Red	50°	16°	25°	45°
Green	27°	15°	20°	20°
Blue	25°	15°	25°	30°

How do your visual fields compare with the fields for a normal eye?

How does the retinal distribution of rods compare with that of cones?

Rods are located in periphery, cones more central near fovea. Green more confined to fovea, red less confined, + blue have greatest distribution of cones and appear to have greater horozontal distribution

Why is the visual field not circular?

Nose, eyebrows, + cheeks obstruct light passage to retina

Anatomy of the Eye and Ophthalmoscopy

1. Name, in order, the refractive parts of the eye through which light passes before it reaches the retina. Which structure has the greatest refractive power?

2. Where is the aqueous humor produced, and how does it leave the interior of the eye?

3. What is the function of the melanin pigment found in the choroidal and retinal layers?

4. Why is visual acuity better when light is focused on the macula lutea and fovea centralis than when focused on other areas of the retina?

5. Name the extraocular muscles and the movements they control. What is strabismus?

6. Outline the role of the autonomic nervous system in controlling pupil size and lens shape.

7. What is the principal cause of night blindness? Explain.

9 Reproductive Physiology

INFLUENCE OF HORMONES ON REPRODUCTION

The interplay between the hypothalamus, pituitary, gonadotropic hormones, sex hormones, and sex organs is highly fascinating and quite complex. In this experiment you will examine the interrelationships of these hormones and organs using rats that are castrated or ovariectomized, maintained for 2 weeks (some injected with replacement hormones), and then sacrificed to examine the effects on certain reproductive organs. You will also have the opportunity to examine the 5-day cycle of estrus, or heat, in the female rat.

It will be your responsibility to perform the surgical operations, care for the animals and make injections when needed, make the required dissections, and present your results to the rest of the class for discussion on the last day of this experiment.

You are urged to read as much as possible about reproductive physiology so that you can properly explain the results of the experiments. Although the experiments themselves are somewhat dogmatic, their presentation and explanation should open many areas for discussion and thinking. Before you begin, acquaint yourself with the reproductive system in the rat by studying Figures 9.1 and 9.2 and observing any available demonstration specimens.

General Directions for Aseptic Operations

These experiments on the endocrine system are performed on "chronic" animals, that is, animals that are kept alive for an indefinite period following the operation. Rats are quite resistant to infection, so successful operations can be performed with a minimum of aseptic techniques.

1. Instruments used in the operation are kept in a tray containing 70% alcohol. They are rinsed in boiled saline solution before being used in an open incision.
2. It is best to work in teams of two during the operation, one person being the anesthetist and the other the surgeon. The anesthetist anesthetizes the rat with ether, clips the hair from the operation site, and swabs the skin with alcohol. It is the anesthetist's job to watch the animal closely during the operation and regulate the depth of anesthesia (we want only live patients). The surgeon scrubs his or her hands thoroughly before the operation, performs the operation, and closes the incision with sutures or wound clips.
3. The animals are anesthetized by placing them in a desiccator containing a wad of cotton soaked with ether. As soon as the rat becomes unconscious, remove it from the desiccator and use an ether cone to maintain the depth of anesthesia. The ether nose cone is simply a large test tube containing a small ether-soaked piece of cotton.

Caution! *Watch the depth of respiration continuously so it does not become too shallow. Do not apply the ether cone continuously.*

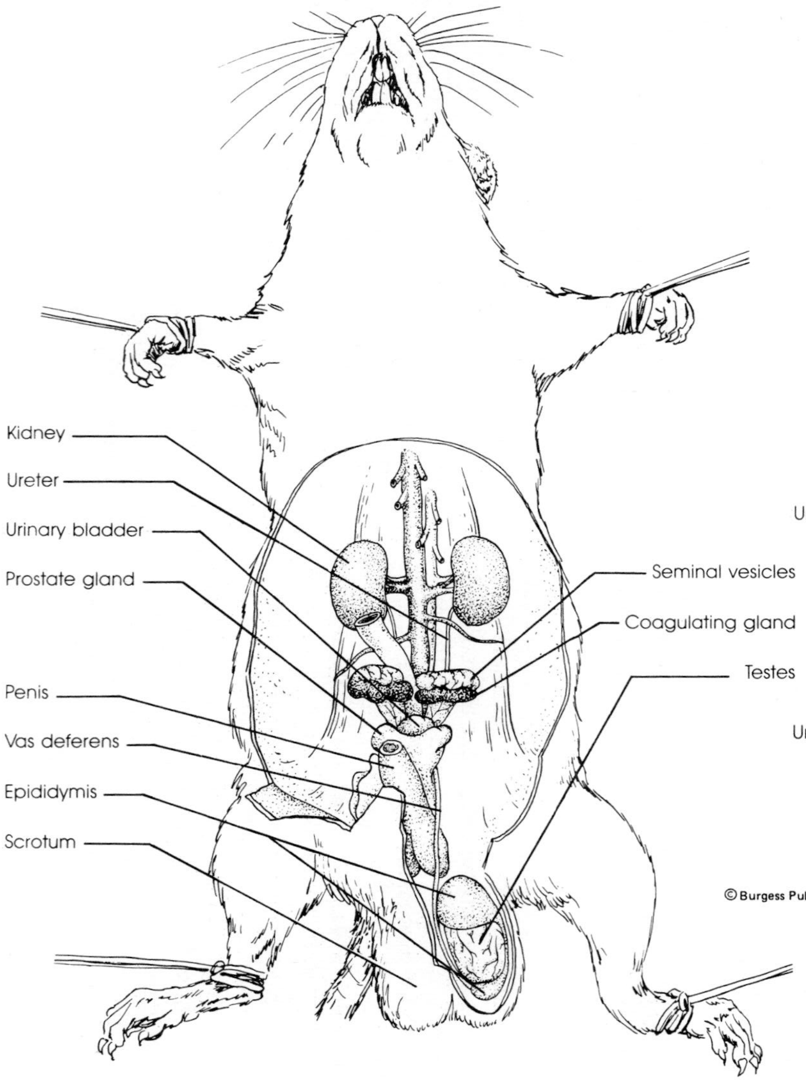

FIGURE 9.1. Male rat urogenital system.

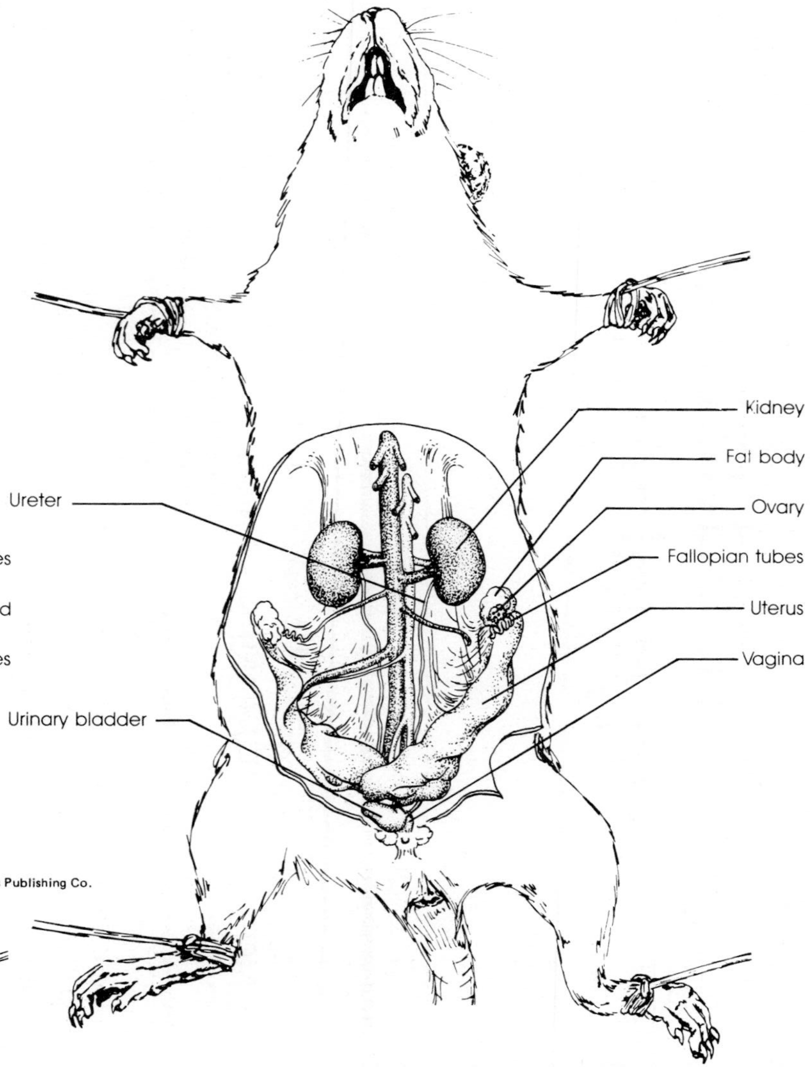

FIGURE 9.2. Female rat urogenital system.

4. After the operation, the animals are placed in a box containing clean towels until they are fully conscious. A lamp placed over the box will help them regain the body heat they lost during the operation. Later they will be returned to their clean "home" cage.
5. While still anesthetized, rats should be numbered for positive identification later. Tail markings or ear tags are the best means of identification.

TESTICULAR AND GONADOTROPIC HORMONES

Male rats weighing 75 to 100 g are divided into the following categories:

Normal rat (control).

Rat castrated 14 days prior to the last lab.

Rat castrated 14 days prior to the last lab and given a subcutaneous injection of 0.1 mg of testosterone daily for the last 10 days.

Normal rat given a subcutaneous injection of chorionic gonadotropin (20 units) daily for the last 10 days.

Rat castrated 14 days prior to the last lab and given a subcutaneous injection of chorionic gonadotropin (20 units) daily for the last 10 days.

Initial and final body weights for all rats will be recorded in the table in the Laboratory Report. On the final day the rats will be sacrificed using an overdose of ether in a desiccator. The seminal vesicles will be carefully dissected out and weighed. Their weight will be recorded in milligrams per 100 g of body weight. Students will then present their findings to the class and discuss the results.

Directions for Removal of Testes (Castration)

1. Anesthetize the rat to a depth where the animal is unaware of the operation and does not move when the skin is cut. Slow, shallow breathing or gasping indicates that excessive ether is being given. Remove the ether cone.
2. Clip the hair along the ventral midline of the scrotum with scissors and swab the area with alcohol. Using fine-pointed scissors, make a midline incision about 1 ½ cm long through the scrotal skin.
3. Sometimes the testes retract into the abdominal cavity and therefore are not visible in the scrotal sac. Slight pressure on the lower abdominal area will force the testes back into the scrotum. You will note that each testis is surrounded by a translucent membrane called the tunica (Figure 9.3a). Grasp the tu-

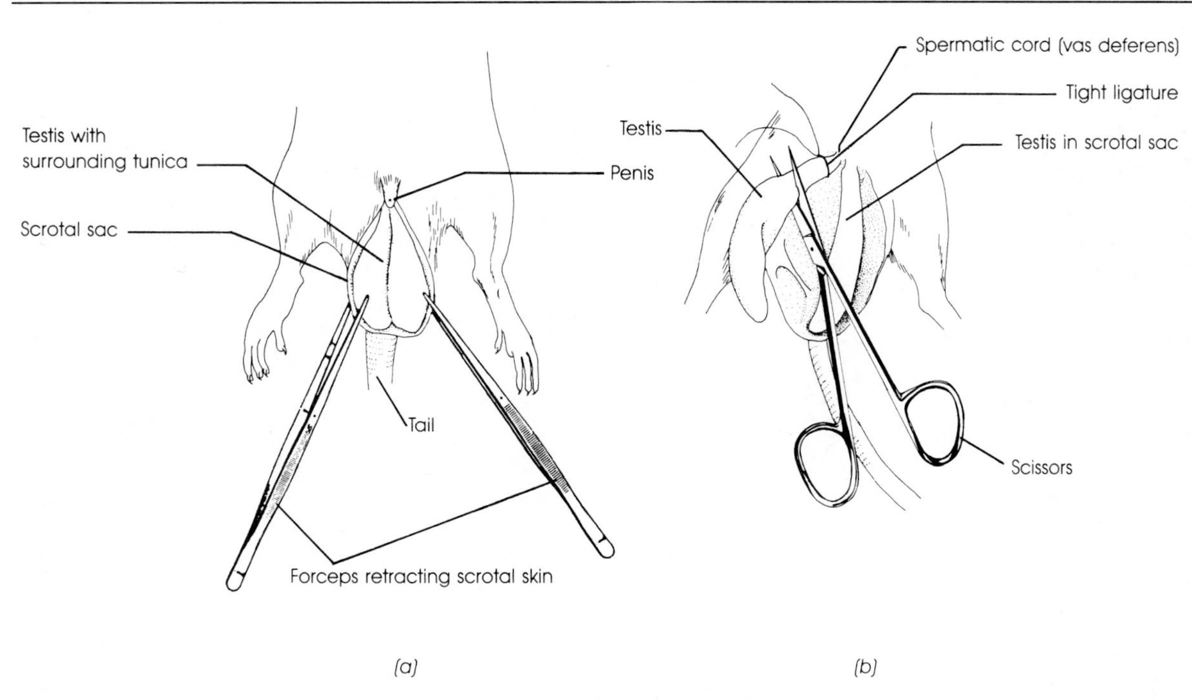

FIGURE 9.3. Castrating a rat. (a) Exposing the testes. (b) Removing the testes.

nica with forceps and slit it with the scissors to free the testis.

4. Tie a heavy thread tightly around the vas deferens (spermatic cord) and surrounding fat tissue; then cut the vas deferens between the knot and the testis and remove the testis (Figure 9.3b). Remove the other testis in the same manner.

5. Close the skin incision using two or three wound clips. Place the rat in the "recovery room" box until it regains consciousness.

OVARIAN HORMONES AND ESTRUS CYCLE

Female rats weighing 75 to 100 g are divided into the following categories:

Normal (control).

Rats ovariectomized 14 days prior to the last lab.

Rats ovariectomized 14 days prior to the last lab and given a subcutaneous injection of 0.1 mg of estrogens daily for the last 10 days.

Initial and final body weights will be recorded for all rats in the table in the Laboratory Report. On the final day the rats will be sacrificed using an overdose of ether in a desiccator. The entire uterus will be dissected out and weighed. The uterus weight will be recorded in milligrams per 100 g of body weight.

Starting 3 days after surgery, vaginal smears will be taken daily on each rat to ascertain the condition of the estrus cycle in each animal. In the last lab, students will present their findings to the class and discuss the results.

Directions for Removal of Ovaries (Ovariectomy)

1. Anesthetize the rat and make a ventral midline incision into the abdominal cavity. If this incision is made through the linea alba (connective tissue in abdominal midline), a bloodless field will be obtained. Be careful not to cut the diaphragm.

2. Push aside the intestines and locate the two horns of the bicornate uterus. Follow the uterus forward until the ovary comes into view.

3. Place a tight ligature around the uterus just below the ovary (Figure 9.4); then cut through the uterus on the ovarian side of the tie and remove the ovary. Repeat this procedure for the other ovary.

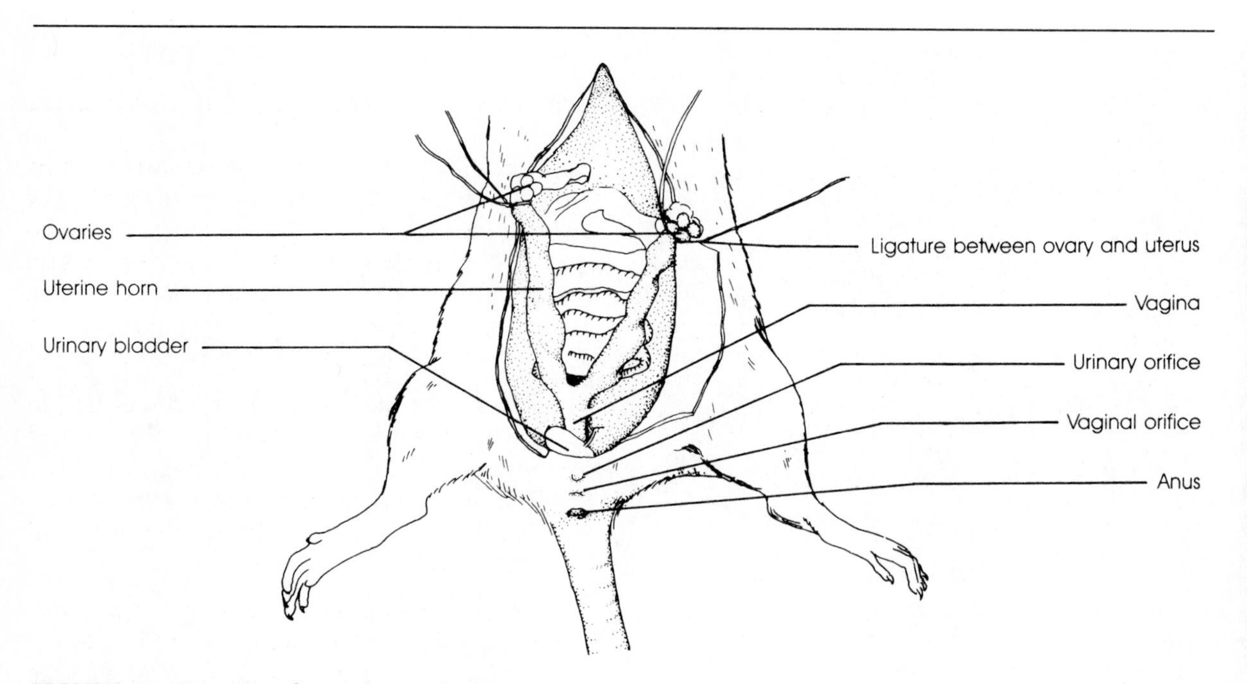

FIGURE 9.4. Preparing the rat for ovariectomy.

4. Close the muscle incision with silk-thread sutures and the skin incision with wound clips.
5. Allow 3 days for recovery from the operation, following which vaginal smears may be taken daily for detection of the estrus cycle.

Determination of the Estrus Cycle Using Vaginal Smears

The term **estrus** means "frenzy." This is the time in the rat's reproductive cycle when she is receptive to sexual copulation and ovulation occurs. We sometimes say the animal is in "heat" in this period. The estrus cycle in the rat is about 5 days in length and repeats itself throughout the year (polyestrus). The cycle is typically divided into four stages that can be identified by examining changes in the types of cells lining the vagina.

Proestrus. This is a period of increasing levels of **follicle-stimulating hormone** (FSH) and **luteinizing hormone** (LH), which stimulate follicle growth and secretion of estrogens. Vaginal smears contain mainly nucleated epithelial cells. Proestrus lasts 8 to 12 hours.

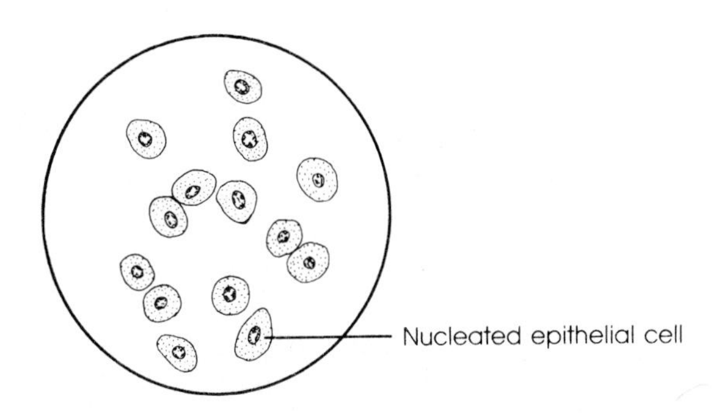

Estrus. This is the period of "heat" and copulation. High levels of estrogens stimulate mitosis of cells in the uterus and vagina. Vaginal smears contain many cornified cells and few leukocytes or nucleated epithelial cells. Estrus lasts 9 to 15 hours.

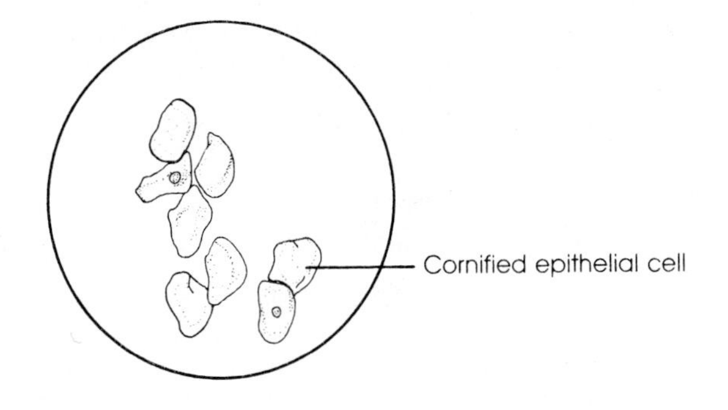

Metestrus. LH and **luteotropic hormone** (LTH) promote the formation of the corpus luteum in this period, which lasts for 10 to 14 hours. The secretion of both progesterone and estrogens increases. Vaginal smears contain many leukocytes and some cornified cells.

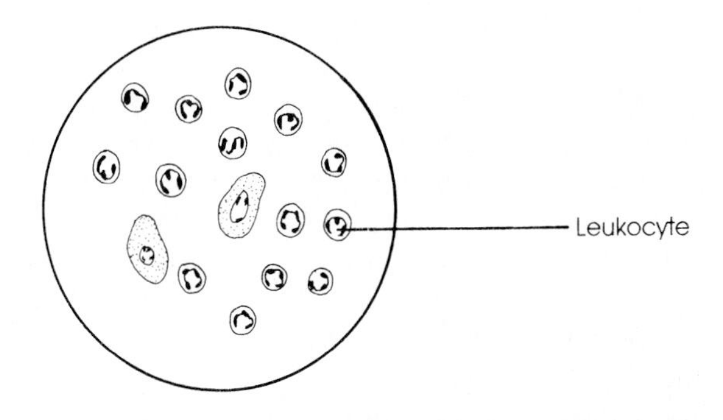

Diestrus. This is the longest stage, lasting 60 to 70 hours. The corpus luteum regresses and the uterus is small and poorly vascularized. Levels of gonadotropic and sexual hormones are at low levels. Vaginal smears contain mainly leukocytes.

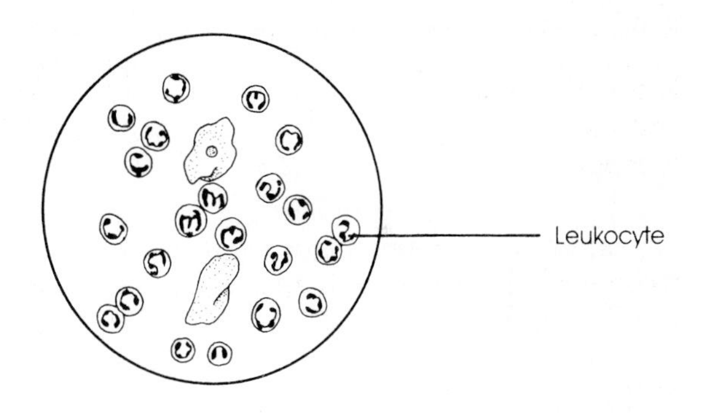

Experimental Procedure

1. Grasp the rat behind its jaw, holding it on its back in the palm of your hand. Wrap a tuft of cotton tightly around a toothpick, moisten the swab with saline, and *gently* insert and rotate the swab within the vagina. Then press the swab in a drop of saline on a microscope slide and smear the cells evenly over the slide.
2. To stain the cells, place the air-dried slide in methyl alcohol for 5 seconds, remove it, and air dry it. Then place the slide in Giemsa stain for 15 minutes, remove it, rinse it with distilled water, and air dry it before examining it under the microscope.
3. Determine the stage of the estrus cycle each rat is in for each of the 10 days prior to the last lab. Enter your results in the table in the Laboratory Report.

PREGNANCY TESTS

Tests for human pregnancy are good demonstrations of some endocrinologic principles and are relatively easy to do. Several types of pregnancy tests are used, all of which are based on the detection of **chorionic gonadotropin,** which is produced in the female during early pregnancy.

After the fertilized ovum is implanted in the uterine wall, trophoblastic cells around the ovum begin producing chorionic gonadotropic hormones. (Trophoblastic cells differentiate into a part of the placenta called the chorion, hence the name *chorionic* gonadotropin). These hormones closely resemble the anterior pituitary gonadotropic hormones in activity, having both luteinizing and luteotrophic properties. Hence, they promote the growth and maintenance of the corpus luteum to ensure adequate levels of estrogen and progesterone in early pregnancy. The production of **human chorionic gonadotropin** (hCG) increases sharply as the placenta develops and reaches a peak level approximately 8 weeks after the first day of the last menstrual period. In early pregnancy, then, the hCG level becomes so elevated that it spills over into the urine, and the detection of hCG in the urine becomes the basis for the pregnancy tests.

Two classes of pregnancy tests are used: **biological** and **immunological.** The biological tests depend on the effect of the chorionic gonadotropin from the woman being tested on the reproductive organs of test animals. The Ascheim-Zondek test is based on the fact that hCG stimulates immature ovaries to cause ovulation and to secrete hormones that stimulate uterine growth.

Gonadotropin can also be detected by immunologic reactions. Chorionic gonadotropin is a protein and behaves as an antigen when injected into an animal. The animal becomes immune to the chorionic gonadotropin by producing antichorionic gonadotropin **antibodies.** The blood serum of the injected animal will thus contain these antibodies and is therefore called the **antiserum.**

Experimental Procedure

1. Preparation of Pregnancy and Nonpregnancy Urines

Urine from pregnant and nonpregnant women can be obtained from hospitals. The urine used should be the first voided by the donors after they arise in the morning. Because human urine contains estrogenic substances toxic to the animals used for the biological assays, urine from both pregnant and nonpregnant women should be extracted with ether to remove these substances. Filter the urine through filter paper, and then place the urine and about one fourth as much ether in a separatory funnel. Stopper the funnel and shake it thoroughly for at least 5 minutes; then allow the urine and ether to layer out. Open the stopcock of the separatory funnel and collect the extracted urine. Discard the ether. The urine can be kept in a stoppered bottle in a refrigerator for several days without loss of hCG potency.

Synthetic Pregnant Urine. The pregnant urine obtained from hospital laboratories varies greatly in its potency from one woman to another. This variation may adversely affect the pregnancy test run in the physiology lab. Consistent, positive tests can be obtained by preparing a synthetic pregnant urine made of 0.9% NaCl containing 100 IU/ml of chorionic gonadotropin.

2. Immunologic Test for Pregnancy (UCG-Slide Test; Wampole)

This test is based on the principle of latex agglutination inhibition. The hCG antigen is bound to latex particles, which become visibly agglutinated when mixed with rabbit antiserum containing hCG antibody. When urine containing hCG is first mixed with the antiserum reagent, the antibodies will be bound to the urine hCG. Then, when the latex-bound hCG (antigen reagent) is added, agglutination of the latex particles will not occur.

a. Make sure the reagents and urine are at room temperature. (They are stored under refrigeration).

b. Carefully clean a glass slide by washing it with a detergent and rinsing it several times with tap water and then with distilled water. No traces of detergent or reagents from a previous test should remain on the slide.

c. Using the disposable pipet provided, place 1 drop of urine on the clean slide. Add 1 drop of antiserum reagent to the drop of urine and mix with a toothpick for 20 seconds.

d. Add 1 drop of well-shaken antigen reagent to the urine and antiserum mixture and mix with a different toothpick.

e. Gently rock the mixture for 2 minutes and immediately observe for agglutination, using a light source directly above the slide.

f. If no hCG is present in the urine, agglutination will be visible. No agglutination means there is sufficient hCG in the urine to indicate that pregnancy has occurred.

3. Ascheim-Zondek Pregnancy Test

The Ascheim-Zondek test is a biological test for pregnancy that was widely used before the development of the immunologic test. It is based on the effect of chorionic gonadotropin on the ovaries and uterus of mice, rats, and rabbits.

Select immature female rats or mice, 15 to 25 days old (weighing 30–45 g). Half of the animals will be given a subcutaneous injection of **pregnant urine (PU)** twice daily (morning and evening) for 3 days prior to lab. The other animals will receive injections of **normal (nonpregnant) urine** on the same schedule. For rats, inject 1 ml of urine each time; for mice, inject 0.5 ml of urine.

Four to five days after the first injections, sacrifice and examine the rats.

First, examine the vaginal orifice (opening) for any differences in size or patency between the experimental groups. Explain any differences.

Next, open the abdominal cavity and examine the uterus and ovaries of the animals. Compare the uteri of the PU-injected and control animals for differences in size and vascularity. Examine the ovaries carefully, using a hand lens or a dissecting microscope. The presence of hemorrhage spots (corpora hemorrhagica), or yellow bodies (corpora lutea), indicates that eggs have been released—their presence is considered a positive test for pregnancy.

LABORATORY REPORT

Name ______________________

Date ____________ Section ____________

Score/Grade ______________________

9. Reproductive Physiology

Testicular and Gonadotropic Hormones

EXPERIMENTAL ANIMAL	BODY WEIGHT (G)		SEMINAL VESICLE WEIGHT (MG)	SEMINAL VESICLE WEIGHT (MG/100 G BODY WT)
	INITIAL	FINAL		
Normal-control				
Castrated				
Castrated plus testosterone				
Normal plus gonadotropin				
Castrated plus gonadotropin				

Ovarian Hormones and Estrus Cycle

EXPERIMENTAL ANIMAL	BODY WEIGHT (G)		UTERUS WEIGHT (MG)	UTERUS WEIGHT (MG/100 G BODY WT)
	INITIAL	FINAL		
Normal-control				
Ovariectomized				
Ovariectomized plus estrogens				

1. Record the stage of estrus for each rat for each of the last 10 days. Let P = proestrus, E = estrus, M = metestrus, and D = diestrus.

EXPERIMENTAL ANIMAL	DAY									
	1	2	3	4	5	6	7	8	9	10
Normal-control										
Ovariectomized										
Ovariectomized plus estrogens										

Pregnancy Tests

1. Make a labeled, colored sketch of the uterus and ovaries of the rats injected with pregnant urine and nonpregnant urine. Indicate the major differences between these organs in the two animals.

2. Explain the role of chorionic gonadotropin in gestation. Why does its presence in the urine indicate that a woman is pregnant?

3. Outline the sequence of events responsible for the effect of gonadotropic hormones on the seminal vesicles.

4. Why are immature rats used in the Ascheim-Zondek pregnancy test?

5. Explain in detail exactly why the Ascheim-Zondek test is a valid test for pregnancy.

6. Why does the presence of cornified cells in the vaginal smear indicate that the rat is in estrus?

10 Digestion

Living organisms run on energy, and it is the job of the digestive system to reduce the foods we eat to small molecules that can be used by the cells to release adenosine triphosphate (ATP). This degradation process is catalyzed by hydrolytic enzymes, which split large molecules into smaller units by combining with water. The end result of digestion is the reduction of carbohydrates to monosaccharides, proteins to amino acids, and fats to fatty acids and glycerol. Hydrolytic reactions are made more efficient by the division of the digestive tract into compartments where specific enzymes can operate at their optimum pH. Release of these enzymes at the proper time is controlled by neural reflexes and endocrine hormones such as gastrin, secretin, cholecystokinin, and gastric inhibitory peptide. In this exercise, we will examine the action of some of the key digestive enzymes and the factors that alter their activity. As you work, you should use your text and lecture notes to become better acquainted with the enzymes and hormones operating along the digestive tract.

SALIVARY DIGESTION OF CARBOHYDRATES

Digestion of carbohydrates begins in the mouth where the salivary glands (parotid, sublingual, submandibular) secrete an **amylase** called **ptyalin** that begins the hydrolysis of complex polysaccharides:

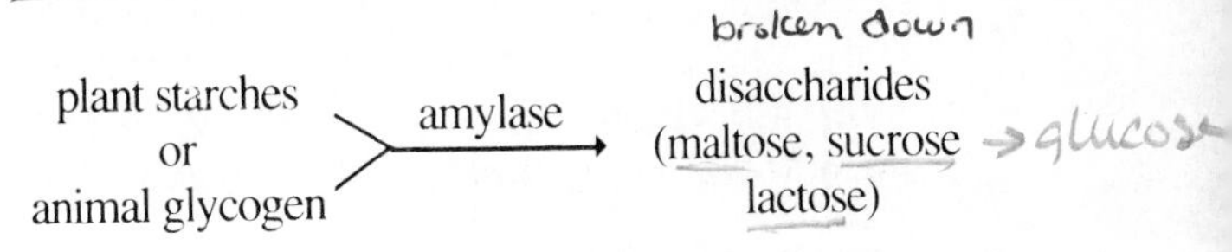

Ptyalin has an optimum pH of around 6.8, which is roughly the pH found in the mouth. In the following experiment we will examine ptyalin digestion of starch, using **Benedict's** test to measure maltose formation and **Lugol's** (iodine) solution to test for starch.

Experimental Procedure

1. Collect 10 ml of your own saliva in a graduated cylinder. Chewing paraffin will increase salivary secretion without altering the chemical composition of the saliva. Dilute the saliva with an equal amount of water if you are unable to collect 10 ml.

2. Test the saliva with pH paper. Is it acid or alkaline?

3. Place a small amount of saliva in a spot plate and add a few drops of 1% acetic acid. A precipitate indicates that mucin (a glycoprotein) is present.

4. Prepare and label four test tubes as follows (use a 0.5% starch paste):

Tube 1	Tube 2
3 ml starch	3 ml starch
+	+
3 ml water	3 ml saliva
↓ in	↓ in
37 °C water bath	37 °C water bath

Tube 3	Tube 4
3 ml starch (cooled)	3 ml starch
+	+
3 ml saliva (cooled)	3 ml saliva
↓ in	+
ice bath	5 drops conc. HCl
	↓ in
	37 °C water bath

5. After the tubes have incubated for 1 hour, pour half of each tube's contents into a new test tube. Test one set of tubes for starch using Lugol's solution and the other set for maltose using Benedict's solution.
   a. Starch Test: Add 3 drops of Lugol's solution to each tube. A dark purple color indicates the presence of starch. Shades of reddish brown indicate lesser amounts of starch. Rate the amount of starch (+ + +), (++), (+), or (−).
   b. Maltose Test: Add 4 ml of Benedict's solution to each tube and place in a boiling water bath for 2 minutes. Remove the tubes using a clamp and compare the concentration of maltose using the following scale: (+ + +) red, (++) orange-yellow, (+) green, (−) blue.

GASTRIC DIGESTION OF PROTEIN

Protein digestion begins in the stomach where the enzyme pepsin splits proteins to shorter polypeptide chains containing amino acids. Secretion and activation of pepsin occurs as follows:

Chief cells in gastric pits → Pepsinogen —HCl→ Pepsin
+

The strong hydrochloric acid secreted by the parietal cells has two functions in gastric digestion: It activates pepsin and it produces a stomach pH of around 2, which is optimal for pepsin activity. The following experiment will illustrate some of the features of protein digestion by pepsin.

Experimental Procedure

1. Place thin slices of cooked egg white in four test tubes. It is important to make these slices the same size (about 0.5 cm²) and as thin as possible.
2. Add the following solutions to the tubes and determine the pH of each tube:

Tube 1	Tube 2
5 ml pepsin (5% soln)	5 ml pepsin (5% soln)
+	+
5 ml HCl (0.5%)	5 ml water

Tube 3	Tube 4
5 ml HCl (0.5%)	5 ml pepsin (5% soln)
+	+
5 ml water	5 ml NaOH (0.5%)

3. Allow the tubes to incubate in a 37 °C water bath for 1 hour. Test the final pH of the solutions and estimate the amount of protein digestion using a scale of (+ + +), (++), (+), and (−) to compare the four tubes.

DIGESTION OF FAT WITH PANCREATIC LIPASE AND BILE SALTS

Pancreatic lipase has a major role in fat digestion, but by itself lipase is ineffective, because it is a water-soluble enzyme trying to act on large lipid droplets, which are water insoluble. Bile salts help overcome this problem by acting as emulsifying agents, which break the fat into smaller droplets so that lipase has a larger surface area for the hydrolysis of fats.

The pancreas also aids digestion by secreting sodium bicarbonate. This compound provides a pH of around 7.8 in the small intestine, which is optimal for the action of the pancreatic enzymes. In the following exercise, we will examine some aspects of the action of pancreatic lipase and bile salts on lipids.

Experimental Procedure

1. In each of two test tubes (A and B) place 3 ml of distilled water and 3 ml of vegetable oil. To tube B add a small pinch of bile salts. Shake each tube for 30 seconds and observe it for several minutes.
2. Add litmus powder to dairy cream until a blue color is produced. Preincubate the litmus cream and a 1% pancreatin solution at 37 °C for 5 minutes. Prepare a series of test tubes as follows:

Tube 1	**Tube 2**
3 ml cream + 3 ml pancreatin	3 ml cream + 3 ml water

Tube 3	**Tube 4**
3 ml cream + 3 ml pancreatin + pinch bile salts	3 ml cream + 3 ml water + pinch bile salts

3. Incubate all tubes in a 37 °C water bath for 1 hour, or until a color change occurs in one tube. Blue litmus will turn pink in an acid environment. Test the pH using pH paper, and note the odor of each tube.

LABORATORY REPORT

Name ______________________

Date __________ Section __________

Score/Grade ______________________

10. Digestion

Salivary Digestion of Carbohydrates

1. pH of saliva 7 Mucin present? yes
2. What is the function of mucin in the mouth? causes food to be lubricated & slimy so you can swallow
3. Indicate the relative amounts of starch and maltose after incubation:

TUBE	STARCH	MALTOSE	EXPLANATION
1 Water	+++	(−)	no enzyme to convert it into maltose
2 Saliva	+	+	enzyme converted starch to sugar because cond. were ideal
3 Cooled saliva	++	+	Some conversion, but digestive process inhibited by cold
4 Saliva, HCl	++	−	enzyme inhibited, too much

4. What *in vivo* (in the body) situation is simulated by the conditions in tube 4?
entrance of food and saliva into stomach
5. Does ptyalin hydrolysis of carbohydrate continue in the stomach?
Explain. yes, but only within the bolus itself
6. Where else is amylase secreted in the digestive system?
in the pancreatic juice and villus cells of the small intestine

Gastric Digestion of Protein

1. Record the initial and final pH of the solutions and the estimated amount of egg white digestion in each tube.

TUBE	INITIAL pH	FINAL pH	ESTIMATED DIGESTION	EXPLANATION
1 Pepsin, HCl	3.0	3.0	+++	enzyme working at optimal pH
2 Pepsin, water	6.5	6.5	+	enzyme activity - weak Mom favorable pH
3 HCl, water	3.0	3.0	−	no enzyme & no activity but HCl
4 Pepsin NaOH	8.0	8.0	−	Enzyme not active @ ↓

2. What *in vivo* situation is simulated by tube 4?
entrance of chyme into duodenum
3. Which other enzymes have major proteolytic activities in the digestive tract?
trypsin, chymotrypsin, aminopeptidase, carboxy peptidase
4. A person with achlorhydria has defective secretion by the parietal cells. What is the physiologic effect of achlorhydria in the body?
tube 2 demonstrates lack of Hcl leads to poor activity of pepsin
5. What is the function of the mucous cells in the gastric pits?
protection of gastric mucosa against digestion

Digestion of Fat with Pancreatic Lipase and Bile Salts

1. Which tube (A or B) has the smaller and more dispersed fat droplets? B
2. What are bile salts? excretory products (steroids) emulsify fats
What are bile pigments? Breakdown products of Red blood cells (Bilirubin, Biliverdin)
3. Where is bile secreted? liver & stored in gall bladder
4. Describe the mechanism of bile salts in the emulsification process (a diagram would help).
Breakdown of large fats droplets into many smaller due to hydrophobic portion of bile salt which enters into lipid but other portion is soluable in water, thus allowing droplets to be exposed to water which contains water soluable enzymes

5. Record the final color, pH, and odor of each tube involved in the digestion of cream.

TUBE	COLOR	pH	ODOR	EXPLANATION
1 Pancreatin	pink	5 acid	sour	Some fat digestion releasing fatty acids
2 Water	blue	6 basic	no smell	No lipase or bile salts no digestion
3 Pancreatin, bile salts	Pink	5 acid	sour	Large amount of digestion lipase & bile salts present
4 Water, bile salts	blue	6 basic	none	no lipase present no digestion

6. What produces the acid pH, indicating that fat digestion has occurred?
during fat digestion, fatty acids are released
7. What produces the rancid odor with fat digestion? Butyric acid in cream
8. Which enzymes are present in the pancreatin solution?
lipase, amylose, trypsinogen, chymotrypsinogen, procarboxy peptidase

Name

LABORATORY REPORT

Date Section

Score/Grade

9. Which enzymes are present in the microvilli brush border of the small intestine?

lipase, maltase, sacrase, lactase, peptidases

10. Briefly list the site of origin, stimulus for release, and function of the following gastrointestinal hormones.

HORMONE	SITE OF ORIGIN	RELEASE STIMULUS	FUNCTION
Gastrin	Stomach antrum cells	digestion products in stomach	stimulates pepsin + Hcl secretion
Secretin	mucosa cells of duodenum	acid in duodenum	Stimulates pancreatic duct cells to produce bicarbonate
Cholecystokinin	mucosa cell of duodenum	food products in duodenum	stimulate acinar cells to secrete enzymes + gall bladder
Gastric inhibitory peptide	mucosa of duodenum	food product in duodenum	inhibits gastric secretions + motility

11. Why aren't the acinar cells of the pancreas digested by the proteolytic enzymes they secrete?

the enzymes are secreted in ~~intestine from~~ inactive form + are activated w/i intestinal lumen

11 Smooth Muscle Motility

Locke's Solution - similar to intestine

RESPONSES OF INTESTINAL AND UTERINE SEGMENTS

Smooth muscle is a nonstriated involuntary type of muscle found in the stomach, intestines, bladder, and uterus and in the walls of most arteries and veins. Its activity is controlled by the autonomic nervous system, but all smooth muscle does not respond in the same manner to the same stimulus. For example, sympathetic stimulation produces a decrease in the rhythmic contractility of the gut smooth muscle, but causes the circular smooth muscle of the arteries to contract and constrict the vessels. Parasympathetic stimulation evokes an increase in intestinal contractility, but has little effect on arterial smooth muscle. Another unique property of smooth muscle is its ability to maintain approximately the same amount of tone and rhythmic contractility at different degrees of stretch. Because of this property, the intestines, uterus, or bladder can undergo "receptive relaxation." For example, the uterus can be stretched greatly to receive the growing fetus without its muscular tension on the fetus or its contractility being increased. In this experiment, you will use an isolated segment of the uterus or intestine to demonstrate some of these special properties of smooth muscle.

Experimental Procedure

1. Prepare a muscle warmer setup complete with aeration as shown in Figure 11.1. During the experiment, the muscle warmer will be placed in a 37 °C water bath to keep the smooth muscle at its normal physiological temperature. Reserve flasks containing Locke's solution should also be placed in the water so they will be at the proper temperature when needed.
2. The rats used in this experiment should be given no food for 24 hours prior to the lab period to remove most of the fecal matter from the intestinal lumen.

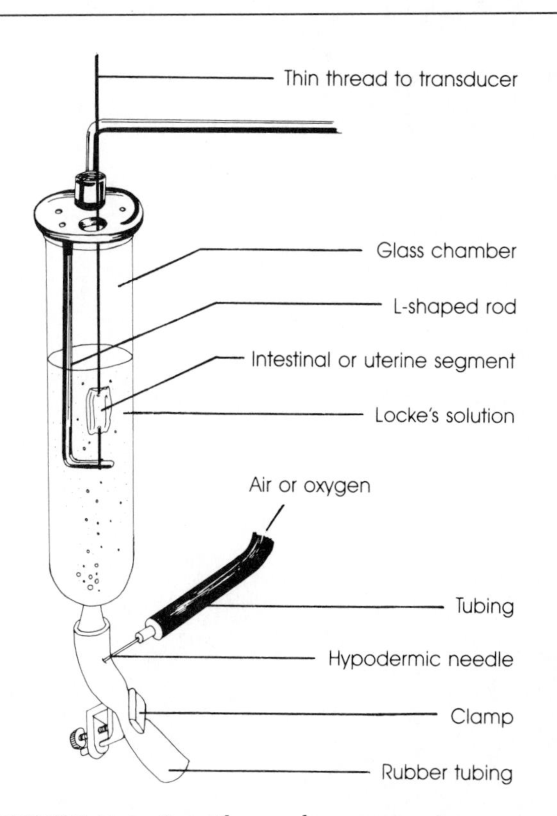

FIGURE 11.1. Smooth muscle warmer setup.

3. Kill a female rat using an overdose of gas (150–200 g) in a desiccator. (Many anesthetic agents such as ether or Nembutal depress intestinal motility, whereas gas has little effect.)

4. Open the abdominal cavity and isolate a 10-cm segment of the jejunum. The jejunum begins about 8 cm below the gastroduodenal junction (pyloric sphincter). Isolation and removal of the segment must be done gently, avoiding excessive stretching or drying of the tissue. Trim off the mesentery and place the segment in a dish containing aerated Locke's solution at 37 °C. Flush out the gut contents with Locke's solution using gentle pressure from a syringe into the lumen of the jejunum segment.

5. Also at this time, remove the entire uterus (both horns) and place it in an aerated Locke's solution (37 °C) for use later in the lab.

6. Cut the jejunum segment into smaller segments about 2 cm long. Tie one end of a segment to the L-shaped rod of the muscle warmer and the other end to the transducer (use a thin thread).

7. Fill the muscle warmer with Locke's solution so that the segment is completely immersed. Turn on the air or oxygen so that it gently bubbles through the solution. Adjust the transducer sensitivity so that normal contractions about 2 cm high are recorded.

8. Record the contractions of the jejunum segment under each of the experimental conditions listed in the table in the Laboratory Report. You will need to drain the muscle warmer and refill with fresh Locke's solution periodically as indicated.

9. After you are finished with the jejunum segment, mount one horn of the uterus in the muscle warmer and record its contractility under the conditions given in the table in the Laboratory Report.

Save your myographs for insertion in the Lab Report.

normal

Epinephrine

LABORATORY REPORT

Name ______________________

Date ____________ Section ____________

Score/Grade ______________________

11. Smooth Muscle Motility

Observations on Contractility of Smooth Muscle

EXPERIMENTAL CONDITIONS	FREQUENCY	CONTRACTILE STRENGTH	TONE	EXPLANATION
JEJUNUM SEGMENT				
Normal activity				
Increased stretch (raise transducer or add counterweights)	↑	↑	↑	Contractility increases w/stretch similar for fillings w/ food
Epinephrine—add 1 or 2 drops of 1:10,000	↓	↓	↓	stimulate B receptors to ↑ motility
Drain muscle warmer and add new Locke's solution				
Acetylcholine—add 4 or 5 drops of 1:10,000	↑	↑ strength	↑ increase	Stimulates muscarinic receptors for ↑ motility
Pilocarpine—add 2 drops of 1:1000				
Atropine sulfate—add 2 drops of 1:4000	↓ frequency	↓ strength of Contraction	decrease	Block Ach receptors
Pilocarpine—5 min after atropine dose				
Drain muscle warmer and add new Locke's solution				
$BaCl_2$—add a few drops of 0.6% soln				
HCl—add several drops of 2% soln				
NaOH—add several drops of 2% soln				
Isosmotic glucose—drain warmer and fill with 5.4% glucose soln				
UTERINE SEGMENT				
Normal activity				
Acetylcholine—add 4 or 5 drops of 1:10,000				
Epinephrine—add 1 or 2 drops of 1:10,000				

Myographs of Smooth Muscle Motility

Place the records obtained in the following spaces.

Intestine—normal

Intestine—epinephrine

Intestine—acetylcholine

Intestine—pilocarpine

Intestine—atropine + pilocarpine

Uterus—normal

Uterus—acetylcholine

Uterus—epinephrine

12 Insulin Regulation of Blood Glucose

Caution! *Parts of this lab may involve working with human blood. You should handle only your own blood. Dispose of all supplies (cotton, gauze, lancets, etc.) that come in contact with blood in properly marked containers. ALL BODY FLUIDS AND SUPPLIES MUST BE TREATED AS POTENTIALLY INFECTIOUS. Please read Appendix A, "Precautions for Handling Blood."*

ACTION OF GLUCOSE

Insulin is an endocrine hormone secreted by the beta cells of the islets of Langerhans in the pancreas. Its principal function is to assist the transport of glucose across the cellular membrane. When insulin is deficient or lacking, only a small amount of glucose can cross the cell membrane and be used in cellular metabolism. This low rate of transport results in excess accumulation of glucose in the blood, called **hyperglycemia.** An excess of insulin causes a decrease in the level of blood glucose, or **hypoglycemia.** The normal concentration for blood glucose is 90 mg% (90 mg/100 ml of blood) but it may range from 60 mg% to 140 mg%, depending on the individual's dietary intake of glucose.

The disease **diabetes mellitus** can be caused by a lack of insulin. This fact can be demonstrated by either removing the pancreas of an experimental animal (difficult in rats) or destroying the beta cells of the islets of Langerhans by injecting the chemical **alloxan** (alloxan is a specific inhibitor of the beta cells). Either of these procedures will produce the typical symptoms of diabetes in the animal: high blood glucose level and excretion of glucose in the urine. Urinary excretion of glucose (**glucosuria**) results when the concentration of blood glucose exceeds the threshold level for total reabsorption by the kidney. The increased osmolarity of the urine also causes abnormally large quantities of water to be excreted (**polyuria**); this increased excretion of water may lead to dehydration, which in turn stimulates excessive water intake (**polydipsia**). Glucosuria, polyuria, and polydipsia are three major characteristics of diabetes. Diabetes mellitus received its name because the body of the diabetic person was formerly visualized as melting and flowing out in the copious, sweet-tasting urine.

When insulin is deficient and the cells cannot metabolize glucose for energy, the cells compensate by increasing their metabolism of fats and proteins. Thus, the diabetic is usually thin, owing to the loss of fats and proteins from the body structure. The increased metabolism of fats releases into the blood large quantities of **ketone bodies** (e.g., acetone), which are intermediate products of fat breakdown. These are excreted in the urine and have the easily recognizable odor of acetone. Also, ketone bodies are acidic and their accumulation will cause a drop in blood pH; the diabetic becomes **acidotic.** Severe acidosis leads to coma and eventually death.

Hyperinsulinism causes weakness, tremors, hunger, irritability, and other symptoms of low blood glucose; insulin shock can occur if blood glucose falls to a very low level.

GLUCOSE TOLERANCE TEST

In diagnosing for diabetes, several tests are used to determine as precisely as possible what metabolic error is causing the disease. Such tests are urinary glucose level, urinary ketone bodies, fasting blood glucose level, insulin sensitivity, and glucose tolerance tests.

The glucose tolerance test assays the ability of the body (especially the pancreas) to respond to an excess ingestion of glucose. The changes in blood glucose level following glucose ingestion (1 g/kg body weight) are markedly different between the normal and the diabetic person. This difference is shown in Figure 12.1. In the normal person, the blood glucose level rises from about 90 mg% to around 140 mg% in 1 hour and then falls back to normal within 3 hours or even below normal due to excess insulin release by the pancreas. The diabetic person shows a hyperglycemic response in which the blood glucose level rises from about 120 to 160 mg% to as high as 300 mg% and then slowly falls to the fasting diabetic level after 5 to 6 hours. The diabetic's abnormal response is caused by the inability of the pancreas to secrete additional insulin in response to elevated blood glucose levels.

Experimental Procedure

1. Select one person from each team for this experiment, or ask for at least four volunteers from the entire lab. These subjects should report to the lab in the fasted state (not having eaten for the last 12–18 hours). For our purpose it will be adequate if they just skip the meal preceding this lab.
2. Determine each subject's normal blood glucose level, using the Glucostix test. Obtain blood for the test from a finger, using a sterile lancet. Clean the finger with 70% alcohol first. The subject will then obtain a specimen of his or her urine and test it for glucose using Testape.
3. Each subject will then drink a lemon-flavored solution of 25% glucose. The quantity of solution will be based on a quantity of 1 g of glucose per kilogram of body weight. If the glucose solution cannot be made palatable, a good substitute is 30 ml of honey per test. Also, several commercial flavored drinks containing 50 to 100 g of dextrose in 10 oz are available for this test (e.g., Dextol, Tritol).
4. After ingesting the glucose or honey, the subject will repeat the Glucostix test every 30 minutes. As soon as each blood sample has been taken, the subject will obtain another urine sample and repeat the Testape test for urinary glucose. Testing will continue in this manner for 2 hours or until the end of the lab period.
5. Record and graph the results of the blood glucose tests in the Laboratory Reports. Also note the time when glucose appears in the urine. How do the results compare with the normal glucose tolerance test curve?

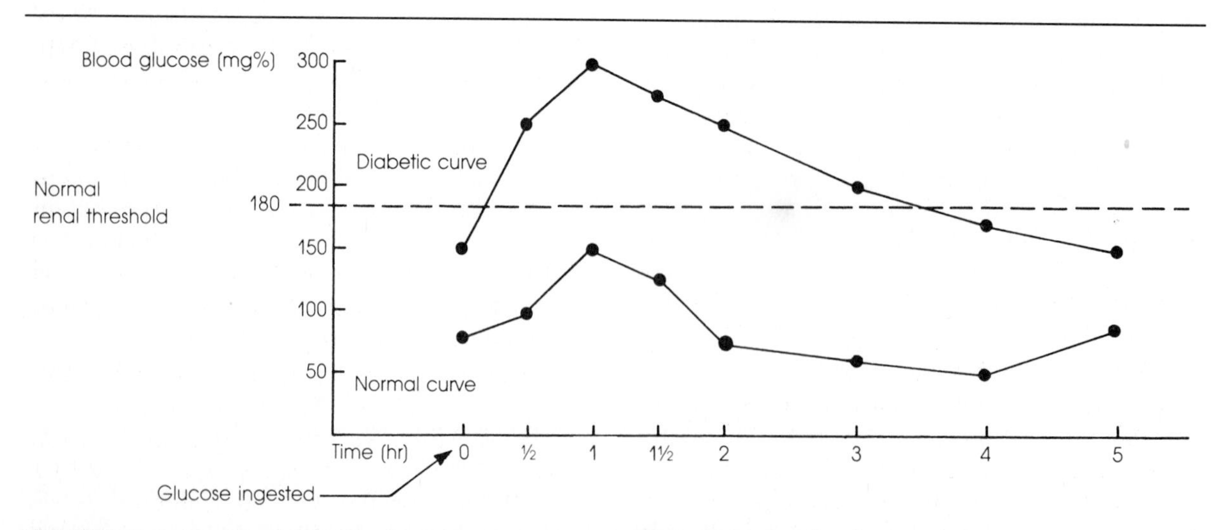

FIGURE 12.1. Changes in the level of glucose in blood following the ingestion of glucose.

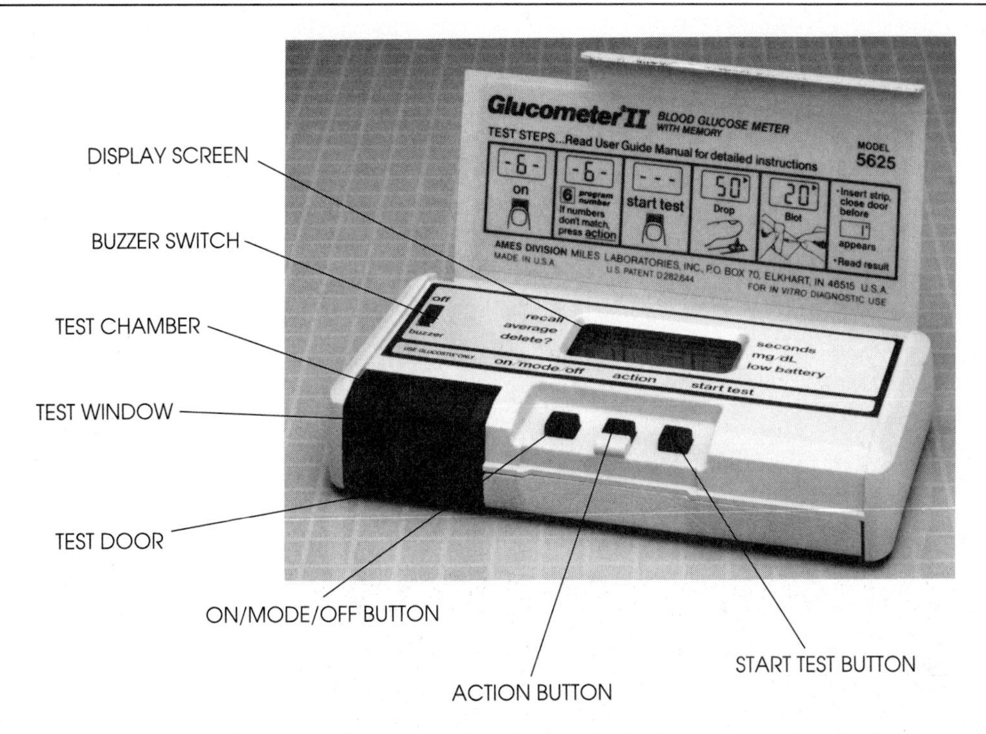

FIGURE 12.2. Glucometer II. (Courtesy of the Ames Division, Miles Laboratories, Elkhart, IN.)

OPERATION OF THE GLUCOMETER II

A more accurate measurement of blood glucose can be obtained using an instrument such as the Glucometer II (Figure 12.2). Many diabetics use these glucose meters to monitor the effect of diet, exercise, and so forth on their blood glucose so that adjustments can be made in their insulin injections or regimen of oral medication.

1. Lift the instrument lid and press the *on/mode/off* button. A program number will appear in the display window. If the number does not match the program number for the Glucostix reagent strips, press the *action* button repeatedly until the numbers match.
2. Remove a Glucostix strip from the bottle and immediately replace the bottle cap. Fold a facial tissue in half and then in quarters so it is ready for the blotting procedure.
3. Clean your finger with 70% alcohol, dry it, and prick your finger to obtain a large drop of blood. Press the *start test* button; at the sound of the beep (50-second display), quickly apply a large drop of blood to the Glucostix test pads. Cover both yellow test pads with blood.
4. At 20 seconds (long beep), immediately blot the test pads by placing the pad side up between the tissue folds on a firm surface. Repeat immediately on a clean area of tissue. (A warning beep will sound at 22 and 21 seconds to alert you to blot at the right time.)
5. After blotting the reagant strip, immediately open the test door and insert the strip into the strip guide with the test pads facing the test window and the end of the strip positioned behind the strip guide tab. Immediately close the test door.

   Note: The reagent strip must be inserted and the test door closed before the instrument timer reaches 1 or inaccurate results (or ERR) will be displayed.
6. The test result will be displayed with a final beep and arrows pointing to *delete?* and mg/dl (mmol/L). To delete the test result, press the *action* button. If it is not deleted, the result will be automatically stored and used to calculate an average when the instrument is turned off or the *start test* button is pressed.
7. Open the test door and throw away the used reagent strip. If needed, clean the reflectance disc and test window, using a cotton swab or lint-free tissue moistened with water. Dry with lens tissue or a lint-free cloth.

   To begin a new test, proceed as before and press the *start test* button.

LABORATORY REPORT

12. Insulin Regulation of Blood Glucose

Name ______________________

Date __________ Section __________

Score/Grade ______________________

Glucose Tolerance Test

1. Record the blood and urine glucose data for the subject in your team and the average values for all team subjects in the laboratory. Plot the blood glucose data on the following graph.

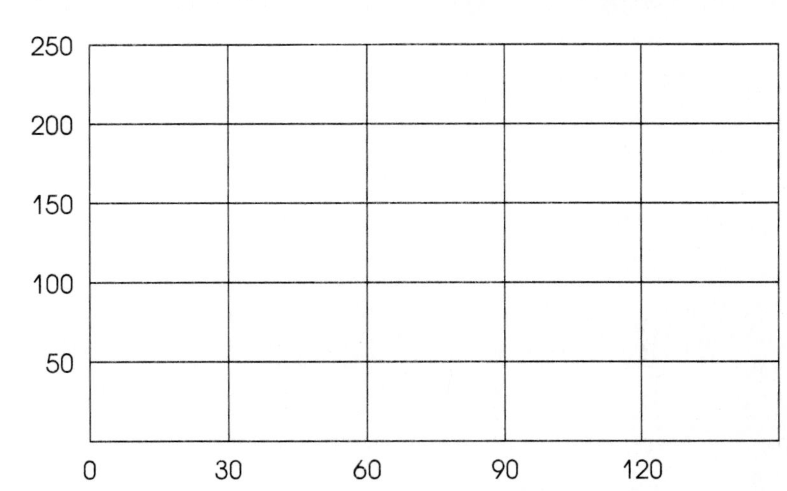

Blood Glucose Level	Team Subject					
	Class Average					

Urine Glucose Level	Team Subject					
	Class Average					

2. List the effect of each of the following hormones on blood glucose, and the mechanism producing the effect.

HORMONE	BLOOD GLUCOSE EFFECT	MECHANISM
Insulin		
Epinephrine		
Glucagon		
Growth hormone		
Cortisol		

3. How are the levels of insulin and glucagon regulated in the body?

4. What is meant by the renal threshold for glucose?

5. Why is there an increase in urine output (diuresis) in diabetes mellitus?

6. Why does a person who has diabetes mellitus have a more acidic urine?

7. Some diabetics control their blood glucose level by ingesting tablets rather than by receiving injections of insulin. How do these tablets work, and who may use them?

8. Define the following terms:

   Glycogenolysis:

   Gluconeogenesis:

   Ketonemia:

   Hyperglycemia:

   Glycogen:

   Glycosuria:

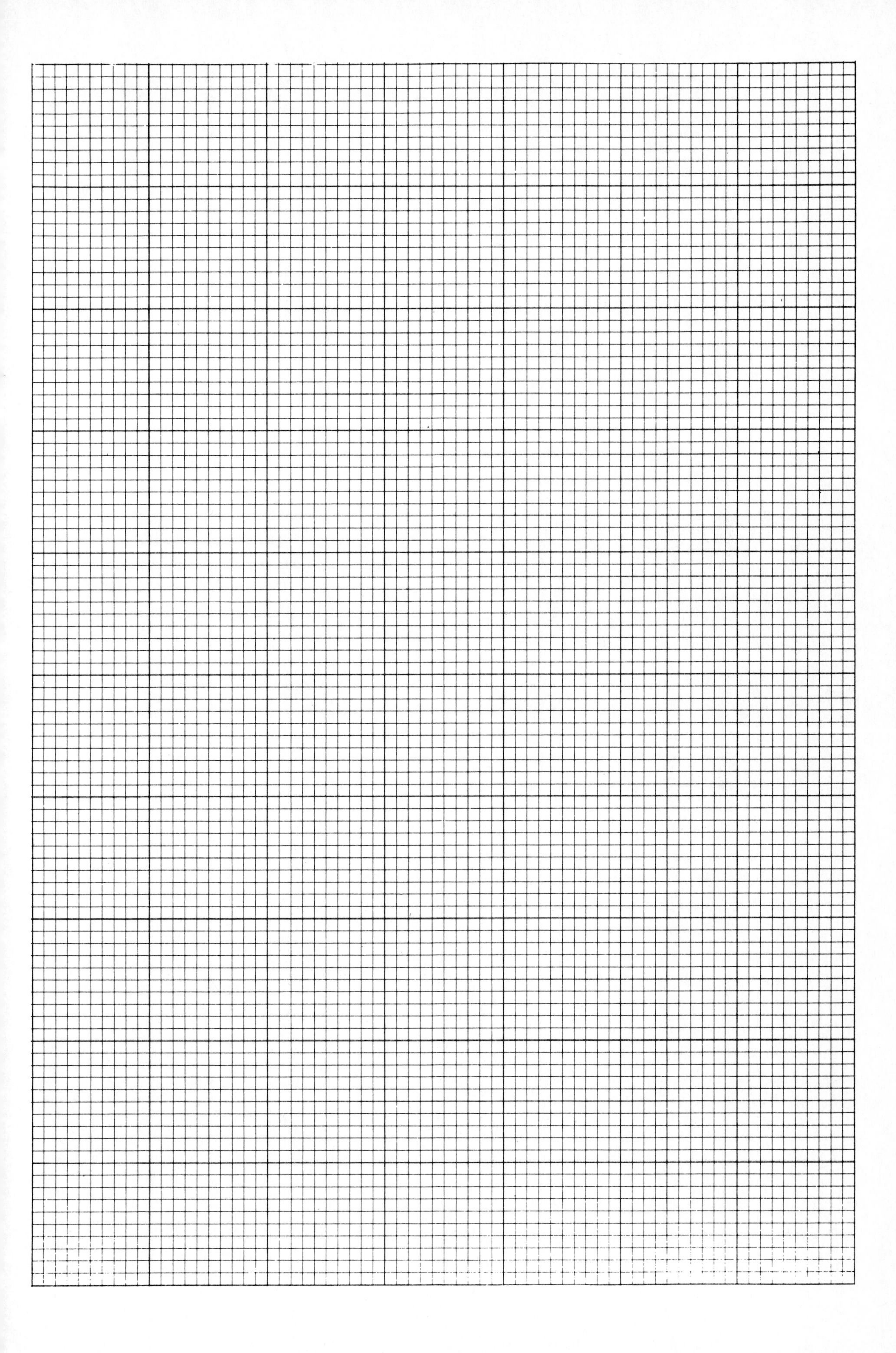

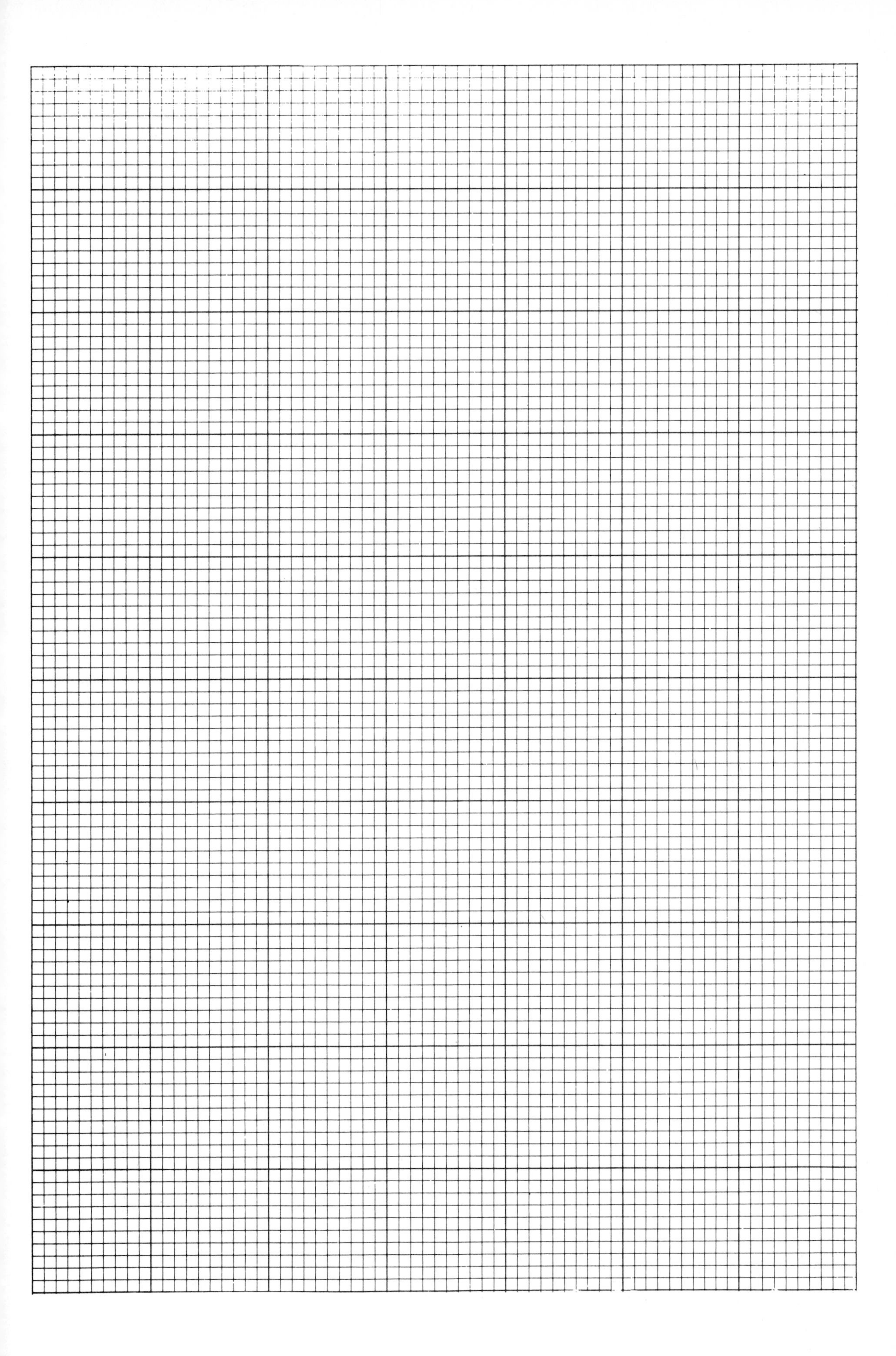

13 Measurement of Metabolic Rate

HUMAN METABOLISM: CALORIMETRY

Metabolism is a broad term that refers to all of the chemical reactions that occur in a biological system, that is, all the reactions of **anabolism** (synthesis of complex molecules from simple molecules) and **catabolism** (breakdown of complex molecules to simple molecules and release of energy). A measurement of metabolism provides information on how the organism obtains its energy and how quickly and efficiently it uses it.

Metabolism would be extremely difficult to measure but for the fact that nearly all of the energy the body uses is eventually converted to **heat.** Only external work, such as lifting of a weight, is not converted to heat. Thus, if we can measure the amount of heat produced by the organism when it is not doing work, we will have a valid measure of its metabolic rate. Such a measurement is called **calorimetry** and is usually expressed as calories or kilocalories of heat produced. To make comparisons between animals of different sizes and weights, the metabolic rate is usually expressed as kilocalories per unit of body weight per hour, or kilocalories per square meter of body surface area per hour.

Direct Calorimetry

At first glance it would appear that the easiest way to measure the heat production (metabolism) of an organism would be to measure directly the amount of heat evolved from the body over a certain period of time. This measurement can be done in a special instrument called a **calorimeter,** an insulated chamber containing a water jacket that absorbs the evolved body heat with consequent increase in the temperature of the water. From the amount of water in the jacket and the temperature rise, we can calculate the heat evolved, using the fact that 1 cal is the amount of heat needed to raise 1 g of water 1 °C. However, this direct method is quite tedious and difficult and has been largely replaced by simpler indirect methods.

Indirect Calorimetry

In the indirect method we make use of the fact that various foodstuffs (proteins, fats, carbohydrates) will produce nearly the same amount of heat whether they are burned in the body (*in vivo*) or outside the body (*in vitro*). Also, the same amount of oxygen must be used *in vivo* or *in vitro* for the complete oxidation of these foodstuffs. The heat of combustion and the oxygen consumed when 1 g of each foodstuff is metabolized are shown in Table 13.1. From the kilocalo-

TABLE 13.1. Metabolic Values for Foodstuffs.

	CARBOHYDRATES	FATS	PROTEINS
Kilocalories per gram	4.1	9.3	4.3
Liters of O_2 per gram	0.75	2.03	0.97
Kilocalories per liter of O_2	5.0	4.7	4.5

ries per gram and the liters of oxygen per gram, we can calculate the **caloric equivalent of oxygen** in kilocalories per liter of oxygen consumed. Thus, we are able to determine the metabolic rate (heat production) indirectly by measuring the oxygen consumption of the organism. This is a much simpler method than the direct calorimetry method. For a person on an average balanced diet of carbohydrates, fats, and proteins, we use an average figure of **4.825 kcal/L O_2 consumed.**

Basal Metabolic Rate (BMR)

The basal metabolic rate (BMR) is a measure of the rate of energy use in a subject in the resting state and awake. It is a measure of the minimal amount of energy needed to maintain just the vital vegetative processes of the body. To be valid, the BMR must be measured under the following rigid conditions:

1. The test is conducted 12 hours or more after the last food is eaten.
2. It is taken after a restful night's sleep, when the activity of the sympathetic nervous system is at its lowest.
3. The subject has been awake and at rest for 30 to 60 minutes prior to the test, with no exercise during this period.
4. The subject is in a reclining position in a quiet room whose temperature is between 62 °F and 87 °F.

Obviously, these conditions for an exact BMR cannot be met in the laboratory; therefore the values obtained in lab should more correctly be labeled metabolic rate (under the existing conditions) rather than basal metabolic rate. The human BMR is commonly expressed as a percentage of the normal standard metabolic rate as found in tables based on age, sex, and body surface area. A BMR of ±10% is considered normal.

Experimental Procedure

Human metabolism is measured using a **respirometer** or **metabolator** (Figure 13.1). This instrument resembles a spirometer but is modified so that it can be filled with 100% oxygen, and it contains a canister of soda lime for the absorption of carbon dioxide. As the sub-

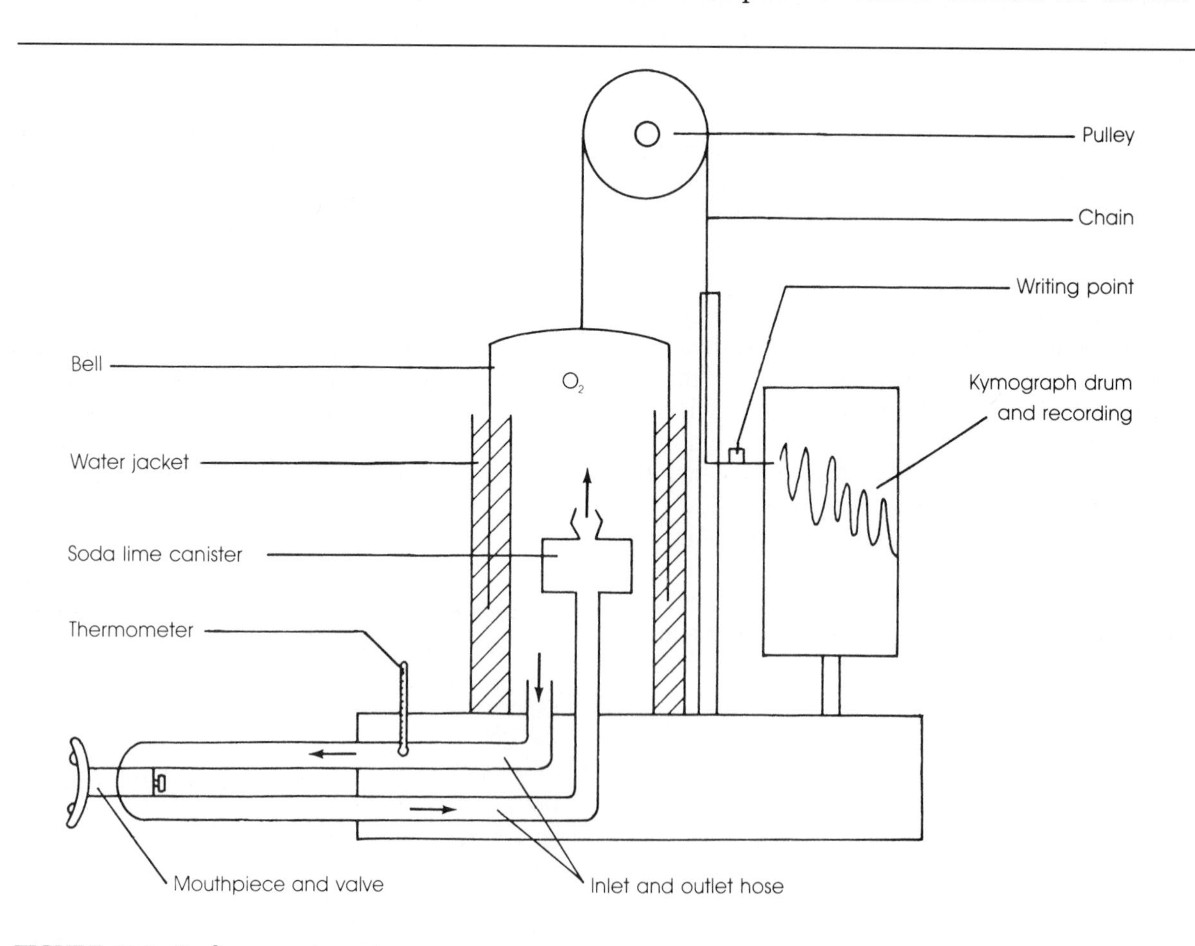

FIGURE 13.1. Sanborn respirometer.

ject breathes in and out of this closed system, oxygen is used from the tank, and the expired carbon dioxide is immediately absorbed by the soda lime. Thus, any decrease in the tank volume represents the volume of oxygen consumed by the subject.

1. Calculation of Oxygen Consumption

a. Let the subject rest in a reclining position for 10 to 15 minutes. In this time, fill the respirometer with 100% oxygen and adjust the instrument for recording.

b. In the last 5 minutes of rest, connect the respirometer to the subject using a clean mouthpiece and a nose clamp on the nostrils. Adjust the valve near the mouthpiece so the subject is breathing atmospheric air.

c. After the subject has breathed room air for 5 minutes, close the mouthpiece valve. Let the subject breathe pure oxygen as naturally as possible for 9 to 10 minutes. The first 1 to 2 minutes should not be used in your calculations, because while the body is adjusting to breathing pure oxygen, the respiratory pattern is abnormal. The normal record will resemble that shown in Figure 13.2.

d. As oxygen is used from the tank, the tank drops and the writing pen moves up on the recording paper. The distance between each numbered vertical line on the record shown in Figure 13.2 represents 1 minute of time. Draw an average slope line by connecting the expiratory excursions as shown. Draw a horizontal line from the beginning of the second or third minute. In Figure 13.2, the distance X represents the distance the tank drops in 4 minutes. Measure the tank drop in millimeters and multiply this distance by the **tank constant** (ml/mm) for your respirometer. (The tank constant is stamped on the respirometer.) Dividing this volume by the minutes of respiration during this tank drop gives the gross milliliters of oxygen consumed per minute. Place this value on your metabolism data sheet and attach the record to the Laboratory Report. With some respirometers, the oxygen consumption can be obtained directly from the recording paper.

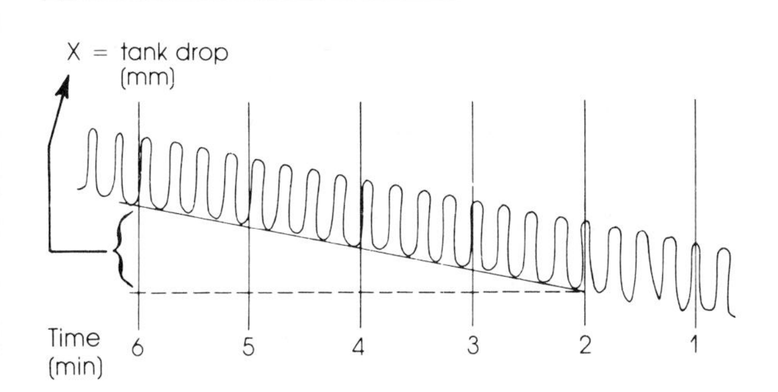

FIGURE 13.2. Normal respiratory record.

2. Calculation of Metabolic Rate

The metabolism data sheet is arranged so that the metabolic rate can be calculated in an orderly, stepwise manner. This sheet is for your convenience, but you are cautioned not to make your calculations from rote memory alone; try rather to understand why each step is performed. Use the following overall procedure to calculate metabolic rate.

a. Determine gross oxygen consumption (from your record).

b. Determine the room barometric pressure (from a barometer) and the temperature of the oxygen during the run (from the thermometer on the respirometer).

c. Calculate the **STPD (standard temperature pressure dry) factor.** This factor is used to convert the volume of oxygen consumed under the experimental conditions to the volume it would occupy as a dry gas at standard temperature (0 °C or 273 K) and standard pressure (760 mm Hg).

$$\text{STPD} = \frac{\begin{matrix}\text{Room barometric pressure (mm Hg)}\end{matrix} - \begin{matrix}\text{Water vapor pressure (mm Hg)}\end{matrix}}{\text{Standard pressure (760 mm Hg)}} \times \frac{\text{Standard temp. (273 K)}}{\text{273 K} + O_2 \text{ temp. (°C)}}$$

The vapor pressure of water at the oxygen temperature is obtained from a standard table (Table C.2, Appendix C).

d. Calculate the corrected oxygen consumption volume.

Corrected volume = Gross volume × STPD

e. Convert the corrected oxygen volume to heat produced.

Heat production =

Corrected L O_2/hr × 4.825 kcal/L O_2

f. Calculate the metabolic rate (heat production per unit of surface area).

$$\text{Metabolic rate (Kcal/m}^2\text{/hr)} = \frac{\text{kcal/hr}}{\text{m}^2 \text{ of body surface area}}$$

The body surface area is obtained from a standard table based on height and weight (Table C.3, Appendix C).

g. Calculate the BMR.

$$\text{BMR (\%)} = \frac{\text{Measured metabolic rate} - \text{Normal metabolic rate}}{\text{Normal metabolic rate}} \times 100$$

The normal metabolic rate is obtained from the Mayo standards table (Table C.4, Appendix C).

Determine the resting metabolic rate with the subject reclining. If desired, you can obtain the exercise metabolic rate after having the subject run in place for 200 steps or do a similar exercise.

RELATIONSHIP OF METABOLISM TO SURFACE AREA AND BODY WEIGHT

It has long been recognized that metabolism is related to body size, and when different-sized animals are compared we usually express their metabolism per unit of body weight (kcal/kg) or surface area (kcal/m^2). The relationship of metabolism to body weight and surface area is not, however, as simple as was once believed, and it has generated considerable controversy and numerous research studies in the last 150 years.

Perhaps the most famous generalization that came from these studies was the **surface area law.** Proponents of this law noted that the rate of heat loss of a body is proportional to its surface area, and because heat production equals heat loss in resting animals, they argued that heat production must also be proportional to surface area. Surface area is roughly proportional to Weight$^{0.67}$ for objects of similar geometry and specific gravity. DuBois's studies, which provided us with the human body surface area tables (Table C.3, Appendix C), were based on the surface area law. Although the kilocalorie-per-meter-squared relationship has been useful as an empirical approximation, it is not founded on strong physiological concepts. The surface area of an animal changes with changes in body position, heat loss varies greatly over various parts of the body, and heat loss and metabolic rate are governed by complex neural and hormonal mechanisms. An added complication is that it is almost impossible to measure the surface area of animals accurately.

The relationship of metabolism to body weight also seemed obvious to early investigators, but it was soon evident that this is not a simple linear relationship. On a kilogram basis, the metabolic rate was found to be higher in small animals than in large animals; hence the metabolic rate is actually related to a power of body weight as expressed in the following formula:

$$M = aW^b$$

$$\log M = \log a + b \log W$$

M = Metabolic rate

W = Body weight

a = Metabolic rate/unit weight

b = Rate at which metabolism changes with size

When the log of metabolic rate is plotted against the log of body weight, a constant linear relationship is found (Figure 13.3); this relationship has been demonstrated for fish, amphibians, reptiles, birds, and mammals. Over the years, the most heated controversy has arisen over the exact value of the exponent b, whose average value is around 0.75 but may vary from 0.7 to 0.8, depending on the species.

In this experiment you will determine your own value for b by constructing a simple version of Benedict's famous "mouse-to-elephant" curve for mammals, which we will call a "mouse-to-human" curve. You will also relate metabolic rate to body surface area for the three animals (mouse, rat, and human) to examine the validity of the surface area law.

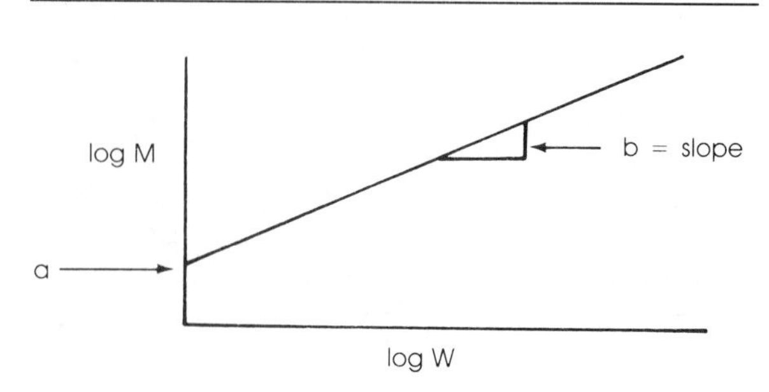

FIGURE 13.3. Plot of log of metabolic rate against log of body weight.

Experimental Procedure

1. Rate of Oxygen Consumption of Human

Determine the rate of oxygen consumption of the human as you did previously or use the data you obtained them.

2. Rate of Oxygen Consumption of Rat

Determine the rate of oxygen consumption of the rat by using the technique described in Experiment 14.

3. Rate of Oxygen Consumption of Mouse

To measure oxygen consumption of the mouse, use either a smaller desiccator than you used for the rat (Figure 14.1) or make a simple mouse metabolator (Figure 13.4).

a. Determine the oxygen consumption of your mouse by following the steps that you used to determine it for the rat.

b. Convert the oxygen consumption to the volume of oxygen under standard conditions (STPD) and calculate the heat production in kilocalories per hour and per day using the conversion factor of 4.825 kcal/L O_2.

c. Express the metabolic rate as kilocalories per square meter per hour. Use Tables C.3 and C.5 in Appendix C to ascertain the body surface areas of the human and the rat tested. Calculate the surface area of the mouse from the following equation:

$$\text{Surface area (m}^2) = \text{Weight}^{0.425}\ \text{(kg)} \times \text{Length}^{0.725}\ \text{(cm)} \times 0.007184$$

$$\log \text{area} = 0.425 \log \text{(kg)} + 0.725 \log \text{(cm)} + \log 0.007184$$

The length of the mouse is measured from the tip of the nose to the base of the tail. Are the metabolic rates of the three mammals comparable when based on square meters of body surface area?

4. Plotting the "Mouse-to-Human" Curve

Using 3 × 3- or 4 × 4-cycle logarithmic graph paper, plot the metabolism (kcal/day) against the body weight (kg) and draw a "mouse-to-human" curve. From this plot, determine the a and b values for the formula $M = aW^b$.

How does your exponent b compare with the values usually cited in the literature of around 0.65 to 0.75?

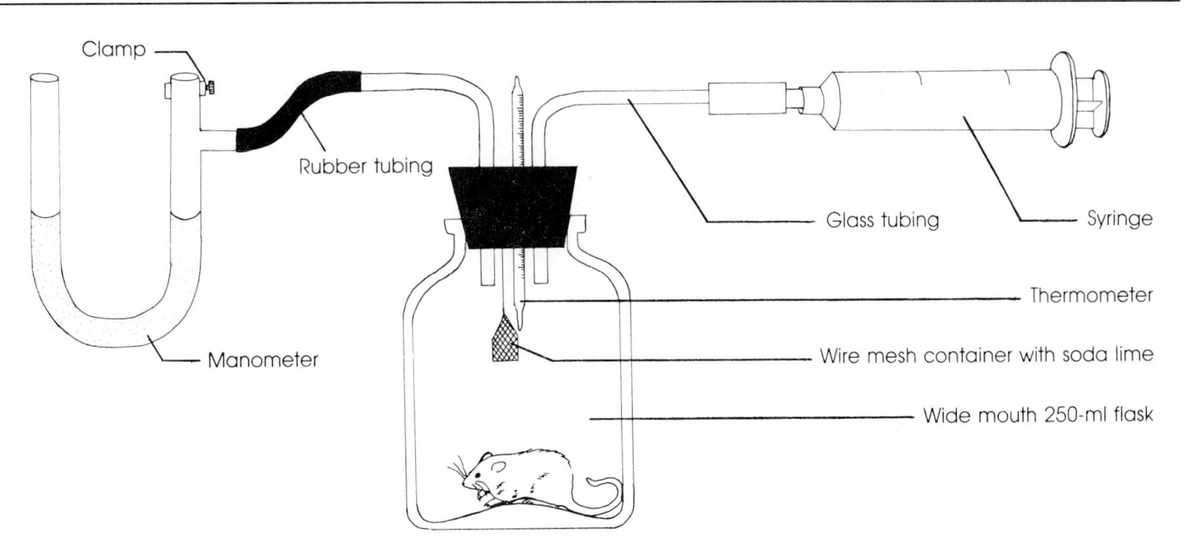

FIGURE 13.4. Mouse metabolator.

LABORATORY REPORT

Name ______________________

Date __________ Section __________

Score/Grade ______________________

13. Measurement of Metabolic Rate

Human Metabolism—Calorimetry

Metabolism Data Sheet (Humans)

Name ______________________ Age __________ Sex __________

Height (cm) __________ Weight (kg) __________ Surface area (m^2)* __________

STEP	EXAMPLE CALCULATION FOR A 70-KG MALE 180 CM TALL, AGE 20	EXPERIMENTAL SUBJECT
1. Oxygen temperature, °C	27	
2. Barometric pressure, mm Hg	700	
3. Vapor pressure of H_2O, mm Hg	26.7	
4. STPD factor (standard temperature, pressure, dry)	0.806	
5. Gross oxygen consumption, ml/min	400	
6. Corrected oxygen consumption, ml/min (gross × STPD)	322.4	
7. Oxygen consumption, ml/hr	19344	
8. Oxygen consumption, L/hr	19.344	
9. Heat production, kcal/hr (L O_2/hr × 4.825)	93.33	
10. Metabolic rate per unit surface area, kcal/m^2/hr	49.38	
11. Normal metabolic rate, kcal/m^2/hr	41.43	
12. Metabolic rate expressed as percent of normal $\frac{(\text{step 10} - \text{step 11})}{\text{step 11}} \times 100$	+19.19%	

*See Table C.3, Appendix C.

Respirometer record used in calculating your oxygen consumption (place your record in the following space).

1. What is the purpose of correcting the oxygen volume using the STPD factor?

2. Why is metabolism expressed in relation to body surface area?

3. List five factors that alter metabolic rate and how they affect it.

4. Is the determination of metabolic rate a valid test for hyperthyroidism or hypothyroidism? Why? What other tests are used to detect alterations in thyroid function?

5. Assume that your resting gross carbon dioxide production for 5 minutes is 1.6 L. Calculate your resting respiratory quotient (RQ). (RQ = CO_2 produced/O_2 consumed.)

RQ = ___________

What is the importance of the RQ in physiology?

Name ______________________

Date __________ Section __________

Score/Grade __________________

Hormone Regulation of Metabolism

The capture and consumption of energy by the body is regulated by the action of several hormones that promote anabolism or catabolism of the body's nutrients and minerals. Briefly list the major effects of the following hormones on nutrients and minerals using (+) for stimulatory and (−) for inhibitory actions. Also include any other physiological effects associated with these hormones that are unrelated to their metabolic actions.

Insulin:

Glucagon:

Epinephrine:

Growth hormone:

Cortisol:

Thyroxine:

Parathyroid hormone:

Calcitonin:

Relationship of Metabolism to Surface Area

MEASUREMENT	MOUSE	RAT	HUMAN
Weight (kg)			
Surface area (m^2)			
Gross O_2 consumption (L/hr)			
Corrected O_2 consumption (L/hr)			
Heat production (kcal/hr)			
Heat production (kcal/day)			
Metabolic rate (kcal/m^2/hr)			

1. What is the theoretical basis for relating metabolism to surface area?

Relationship of Metabolism to Body Weight

1. Calculate the metabolic rate of the three animals as related directly to body weight (kcal/kg/hr).

	MOUSE	RAT	HUMAN
kcal/kg/hr			

Attach your log-log plot of metabolism (kcal/day) against body weight (kg) to the lab report.

2. What exponent b value did you obtain? __________
What does this b value tell you about the relationship between metabolism and body weight?

3. Which of the three means of expressing metabolic rate do you think gives the best comparison between different-sized animals? Why?

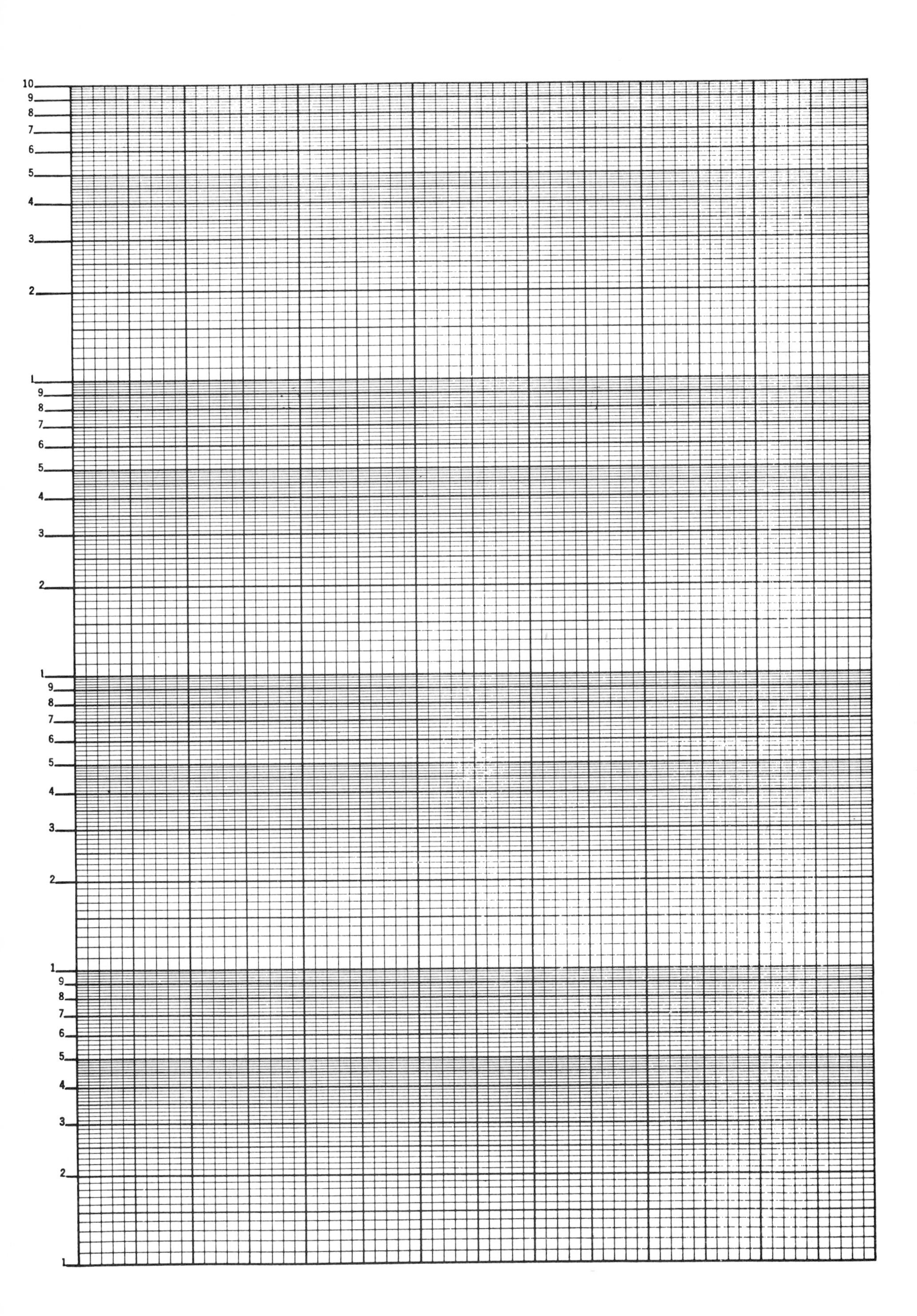
10
9
8
7
6
5
4
3
2
1
9
8
7
6
5
4
3
2
1
9
8
7
6
5
4
3
2
1
9
8
7
6
5
4
3
2
1

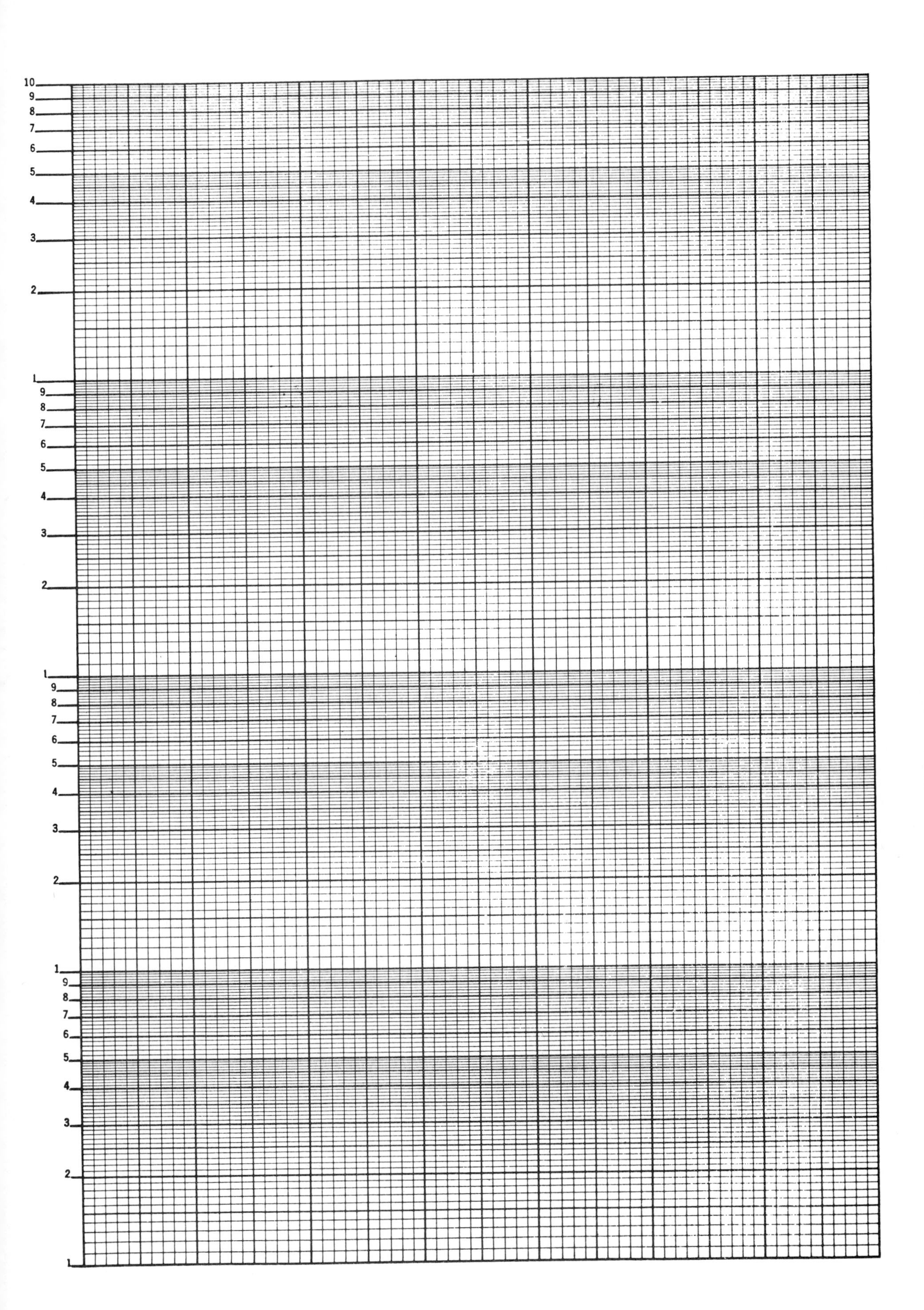
10
9
8
7
6
5
4
3
2
1
9
8
7
6
5
4
3
2
1
9
8
7
6
5
4
3
2
1
9
8
7
6
5
4
3
2
1

14 Thyroid Function

THYROID EFFECTS ON METABOLISM

The thyroid is a shieldlike gland located on the ventral side of the trachea. Under the influence of **thyrotropin,** or thyroid stimulating hormone (TSH), from the anterior pituitary, the thyroid secretes several thyroid hormones, the chief of which is **thyroxine.** Thyroxine is a powerful hormone that affects the metabolism of every body cell; its principal effects are an increase in oxygen consumption and calorigenesis (heat production) by the cells. The exact mechanism of action of thyroxine is still unknown. The general effects of thyroxine can be observed in the hyperthyroid person. He or she displays an increase in heart rate, stroke volume, and blood pressure and increased peripheral vasodilation to aid loss of excess body heat. The cells of the nervous system become hyperexcitable, so that the person becomes very irritable and nervous. The increased metabolic rate depletes the body's fat reserves and the person becomes quite thin. The hypothyroid person shows opposite symptoms: increased body weight, sluggishness, low metabolism, and low heat production.

The metabolic effects of thyroxine can be observed in animals fed thyroxine in excess for a period of 2 to 3 weeks. A latent period of around 10 days occurs before the effects of excess or deficient thyroxine can be seen. In this experiment you will examine thyroxine's actions by measuring the metabolic rate of normal, hypothyroid, and hyperthyroid rats.

Experimental Procedure

1. Preparation of Animals

The control and experimental animals should be of the same sex and should be of young (150–200 g) stock not used for other experiments. Two weeks prior to the laboratory period, the rats will be divided into four groups (two rats per group):

a. Euthyroid controls.
b. Hyperthyroid. Fed 1% desiccated thyroid in their rat chow for 2 weeks. An alternate technique is to give the rats an intraperitoneal (IP) injection of 15 μg of triiodothyronine or 25 μg of L-thyronine every third day.
c. Hypothyroid. Fed 0.5% thiouracil or 0.02% propylthiouracil in their rat chow for 2 weeks.
d. Hypothyroid. Fed 1% potassium perchlorate in their drinking water for 2 weeks.

The weights of all rats should be recorded weekly.

2. Principles

The animal is placed in a closed system in which carbon dioxide is continuously removed by reaction with soda lime (Figure 14.1). The pressure in the respirometer is kept constant, so that any change in volume represents oxygen consumed by the animal. The volume change is measured by means of a Vaseline-coated syringe. The heat production of the animal is calculated on the as-

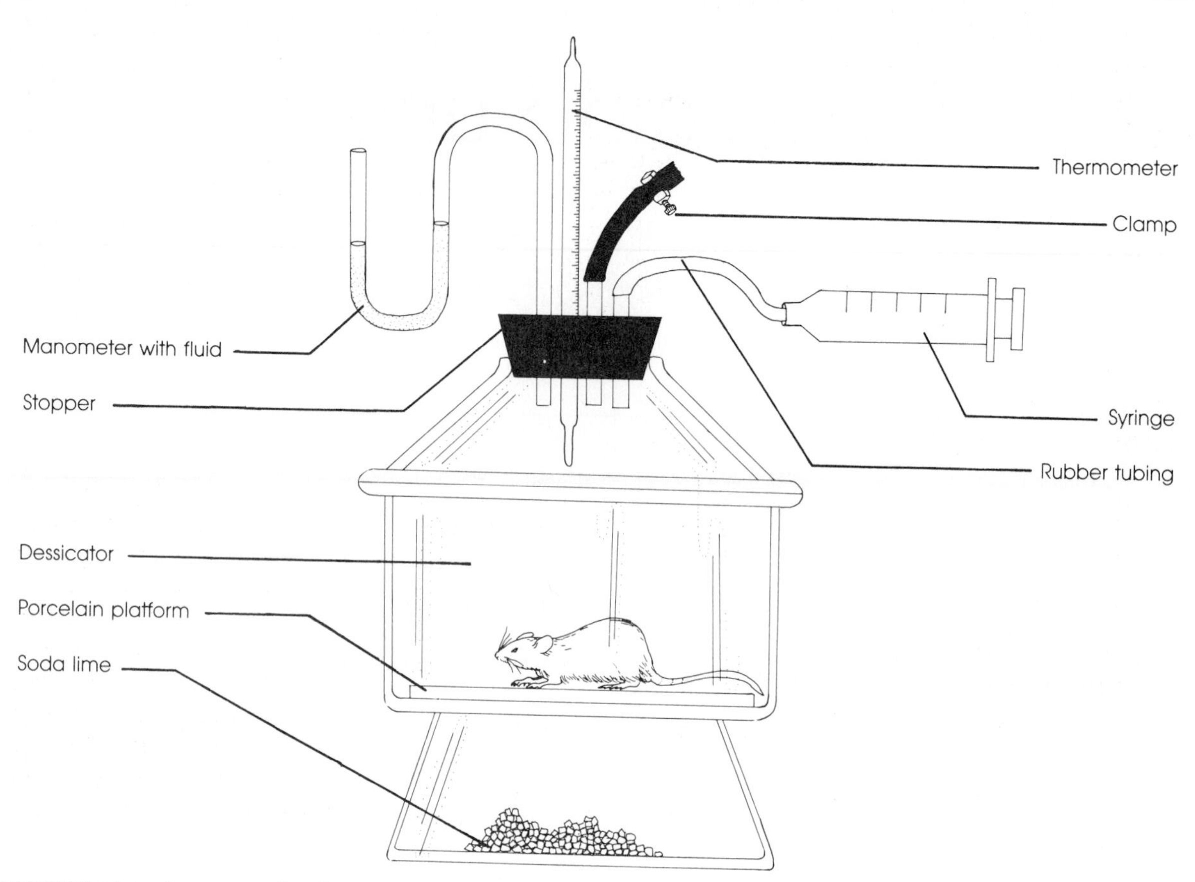

FIGURE 14.1. Apparatus for determining the metabolic rates of small animals.

sumption that for each liter of oxygen consumed, 4.825 kcal of heat is produced. This constant is called the **caloric equivalent of oxygen.**

3. Procedure

You will be assigned animals from one of the four experimental groups. Measure the metabolic rate of your rats before and after the 2-week feeding period. Weigh the rats and obtain their body surface area using Table C.5 in Appendix C.

When measuring oxygen consumption using a desiccator, it is essential that you follow directions exactly. The following precautions will help ensure the success of your measurements: (1) Always keep the clamp on the vertical tube open while making the initial adjustments. (2) Never push abruptly on the syringe, or you will separate the manometer fluid column. (3) During the run, keep the manometer fluid level at all times; this keeps the pressure in the chamber constant. (4) Handle the apparatus and the rat gently at all times.

Each team will now obtain a rat and determine oxygen consumption as follows:

a. Weigh your rat to the nearest gram and record the weight.

b. Remove the top half of the desiccator and place the rat inside. Make certain that adequate fresh soda lime is in the bottom of the desiccator. Be sure the clamp on the vertical tube is open.

c. Replace the top of the desiccator, using Vaseline to provide an airtight seal between the two halves.

d. Test the system for leaks by closing the clamp on the vertical tube and gently pulling back on the plunger of the syringe. The manometer fluid should move toward the desiccator. If there is a leak in the system, the manometer fluid will return to its original level. If a leak is detected, try to seal it by applying Vaseline around the stopper or wherever there is a rubber-to-glass connection.

e. Open the clamp on the vertical tube and adjust the card on the back of the manometer so that the crossline is level with both menisci of the manometer fluid.

f. Wait 4 to 5 minutes to allow the temperature to reach equilibrium in the respirator.

g. Close the clamp on the vertical tube and record the time.

h. Manipulate the plunger of the syringe to keep the two limbs of the manometer fluid constantly on the crossmark.

i. Exactly every minute (or every 2 minutes) read the syringe and record the reading. Also read the thermometer and record the temperature.

j. Continue making observations until the syringe is empty.

k. Open the vertical tube clamp and remove the top half of the desiccator to allow the stale air to be flushed out.

l. Repeat the previous steps two or three times and determine the average milliliters of oxygen used per minute by your rat.

4. Calculation of Metabolic Rate

Using the metabolism data sheet on page 141 as a guide, calculate the oxygen consumption as milliliters of oxygen per minute per 100 g and the heat production as kilocalories per square meter per hour. It is necessary to correct the observed gas volume to the volume it would occupy under standard conditions of pressure and temperature (pressure = 760 mm Hg; temperature = 0 °C or 273 K). To make this correction, the observed milliliters of oxygen per minute consumed are multiplied by the STPD factor. Summarize the data from all the rats in lab to see the effects of the thyroid on metabolism, and record the average values in the table in the Laboratory Report.

THYROID UPTAKE OF IODINE

The production of the hormones thyroxine and triiodothyronine by the thyroid gland depends on the presence of an adequate amount of iodine. Nearly all of our dietary iodine is actively moved into the thyroid by an active transport process (iodine pump), a process so powerful that it produces a concentration of iodine in the thyroid 25 times that in the blood in a euthyroid individual. This uptake of iodine can be demonstrated readily by following the distribution of a radioisotope of iodine, ^{131}I, in the body. The uptake of radioactive iodine is also used as a test to determine the rate of activity of the thyroid gland. In this experiment you will determine the percent incorporation of ^{131}I into the thyroid in 1 hour in three types of rats: euthyroid controls, hypothyroid, and hyperthyroid.

Experimental Procedure

1. Young adult rats will be used. Hypothyroid rats will be produced by adding 1% potassium perchlorate to their drinking water or by feeding them rat chow containing 0.5% thiouracil for 2 weeks prior to the lab. Hyperthyroid rats will be produced by feeding them rat chow containing 1% desiccated thyroid. A euthyroid (control) group will also be used. Each team will be responsible for carrying out the following procedures on one of the four experimental types of rats.

2. Weigh your rat to the nearest gram. Inject it intraperitoneally (IP) with 1 ml of ^{131}I solution (4 microcuries/ml) and record the injection time. It is important that you inject a full 1 ml into the animal. Stagger the injection of the rats by 5 to 10 minutes to provide adequate counting time for each animal at the end of the experiment. It is best to keep the "hot" radioactive substances as far away from the counting apparatus as possible to avoid contamination of the counting area.

3. During the 1-hour waiting period, determine the background counts per minute (cpm) by placing an empty counting planchet under the Geiger-Mueller tube and recording the count for 5 to 10 minutes. The background counts per minute must be subtracted from each gross radiation count to give the net counts per minute of radioactivity.

4. Determine the radioactivity of the ^{131}I injected. Place exactly 1 ml of the ^{131}I injected (4 microcuries/ml) into a 50-ml volumetric flask, dilute to 50 ml with distilled water and mix thoroughly. Measure out exactly 0.5 ml of this diluted ^{131}I, place it in a counting planchet, and dry it under a heat lamp. Place the planchet in the Geiger-Mueller counter and determine the gross counts per minute.

Subtract the background counts per minute to give the net counts per minute. Multiply this value by 100 to obtain the net counts per minute per milliliter of the injected ^{131}I solution.

5. Fifty minutes after injection of the ^{131}I solution, inject your rat IP with 50 mg of Nembutal per kilogram of body weight so that it will be in deep anesthesia at the 60-minute mark.

6. Make an incision in the neck to expose the trachea. Remove the section of the trachea containing the thyroid gland by cutting above and below the larynx. Carefully dissect the thyroid lobes away from the trachea; place them in a counting planchet and determine the net counts per minute of ^{131}I in the thyroid tissue. Weigh the thyroid tissue and record the weight. Calculate the percentage of the injected ^{131}I absorbed by the thyroid during the hour.

LABORATORY REPORT

Name ______________________

Date __________ Section __________

Score/Grade ______________________

14. Thyroid Function

Thyroid Effects on Metabolism

1. Record the metabolic rates (kcal/m^2/hr) for the rats in each experimental group, both pretreatment (before feeding) and posttreatment (after feeding). Use the average values for rats in each group.

METABOLIC RATE KCAL/M^2/HR	EUTHYROID (CONTROL)	HYPERTHYROID (THYROID)	HYPOTHYROID (THIOURACIL)	HYPOTHYROID (PERCHLORATE)
Pretreatment				
Posttreatment				
Difference pre to post				
Percent change				

Thyroid Uptake of Iodine

Background counts per minute (cpm) ______________________

^{131}I injected: Gross cpm/ml ______________ Net cpm/ml ______________

	EUTHYROID (CONTROL)	HYPERTHYROID (THYROID)	HYPOTHYROID (THIOURACIL)	HYPOTHYROID (PERCHLORATE)
Thyroid weight				
Gross cpm				
net cpm				
% of injected ^{131}I absorbed				

1. What are the effects of thyroxine on metabolism and how are these effects brought about?

2. Explain the correlation between your findings for iodine uptake and the mechanism of action of thiouracil and perchlorate.

3. How do you explain your findings for thyroid weight and counts per minute of ^{131}I for the hyperthyroid rats? How is the circulating level of thyroxine regulated in the body?

4. What is a goiter? How is a goiter different in hyperthyroidism compared with hypothyroidism?

5. What role do iodine and the amino acid tyrosine together play in the function of the thyroid gland?

Name ______________________

Date ____________ Section ____________

Score/Grade ______________________

Metabolism Data Sheet (Rats)

Animal number or marking ____________________ Weight (g) ____________________

Treatment ____________________ Surface area (m^2)* ____________________

STEP	EXAMPLE CALCULATION FOR 200-G RAT	EXPERIMENTAL ANIMALS	
1. Chamber temperature, °C	28		
2. Barometric pressure, mm Hg	740		
3. Vapor pressure of H_2O, mm Hg	28.3		
4. STPD factor (standard temperature, pressure, dry)	0.845		
5. Average gross oxygen consumption, ml/min	5		
6. Corrected oxygen consumption, ml/min (gross × STPD)	4.225		
7. Oxygen consumption per unit body weight, ml/100 g body wt/min	2.1125		
8. Heat production, kcal/min (ml/min × 0.004825)	0.0203856		
9. Heat production, kcal/hr	1.223136		
10. Metabolic rate per unit surface area, kcal/m^2/hr	40.63		

*See Table C.5, Appendix C.

15 Nerve-muscle Activity

In the body, skeletal muscle contractility is controlled by the conduction of action potentials over motor neurons and the subsequent release of acetylcholine at the neuromuscular junction. Each motor neuron innervates (makes synapse with) several individual muscle fibers (cells) and each fiber contracts in an all-or-none fashion when its threshold is reached. One motor neuron and all the muscle fibers it innervates is called a *Motor Unit.* The overall contractile strength exerted by a muscle is thus determined by the number of motor units activated by the motor cortex in the brain. We rarely activate all the motor units in a muscle, but during times of extreme anger or fear, humans have performed unusual feats of strength because they were able to activate all their motor units simultaneously.

In this experiment you will examine the operation of the neuromuscular junction and the major characteristics of muscle contractility. Much of our knowledge in this area was derived from the classic studies of the frog gastrocnemius muscle conducted during the last half of the nineteenth century, using the kymograph recording system (Figure 15.1).

DISSECTION OF NERVE-MUSCLE PREPARATION

Pithing the Frog

The frog will be killed by using a pithing needle to destroy both the brain and the spinal cord (double pithed). Hold the frog in one hand, using the index finger to bend the head slightly downward. Insert the pithing needle into the slight depression between the skull and spinal cord. This opening (the foramen magnum) is approximately even with the posterior edge of the tympanic membranes.

Insert the needle quickly into the foramen, move the needle to a position parallel to the top of the skull, push the point forward into the brain, and move the point from side to side to destroy the brain (Figure 15.2a). When the brain is successfully pithed the corneal reflex is lost. Touch the cornea lightly to see if an eye blink can be elicited. Repeat the brain pithing until the corneal reflex is abolished.

A single-pithed frog does not feel pain but still exhibits spinal reflexes. These reflexes can be destroyed by inserting the needle in the foramen and running it down the spinal cord to destroy the cord (Figure 15.2b). When this double pith is complete, the hind legs will become extended and then will relax.

In Vitro Muscle Preparation

Double pith the frog. Then use bone cutters or heavy scissors to cut off a hind leg at the upper end of the thigh. Save the rest of the animal wrapped in a wet paper towel so that the other leg may be prepared if the first muscle fatigues. Using scissors, cut the skin around the ankle and peel back the loose skin from the muscle (Figure 15.3). Avoid touching the nerve or muscle with dry fingers because this will cause depolarization of the membranes and thereby stimulate spontaneous muscle contractions. Cut the skin on the back of the thigh and, using a glass probe,

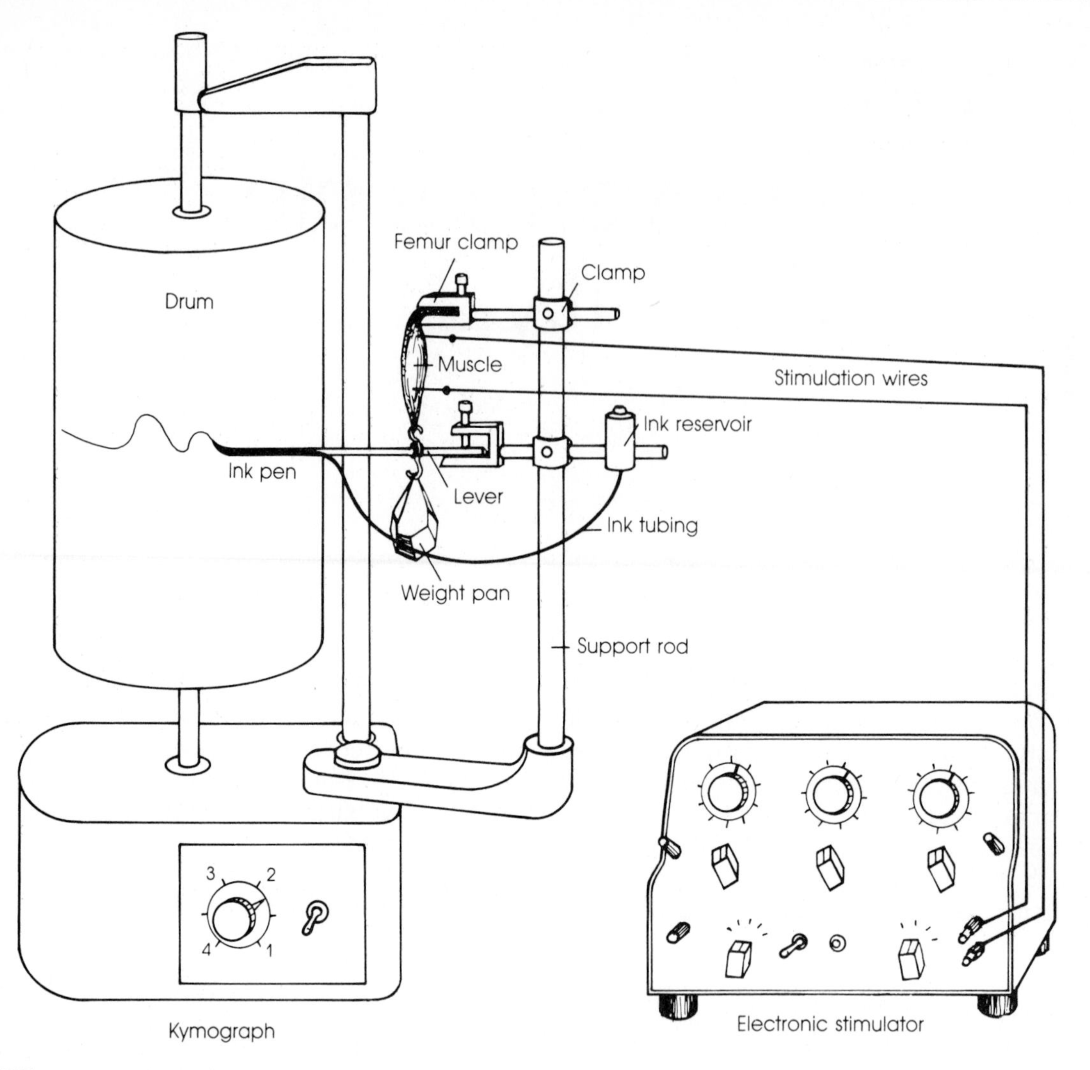

FIGURE 15.1. Kymograph system for recording skeletal muscle contractions.

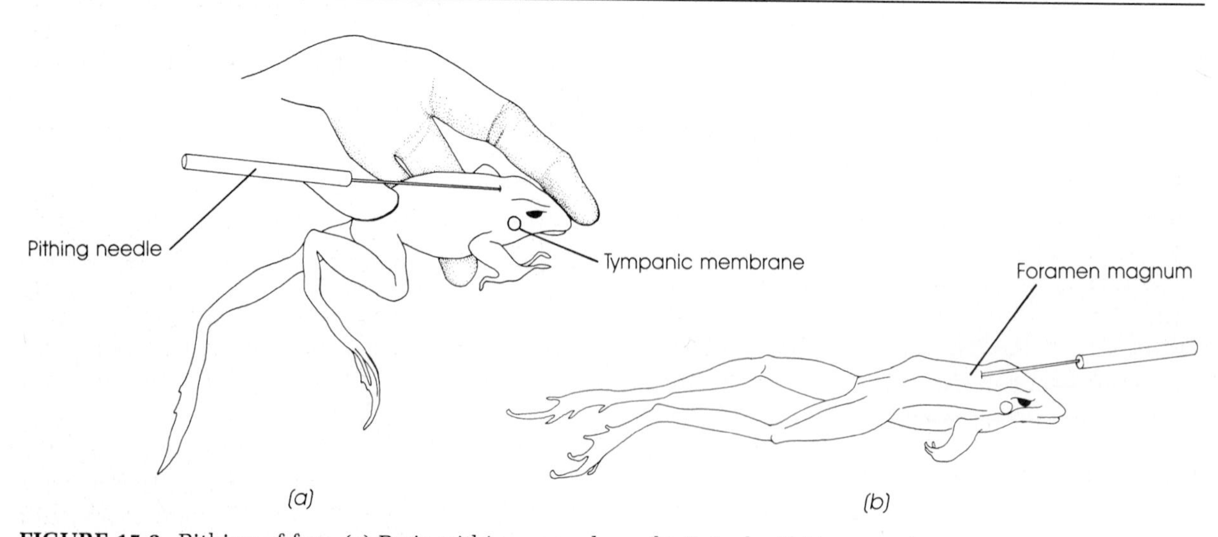

FIGURE 15.2. Pithing of frog. (a) Brain pithing procedure. (b) Spinal pithing procedure.

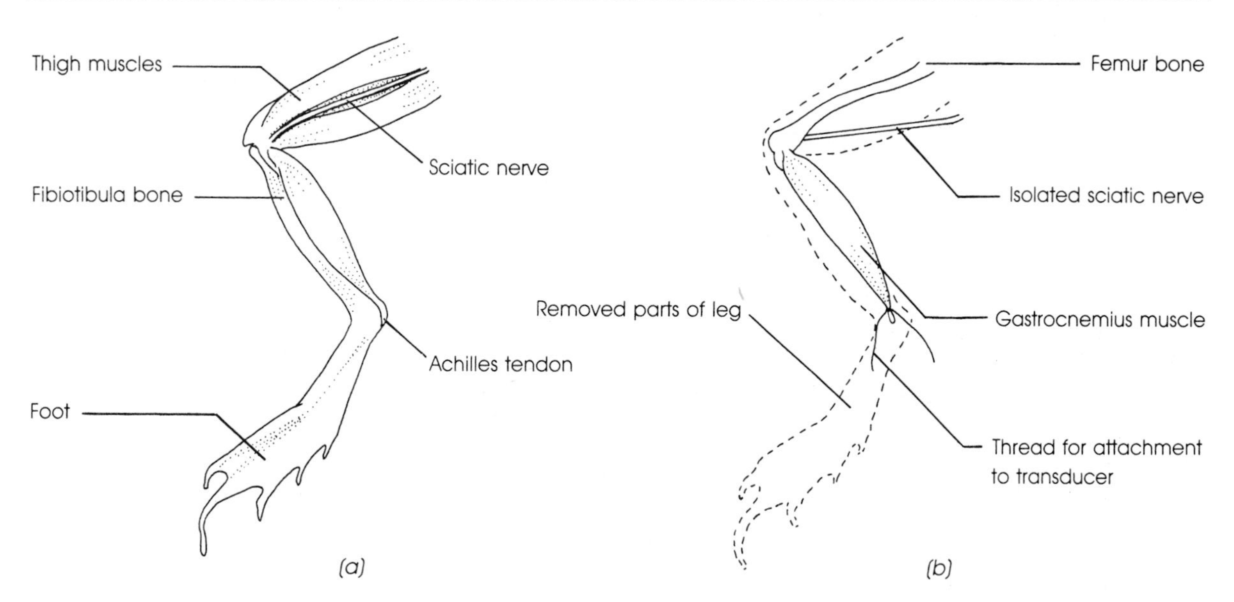

FIGURE 15.3. Dissection of frog gastrocnemius muscle (*in vitro* muscle preparation). (a) Skinned hind leg. (b) Nerve-muscle preparation.

isolate the sciatic nerve from the surrounding muscles in the thigh. Cut away the thigh muscles from the femur, leaving the nerve exposed. Pass a thread around the Achilles tendon at the heel and tie it tightly around the tendon. Sever the tendon below the tie and pull the muscle away from the leg. Cut away the fibiotibula bone just below the knee. This procedure will leave the gastrocnemius muscle attached to the knee, with a section of the femur left to anchor the preparation in subsequent experiments.

The femur is placed between the jaws of the femur clamp and the clamp is tightened securely. The thread attached to the achilles tendon is tied to the lever arm of the kymograph system (Figure 15.1) or to the electronic myograph transducer used in the Physiograph™ system (Figure 15.4). The muscle is positioned so it forms a right angle with the femur and is vertical to the lever arm or transducer. The muscle may be stimulated by positioning an electrode holder so that the electrode prongs are held against the muscle, or pin electrodes may be inserted into each end of the muscle and attached to the stimulator. *The muscle and nerve must be kept moist with Ringer's solution at all times.* Twitching of the muscle when it is not being stimulated indicates that the tissue is drying and needs additional Ringer's solution.

In Vivo Muscle Preparation

In this alternative preparation, the gastrocnemius muscle is left attached to the body with its blood supply intact.

Double pith the frog. Cut the skin on the back of the thigh on one leg and gently retract the gluteus and semimembranous muscles to expose the sciatic nerve deep between the muscle groups. Using a glass probe, slip a thread under the nerve so it can be lifted up and stimulated with electrodes during later experiments. For now, let the nerve retract between the muscle groups and moisten the area with frog Ringer's solution.

Cut and remove the skin over the gastrocnemius muscle. Tie a thread tightly around the Achilles tendon and cut the tendon free of its attachment to the foot. Place the frog prone on the frog board and pin the knee into the holder (Figure 15.5). Tie the free end of the thread to the myograph transducer as shown in the Physiograph setup and insert pin electrodes into each end of the muscle for stimulation. Use the tension adjuster to apply a slight stretch on the muscle so it is at a better length for contraction. *The muscle must be kept moist with frog Ringer's solution at all times.* Cover the rest of the frog with a paper towel soaked in water.

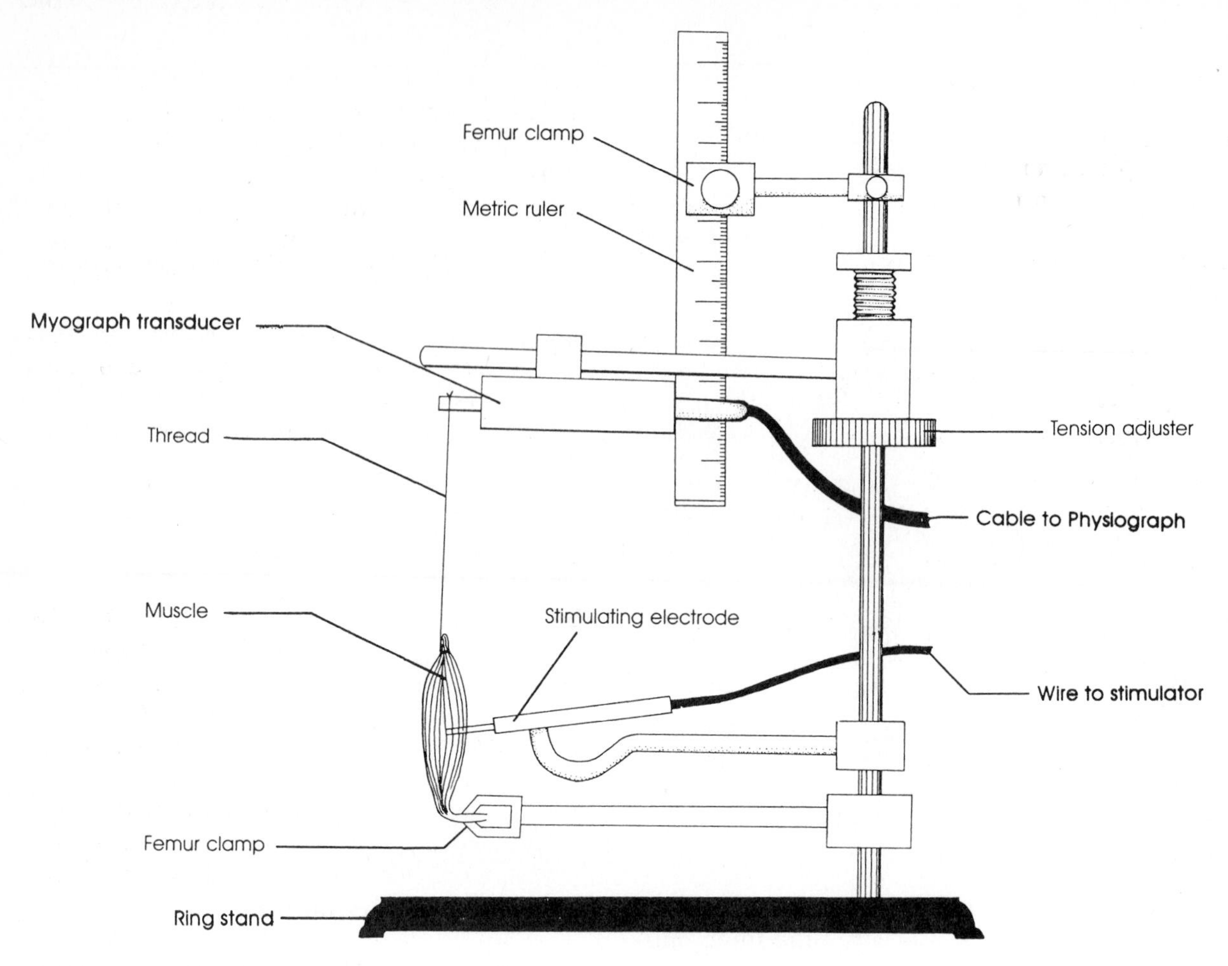

FIGURE 15.4. Physiograph system for recording muscle contractions.

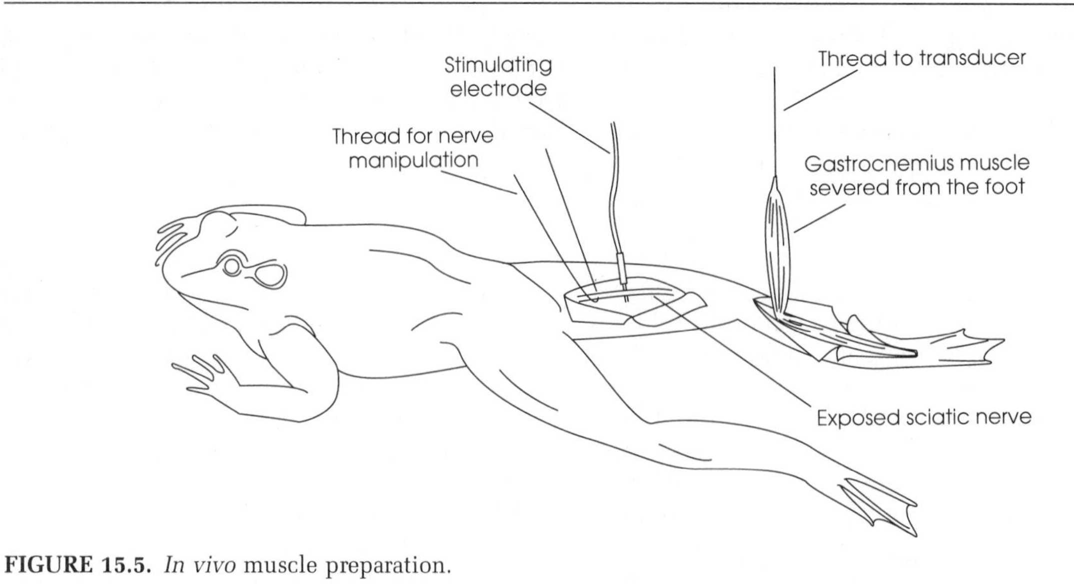

FIGURE 15.5. *In vivo* muscle preparation.

Stimulation of Tissues

If you are unfamiliar with the electronic stimulator, please refer to the section on "Stimulation of Tissues" in the Membrane Action Potentials lab (Experiment 6) before you begin the experimental procedures.

ISOLATED MUSCLE RESPONSES

Experimental Procedures

1. Muscle and Nerve Irritability

Double pith the frog and make an *in vivo* preparation of the gastrocnemius muscle, without attaching the muscle to the transducer.

Lift the sciatic nerve out from between the muscles and lay it over the stimulating electrodes. Using a 1-msec pulse duration, stimulate the nerve with a single pulse and determine the lowest voltage (threshold) that will produce a perceptible (visual—not recorded) muscle twitch.

Replace the nerve between the muscles and place the stimulating electrodes directly on the gastrocnemius muscle. Determine the threshold voltage that produces a perceptible twitch, as was done with the nerve stimulation.

2. Effects of Stimulus Strength

Prepare the muscle for recording contractions, using either the Physiograph or kymograph system. Set the stimulator for 1-msec pulse durations.

Using a very slow paper speed, stimulate the muscle directly with a single pulse and record the first perceptible twitch obtained with a threshold stimulus. Label the recording with the voltage used. Now gradually increase the voltage in a stepwise manner, in small increments, and record the muscle twitches obtained. Continue this process until no further increase in twitch height occurs. Do not exceed this maximal voltage by any large margin or you risk damaging the muscle. Label the record with each voltage applied. You should now be able to categorize stimuli on the basis of the response obtained. Your record should resemble that shown in Figure 15.6.

3. Effects of Stimulus Frequency

Using a moderate paper speed, stimulate the muscle with a supramaximal voltage long enough to obtain five to eight twitches at each of the following frequencies: 0.5/sec, 1/sec, 2/sec, 3/sec, 4/sec, and so on until tetanus is produced. Allow 15 seconds for recovery between each series of stimulations and apply Ringer's solution to the muscle during this time. Your record should resemble that shown in Figure 15.7.

What is the effect of the higher frequencies? What happens to the strength of muscle contraction and the extent of muscle relaxation when the frequency of stimulation is increased?

4. Effects of Stretch or Load on Muscle Contractility

The effects of load or stretch on a muscle prior to its contraction was first described by Fenn in the 1890s. Later, Starling found that the isolated heart responded in the same manner, and this became widely known as Starling's law of the heart.

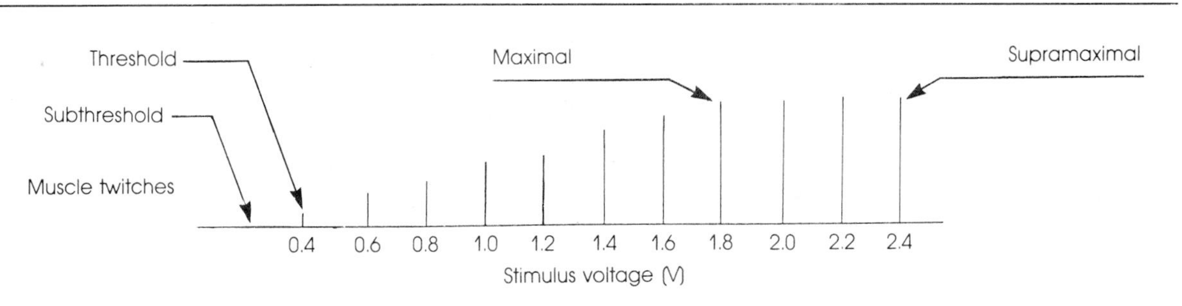

FIGURE 15.6. Muscle contractions at various strengths of stimuli.

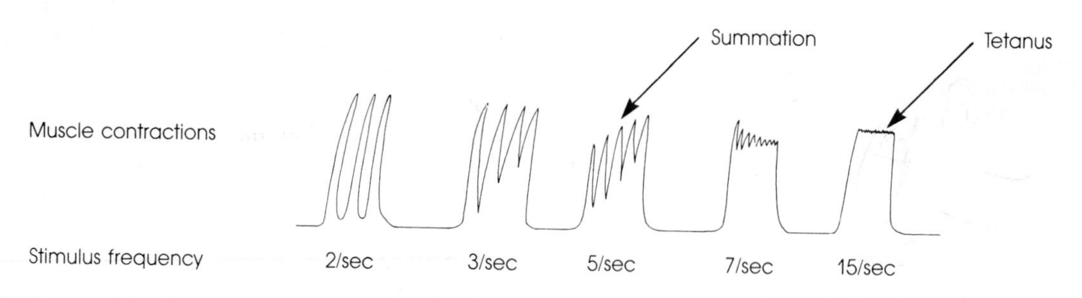

FIGURE 15.7. Muscle contractions at various frequencies of stimuli.

This basic characteristic of muscle response can be demonstrated by increasing the weight that a muscle must lift or by stretching the muscle in a progressive, stepwise manner. The method employed in lab will depend on the recording system available.

a. Effect of Stretch: Length-Tension Curve (Physiograph System)

Mount a metric ruler on the ring stand so that muscle length and stretch can be measured by sighting over the rod that holds the transducer (Figure 15.4). Adjust the amplifier sensitivity so that a single supramaximal stimulus voltage produces a pen deflection of around 2 cm. Adjust the pen so it is at the lowest position in the range of pen movement. Then use the tension adjuster to reduce the stretch on the muscle until a supramaximal voltage produces only a very small pen deflection (1–2 mm) when the muscle contracts. Record this as 0 mm of muscle stretched length. Record also the ruler measurement at this muscle length by sighting over the rod that holds the transducer.

Stimulate the muscle with a single pulse of supramaximal voltage. Record the millimeters of stretch placed on the muscle by writing under each contraction. Use a very slow paper speed for the rest of the experiment.

Stretch the muscle 1 mm and stimulate it as before. Continue in this manner, stretching the muscle in 1 mm increments, stimulating at each new length, and recording the mm of stretch under each contraction. Allow at least 10 seconds of rest time after each stimulation before the muscle is stretched and stimulated again. Continue the experiment until no contraction occurs when the muscle is stimulated. *Keep the muscle moist with frog Ringer's during the entire experiment.* The record produced should resemble that shown in Figure 15.8.

Note: The baseline of the recording will be progressively elevated as the muscle is passively stretched between active contractions. If the pen deflection exceeds the range of pen movement, you will need to reposition the pen lower in the range in order to provide room for recording the muscle contractions.

From your record, determine the *passive* (resting) tension on the muscle in millimeters of pen deflection from the baseline position. Then determine the *active* (contractile) tension developed when the muscle was stimulated.

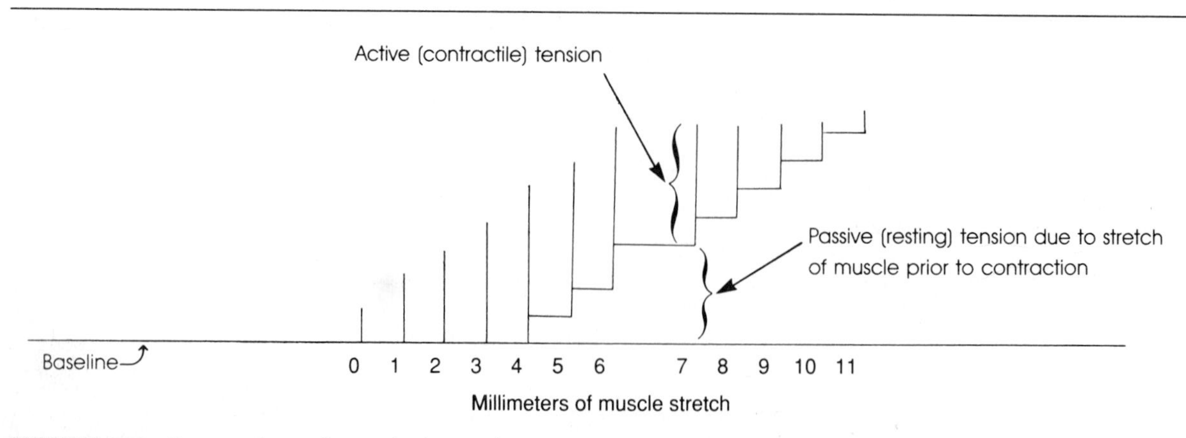

FIGURE 15.8. Contractions of stretched muscle.

In the Laboratory Report, record the mm of *passive* and active tension developed at each muscle length. Then graph the data, with relative muscle length on the abscissa and mm of tension on the ordinate. Plot all three tensions: active, passive, and total (active + passive) on the same graph, using different colors. At what stretched length does the muscle have the greatest active tension? At what stretched length is the active tension reduced to 0?

b. Effect on Load (Kymograph System)

Hang a weight pan from the lever arm directly under the point of muscle attachment as shown in Figure 15.1. The weight of the pan itself will stretch the muscle slightly but will be recorded as zero load when you begin the experiment. Adjust the lever arm and kymograph drum so that the first contractions are recorded in the upper part of the drum. Then, as weights are added, the muscle will be stretched and the contractions will be recorded lower on the drum.

Begin by stimulating the zero-loaded muscle with a single supramaximal voltage. Use a very slow drum speed for the rest of the experiment. After a recovery period of 10 seconds, add 5 g to the weight pan and stimulate as before. Continue in this manner, adding an additional 5 g every 10 seconds and stimulating to record the contractile tension developed. Be sure to mark on the record the weight lifted by the muscle at each contraction. Continue adding weights and recording the muscle contractions until the contraction height is drastically reduced. *Remember to keep the muscle moist with frog Ringer's solution at all times.* The record produced should resemble that shown in Figure 15.9.

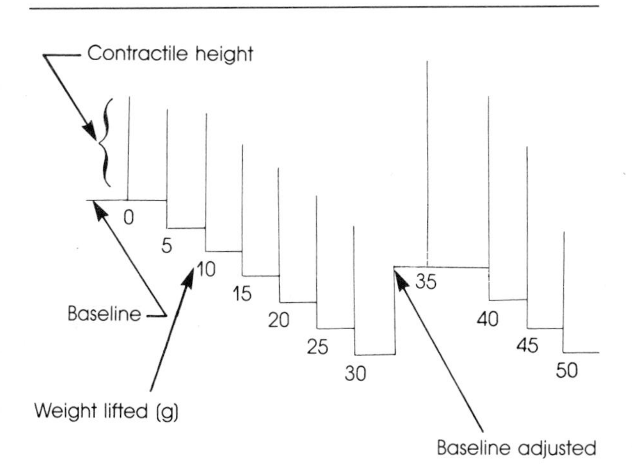

FIGURE 15.9. Contractions of weighted muscle.

In the Laboratory Report, record the weight lifted by the muscle in grams and the relative strength of contraction (mm of pen deflection). Also plot the relative work (g-mm) versus the weight lifted. What happens to the contractile strength as the load is increased?

5. Neuromuscular Fatigue

Lift the sciatic nerve out from between the muscles and lay it over the stimulating electrodes. Stimulate the nerve with a supramaximal voltage and a tetanizing frequency (10–15 /sec). Continue stimulating until contraction fatigue is observed (contraction height is reduced by $\frac{1}{2}$).

Stimulate the nerve with a single shock of supramaximal voltage and record the strength (height) of the muscle contraction. Then place the electrodes directly on the muscle, stimulate it in the same manner, and record the twitch.

Allow the preparation to rest for 5 minutes and then repeat the nerve and muscle single shock stimulations. Has complete recovery taken place?

6. Neuromuscular Blockade

Inject 0.1 ml of tubocurarine (3 mg per ml) into the belly of the muscle near the entrance of the nerve. An alternative method is to inject 1 ml of tubocurarine under the skin on the back of the frog so the drug can be absorbed by the lymphatic system into the blood.

Using a very slow paper speed, stimulate the nerve every 10 seconds, using the same single supramaximal voltage used previously. Within 5 minutes, you should see a change in the strength of the muscle contraction.

When the contractions reach one-half the initial height, stop the nerve stimulation. Stimulate the muscle directly and record the contraction height.

STIMULATION OF MOTOR POINTS

The intact skeletal muscles of humans can be stimulated directly through the skin if a fairly strong stimulus is employed. Such stimulation is used in the diagnosis of certain neuromuscular disorders and to prevent muscle atrophy during temporary paralysis in diseases such as poliomyelitis. Certain spots on the muscle are more sensitive to electrical stimulation than is the rest of the muscle; these spots are called the *motor points*. The motor point usually lies over the point where the nerve enters the muscle; thus, the muscle contraction is produced through stimulation of the innervating nerve. Most motor points are located over the belly of the muscle.

Experimental Procedures

1. Place electrode paste on an electrocardiogram (ECG) plate electrode and secure the electrode to the upper arm with a rubber strap (Figure 15.10). Connect the plate electrode to the negative (−) output terminal of the electronic stimulator. This will serve as the reference electrode.
2. Connect the exploring electrode to the positive (+) output terminal of the stimulator. A banana plug attached to a wire is used as the exploring electrode. Apply a small dab of electrode paste to the tip of the banana plug, and reapply fresh paste as needed during the experiment.

   Note: The best type of stimulator to use for motor point stimulation is one having three terminals: positive, negative, and chassis ground. On stimulators with only two terminals, the negative is common with the ground terminal. If a two-terminal stimulator is used, the subject should not stimulate himself, as he may touch the stimulator, complete the electrical circuit, and receive a strong shock.
3. Have the subject lay his or her forearm on a table with the flexor surface upward (palm up). Set the stimulator at 40 to 60 V, 1-msec duration, and repeat stimulations of one pulse per second. Using the exploring electrode, stimulate various points on the forearm to locate as many motor points as possible. Mark with ink each motor point that produces muscle twitches.
4. On the illustration of the arm provided in the Laboratory Report, indicate the location of the motor points you have stimulated and the threshold voltage of each motor point.
5. Locate the motor points for the flexor muscles of the fingers (e.g., flexor carpi ulnaris or radialis). Increase the voltage stepwise to show gradations of contraction up to a maximum. Then, using a maximal voltage, slowly increase the frequency of stimulation until tetanus is produced, but do not maintain tetanus longer than 2–3 seconds.

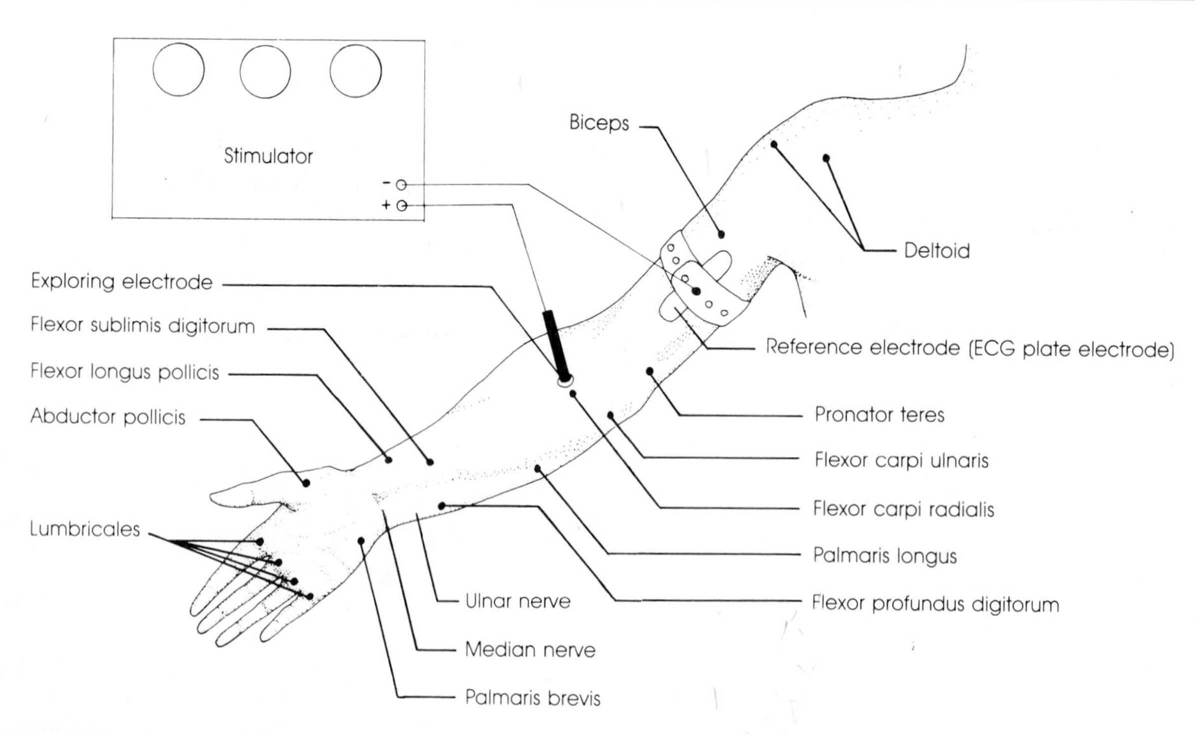

FIGURE 15.10. Stimulation of motor points.

LABORATORY REPORT

Name ______________________

Date __________ Section __________

Score/Grade ______________________

15. Nerve-muscle Activity

1. Muscle and Nerve Irritability

Threshold voltage with nerve stimulation .2

Threshold voltage with direct muscle stimulation 1.1

Why is there a difference in these threshold voltages?

Stimulation via the nerve is lower because it excites the muscle by releasing acetylcholine at the neuromuscular junction while direct muscle stimulation operates via voltage regulated gates

What is meant by the independent irritability of muscle?

the muscle membrane is excitable (capable of being depolarized) directly rather than through nerve stimulation

2. Effects of Stimulus Strength

Place your record in the following space.

9 volts

muscle contractions

stimulus applied voltage 0.5 0.7 1.0 1.5 2.0 2.5 3.0 3.5 4.0 .5 6

sub threshold strength stimuli (no peaks)

threshold stimulus (first peak)

supra threshold or sub maximal stimulus

maximal stimulus

supra maximal stimuli

How is the response obtained related to the all-or-none law of muscle fiber contraction?
each muscle fiber (cell) has a certain voltage at which it depolarizes (becomes excited). When this voltage is ~~not~~ reached the fiber contracts fully (all) If the voltage is not reached, it doesn't contract (none). Increasing stimulus strength causes the threshold of more & more fibers to be reached ∴ stronger contractions (Spatial summation)

3. Effects of Stimulus Frequency

Place your record in the following space.

1/sec 2/sec 3/sec 4/sec 5/sec 10/sec

Explain the mechanism responsible for summation of contractions and the increase in height of contraction when the stimulus frequency is increased. Wave summation
Increasing frequency does not permit Ca++ to be removed from sarcomere back into sarcoplasmic reticulum & muscle cannot relax before next impulse reaches the fiber & the sarcomere begins its next contraction on a partially contracted state.

What is tetanus? Why is it produced?

It is a fusion of muscle twitches with no relaxation, as the stimulus frequency is ↑ by continuous release of calcium from the sarcoplasmic reticulum.

Name ______________________

Date __________ Section __________

Score/Grade ______________________

4. Effects of Stretch or Load on Muscle Contractility

a) Effect of Stretch: Length-Tension Curve (Physiograph System)

Place your record in the following space.

MUSCLE STRETCHED LENGTH (MM)	PASSIVE TENSION (MM)	ACTIVE TENSION (MM)	TOTAL TENSION (MM)

Relative Tension (mm pen deflection)

Muscle Stretched Length (mm)

How is the contractile tension developed with increasing stretch related to the sarcomere structure of the muscle fiber?

Of what practical importance is this response of muscle to stretch in the *in vivo* situation?

b) Effect of Load (Kymograph System)

Place your record in the following space.

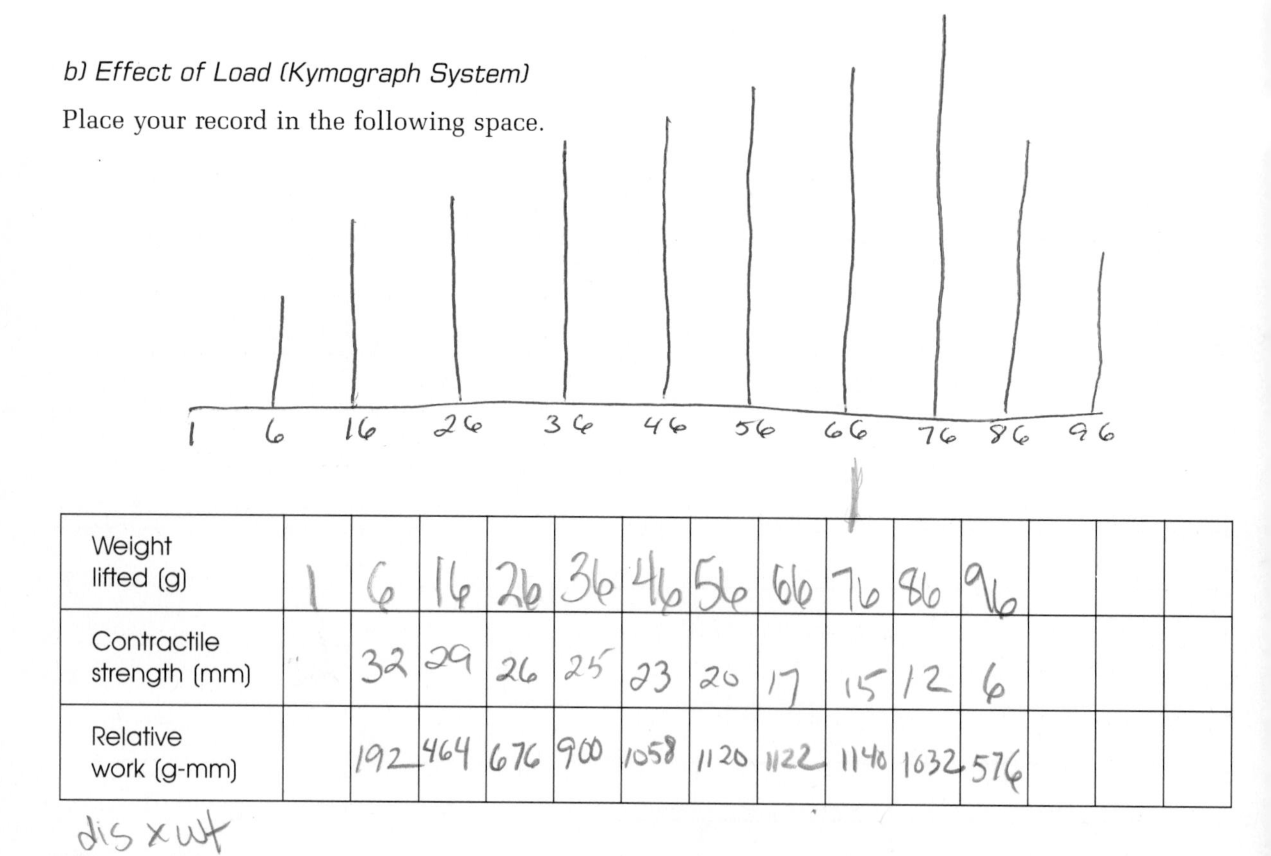

Weight lifted (g)	1	6	16	26	36	46	56	66	76	86	96			
Contractile strength (mm)		32	29	26	25	23	20	17	15	12	6			
Relative work (g-mm)		192	464	676	900	1058	1120	1122	1140	1032	576			

dis x wt

Name ______________________

Date __________ Section __________

Score/Grade ______________________

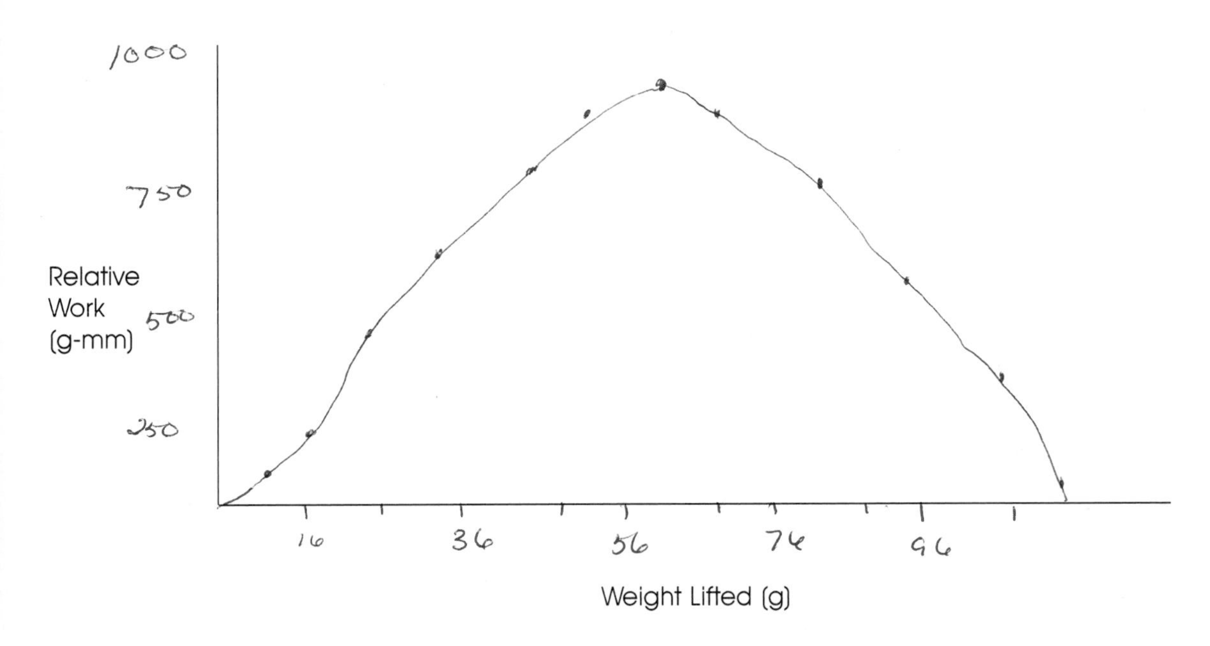

What is meant by optimal load? What is the optimal load for your muscle?

load at which most work is being done

56 grams

Why does the contractile tension and work decrease with heavier loads?

with heavy loads, the myosin cross bridges are pulled away from potential actin binding sites therefore fewer cross bridges will be formed on stimulation

5. Neuromuscular Fatigue

Record the height of the muscle contractions in mm of pen deflection.

Immediately after fatigue		**After 5 minutes of rest**	
Nerve stimulated	Muscle stimulated	Nerve stimulated	Muscle stimulated
__________	__________	__________	__________

Explain the difference in contraction height obtained with nerve and muscle stimulation immediately after fatigue. Where does fatigue occur first? Why?

Which has benefitted most from the rest period—the nerve or muscle? Explain the processes that enable this recovery.

6. Neuromuscular Blockade

Record the height of the muscle contractions before and after tubocurarine injection.

Before Tubocurarine injection		**After Tubocurarine injection**	
Nerve stimulated	Muscle stimulated	Nerve stimulated	Muscle stimulated
________	________	________	________

How do you explain the differences in contraction height? What is the mechanism of action of tubocurarine?

What clinical use is made of drugs such as tubocurarine?

Stimulation of Motor Points

1. On the following diagram, indicate the location of the motor points you have stimulated and the threshold voltage at each point.

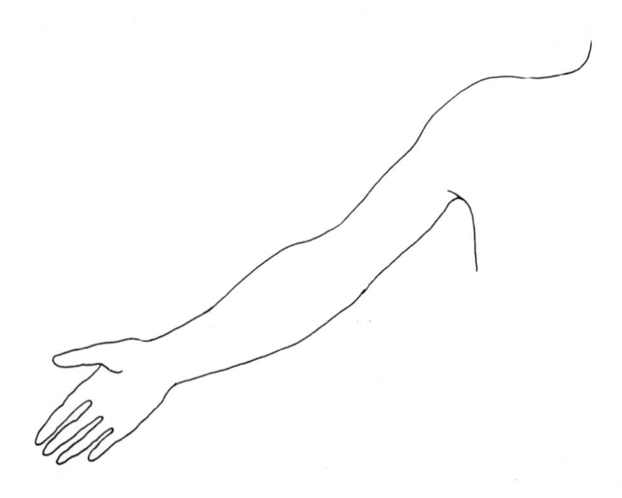

2. At what frequency does tetanus occur when the finger flexor muscles are stimulated?

____

Name ______________________

Date ____________ Section ____________

Score/Grade ______________________

3. How does this frequency compare with the frequency needed to tetanize the frog gastrocnemius muscle you worked with previously?

15-20 for frog muscle in vitro

4. What is a motor unit?

a motor nerve and all muscle fiber which are innerrated by its branchs

5. How does the anatomical composition of motor units controlling different muscle groups vary? What is the purpose of this variation?

ratio of muscle fibers/neuron may be as low as 5:1 in the fingers and eye muscles to as large as 1500:1 in large muscles of leg + back

a low ratio of muscle to neuron indicates greater degree. (fineness) of control - dexterity etc.

a high ratio generally indicates coarse control + usually greater power production

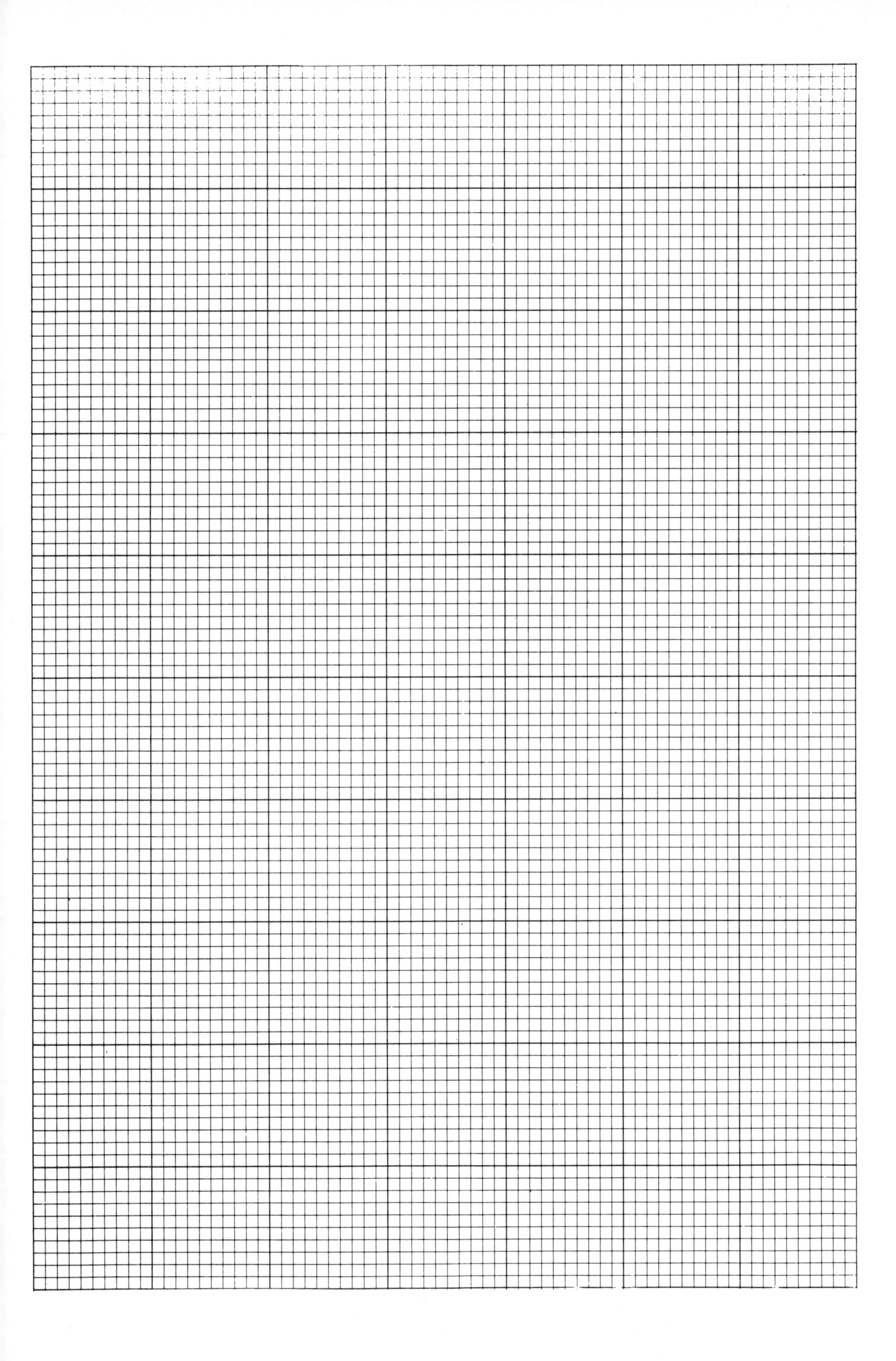

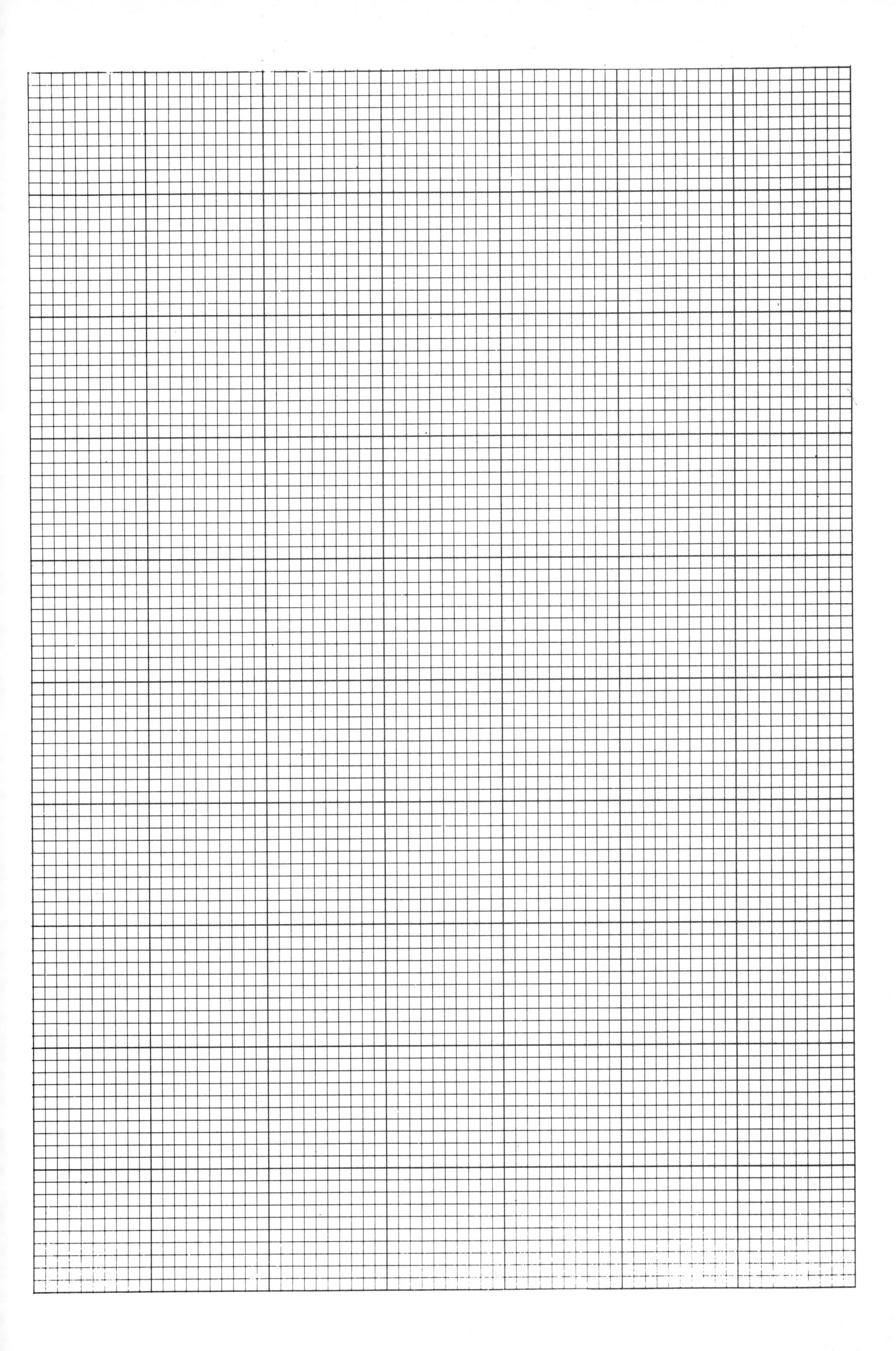

16 Cardiac Function

CHARACTERISTICS OF HEART CONTRACTILITY

The heart's primary function is simply to act as a pump that provides pressure to move blood to its ultimate destination—the tissues. The control of cardiac contractility is complex and represents a balance of **intrinsic** (within the heart) and **extrinsic** (from outside the heart) factors. In the following experiment you will examine some of these intrinsic and extrinsic factors that make the heart such a unique and versatile pump.

Cardiac muscle differs from skeletal muscle both morphologically and functionally. Probably the most striking and fascinating feature of its contractility is its automaticity, that is, its ability to initiate its own rhythmic contractions without requiring a stimulation from outside the heart. This automaticity is believed to be due to "leaky" cell membranes, in which the calcium ions slowly leak into the cells. This leaking causes a slow depolarization to threshold, thus firing an action potential and initiating contractions of cardiac muscle. The cells that are most "leaky" to calcium and that depolarize fastest control the rate of contraction of all other cardiac cells; thus, they act as **pacemakers** for the rest of the heart. In the mammalian heart, the pacemaker is the **sinoatrial (SA) node,** a group of specialized cells near the junction of the vena cava and the right atrium. In the frog or turtle heart, the pacemaker is the **sinus venosus,** an enlarged region between the vena cava and the right atrium. (The mammalian SA node is believed to be an evolutionary remnant of the sinus venosus.) Each region of the heart has its own intrinsic rate of beating: for example, SA node, 72 beats/min; atrium, 60 beats/min; ventricle, 25 beats/min. Only when the faster pacemaker region is blocked is it possible to observe the intrinsic rate of the slower regions. Other intrinsic characteristics of cardiac contraction are shown in the following experiments.

ANATOMY OF AMPHIBIAN OR REPTILIAN HEART

In this series of experiments you will use a frog heart or turtle heart, because it functions well at room temperature and will continue to beat even when excised from the body. Mammalian hearts have the same contractile characteristics, but must be supplied with a constant flow of warm, oxygenated blood to maintain their contractility.

The frog and turtle hearts differ from the mammalian heart anatomically in that they are three chambered rather than four chambered. The pacemaker in the amphibian heart is the sinus venosus, a thin-walled sac that receives blood from the anterior and posterior caval veins and empties blood into the right atrium. The single ventricle receives blood from both atria and pumps blood out through the large artery called the **truncus arteriosus** (Figure 16.1). In contrast, the mammalian ventricle has separate left and right chambers, which prevent mixing of the venous and arterial blood (Figure 16.9).

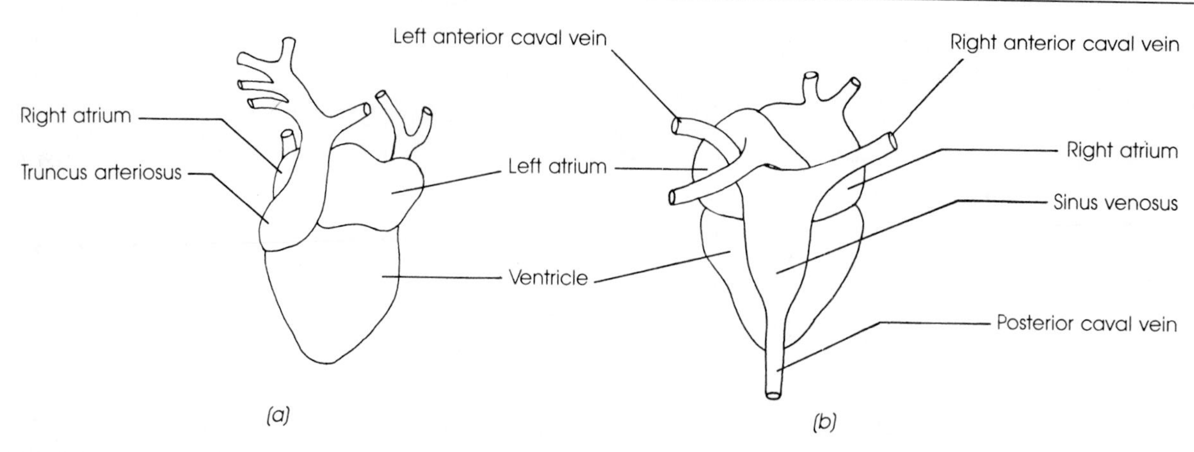

FIGURE 16.1. Frog heart. (a) Ventral view. (b) Dorsal view.

Dissection Procedures

1. Frog Heart Preparation

Double pith a frog and fasten it to a frog board, ventral side up. Use scissors to make a longitudinal incision through the skin and body wall of the thoracic region to expose the heart. Note the pericardial sac surrounding the heart. Hold the pericardium with forceps and carefully cut away the sac from the heart, using scissors. From this point on make sure that the heart is periodically moistened with frog Ringer's solution.

Using forceps, gently lift the apex of the heart upward. Insert a bent insect pin or small fishhook through the tip of the ventricle, being careful not to damage the ventricle. Tie a thin thread to the hook and connect the ventricle to the transducer just as you connected the gastrocnemius muscle to the transducer in Experiment 15. If the kymograph is used, fasten the thread to the heart lever, with modeling clay, near the fulcrum of the lever (Figure 16.2a). If the Physiograph transducer is used, attach the thread to the transducer hook and adjust the tension on the ventricle until the recording pen is raised slightly above the baseline (Figure 16.2b).

2. Turtle Heart Preparation

Using a heavy forceps, grab the turtle's upper beak and pull the head out. Slip a ligature of heavy cord over the head and pull it tightly around the neck. Destroy the brain by pithing it with a heavy probe. It is also possible to anesthetize the turtle by injecting 2 ml of 3% pentobarbital solution (Nembutal) into the peritoneal cavity (insert the needle between the hind limbs and the body).

Cut a circular opening in the plastron (ventral shield) at the level of the heart using a drill press fitted with a hole saw. If a motorized saw is not available, cut away the entire plastron using a hand saw. This latter method is not as desirable because of the chance of hitting major blood vessels and causing excessive hemorrhage.

Attach ligatures to each of the turtle's legs and tie the head and legs to a turtle board. Cut away the pericardial sac around the heart. Tie a thread around the frenulum cordis near the apex of the ventricle, cut the frenulum peripherally, and attach the thread to a heart lever or transducer. Keep the heart moist with frog Ringer's.

PHYSIOLOGY OF AMPHIBIAN OR REPTILIAN HEART

Experimental Procedure

1. Normal Heartbeat

Obtain a recording of the normal cardiac rhythm, using a medium to fast paper speed to distinguish the atrial and ventricular contractions. Run a 1-second time line while recording so that the duration of systole and diastole of the ventricle can be determined. Your record should be similar to that shown in Figure 16.3. Attach a labeled portion of your record to the Laboratory Report.

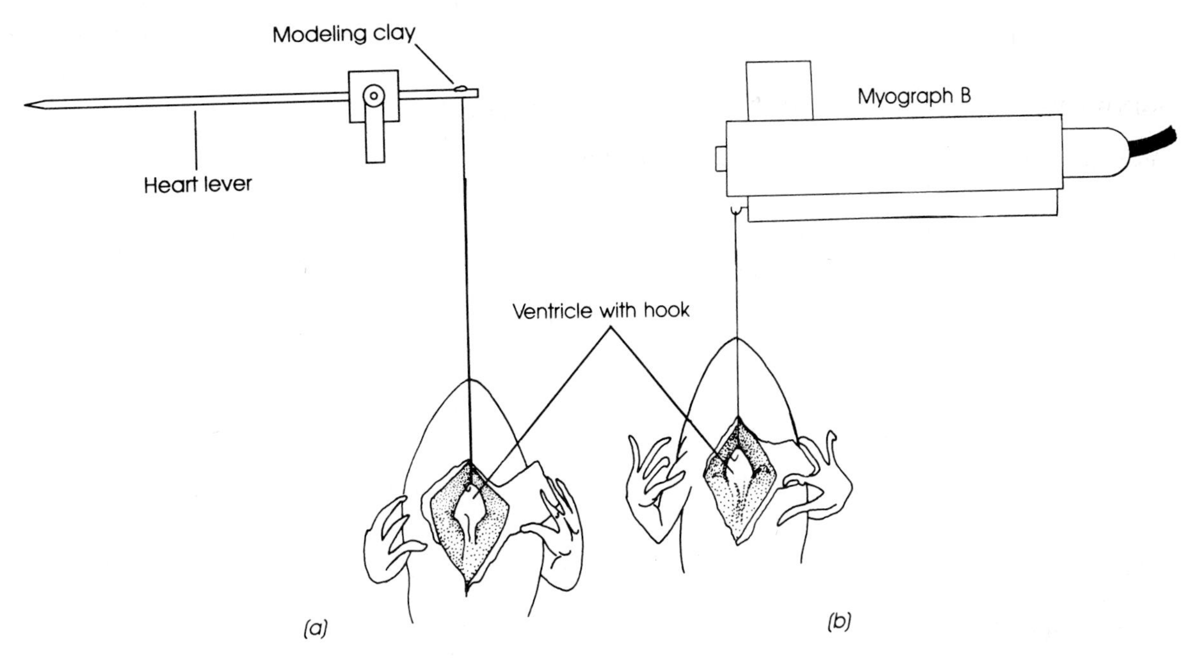

FIGURE 16.2. Frog heart preparations. (a) Kymograph setup. (b) Physiograph setup.

2. Refractory Period of Heart

Position the transducer to eliminate as much as possible the atrial contraction in the recording. Arrange for electrical stimulation of the ventricle by clamping the stimulating electrode so that the points touch the ventricle gently and constantly during the contraction cycle. An alternative method is to connect the stimulator to the heart by means of fine copper wires. Wrap one wire around the base of the ventricle and the other around the apex of the ventricle, thus providing a closed circuit through the length of the ventricle.

Record the ventricular contractions, using a medium paper speed. Using single stimuli of 20 V and 1-msec duration, stimulate the ventricle at different times in the cardiac cycle as shown in Figure 16.4. Begin at time 6 and work backward through the cycle.

What is the result of stimulating during the systolic phase of the cycle? During the diastolic phase? Can a second contraction be elicited before the normal rhythmic contraction occurs? Look for the appearance of an extra systole followed by a compensatory pause as shown in Figure 16.5.

It is often a tricky procedure to obtain an extra systole using single stimuli, because it is difficult to catch the ventricle immediately after its refractory period. If you have trouble, try using repeated stimulation so that one of the stimuli can catch the ventricle at the proper time to produce an extra systole.

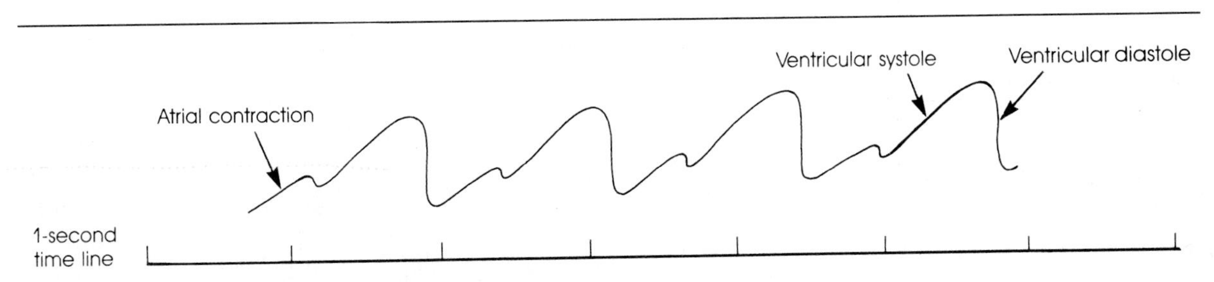

FIGURE 16.3. Record of normal heartbeat.

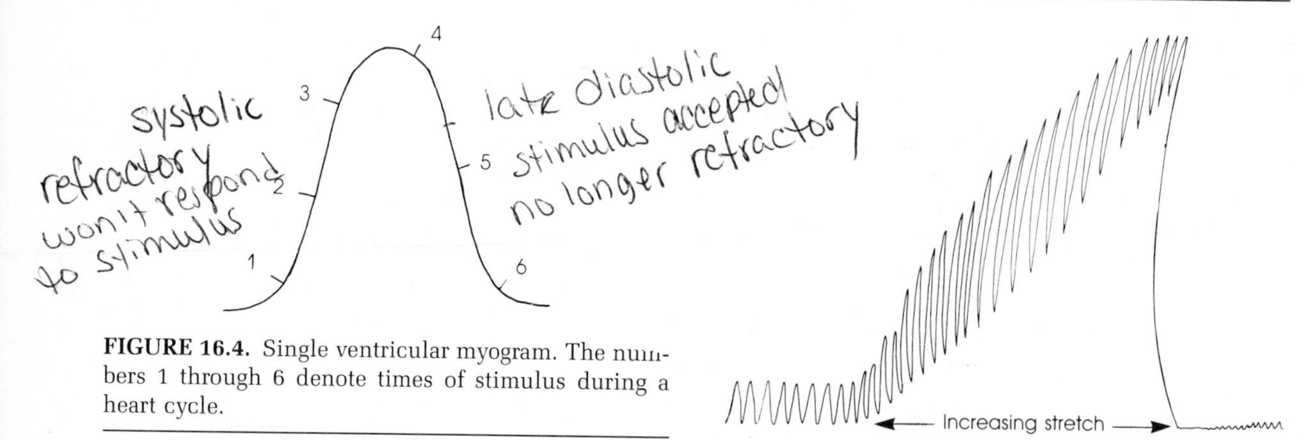

FIGURE 16.4. Single ventricular myogram. The numbers 1 through 6 denote times of stimulus during a heart cycle.

FIGURE 16.6. Contractions of a stretched ventricle.

3. Tetanization of Heart

Record the contraction of the heart while it is being stimulated with a tetanizing frequency (10 to 15 stimuli per second) and 20 V. How do your results compare with tetanization of skeletal muscle? How is this response related to the refractory period?

4. Starling's Law of the Heart

In the early 1900s, Ernest Starling's investigations revealed that "the energy of contraction is proportional to the initial length of the cardiac muscle fiber." This statement became known as Starling's law of the heart, a major concept in cardiovascular physiology. In the intact animal, the length of the cardiac muscle fiber is increased by an increase in diastolic filling of the heart. In this experiment, you will increase the fiber length by simply stretching the ventricle.

Adjust the transducer sensitivity so that the height of the ventricular myogram is 2 cm. Position the recording pen so that contractions are recorded near the bottom of the pen excursion range.

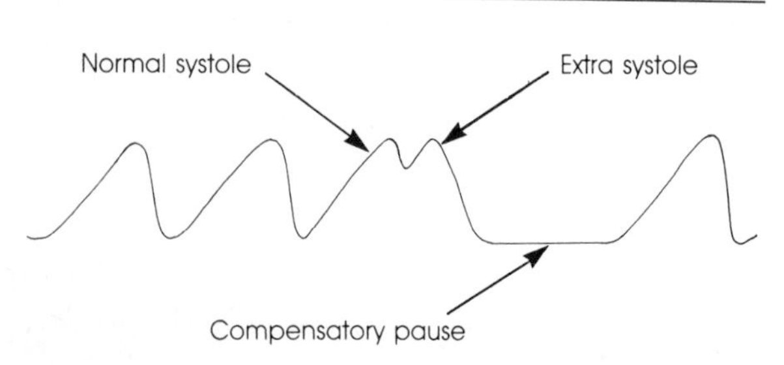

FIGURE 16.5. Extra systole.

Using a medium paper speed, begin stretching the ventricle incrementally by adding 5-g weights to the muscle lever (kymograph system) or turning the tension adjuster (Physiograph system). Continue stretching the ventricle until no additional change is noted. The record obtained should resemble that shown in Figure 16.6.

5. Temperature Effects

Record the heart contractions at room temperature. Then drop warm (40 °C) frog Ringer's solution on the heart until significant changes are seen in rate and contractility. Record contractions at this time. Finally, drop cold (5 °C) Ringer's on the heart and record when changes are observed. Then rinse the heart with room-temperature Ringer's to return the beat to normal before continuing the experiments. Can you see why the poikilothermic animal becomes so sluggish when the temperature drops? Determine the heart rate at each temperature.

6. Drug Effects

Using a syringe or medicine dropper, apply the following drugs to the heart until you see significant changes in rate, contractility, or tone (changes in baseline). Best results are seen if the drug is dropped on the sinus venosus region of the heart. Be very cautious in applying these drugs, because they are very potent and may stop the heart completely if an overdose is given. Acetylcholine and nicotine are especially potent. If the heart does stop, apply epinephrine to restore the beat. After the effect is recorded, rinse the heart with frog Ringer's and allow the heart to return to normal before the next drug is applied.

a. Acetylcholine

0.1 mg/cc (1:10,000). This is the normal transmitter released by the vagus (parasympathetic) nerve innervating the heart.

b. Epinephrine

1 mg/cc (1:1000). This is an analog of norepinephrine, released by the postganglionic sympathetic nerves innervating the heart.

c. Pilocarpine

0.2 mg/cc (1:5000). This is an alkaloid drug from the leaf of *Pilocarpus jaborandi.* It acts as does acetylcholine by directly stimulating muscarinic receptors in effector organs.

d. Nicotine

1 mg/cc (1:1000). This is an alkaloid from the tobacco plant. It acts on the autonomic ganglia, combining with the acetylcholine receptor on the postganglionic neuron. In small doses, nicotine stimulates synaptic transmission in the ganglia, whereas in larger doses it depresses synaptic activity. Because the terminal ganglion (parasympathetic division) lies in the heart muscle, this ganglion can be affected by dropping nicotine directly on the heart. Could the sympathetic ganglia also be affected by this application?

e. Atropine

1 mg/cc (1:1000). This is an alkaloid from *Atropa belladonna* that blocks the receptors for acetylcholine. After applying atropine and observing its effect, drop some acetylcholine on the heart and compare the heart's response with that seen when you applied acetylcholine previously.

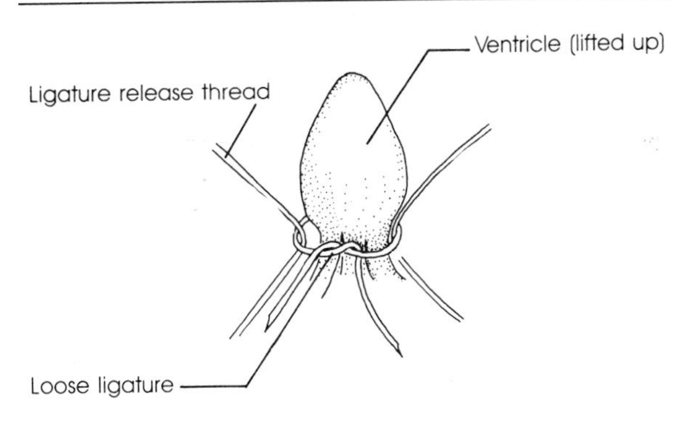

FIGURE 16.7. Heart ligature.

7. Heart Block

Using a heavy thread, tie a loose single-loop ligature around the heart at the junction of the atria and ventricle. Take two short pieces of thread and run each one through the ligature on opposite sides of the loop (Figure 16.7). A Gaskell clamp may also be used to apply pressure on the atrioventricular junction.

Tighten the ligature slowly and observe the beating of the atria and ventricle for changes in rhythm. Do not tighten the ligature too much or you will cut the heart and damage it irreversibly. Can you see any of the following types of heart block?

First-degree block. The interval between atrial and ventricular contraction is prolonged.

Second-degree block. Some impulses fail to reach the ventricle so that the ratio of atrial to ventricular beats is altered. You may be able to see 2:1, 3:1, 5:1, 8:1 types of heart block (Figure 16.8).

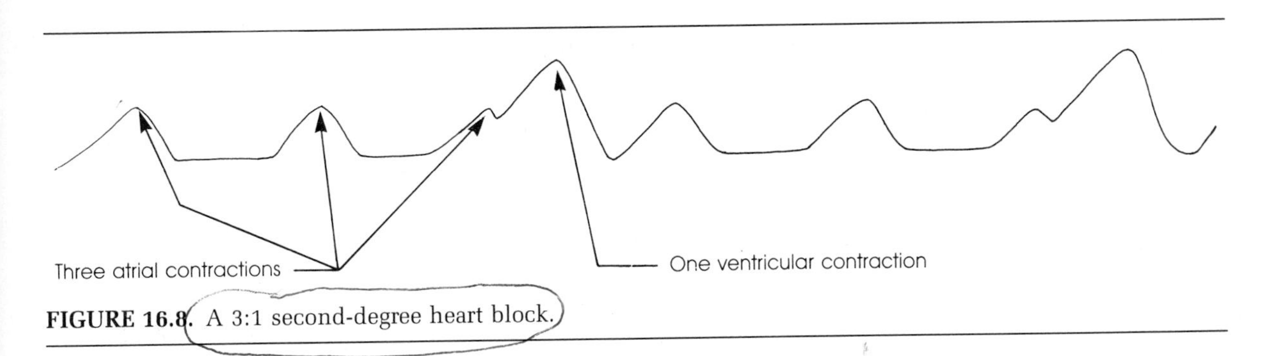

FIGURE 16.8. A 3:1 second-degree heart block.

Third-degree block. Impulses fail to pass through the atrioventricular (AV) node and bundle of His and the ventricle may start its own independent rhythm of beating, or you may see only the atria contracting.

After producing a complete heart block (third degree), release the ligature by pulling on the release threads and see if a normal AV beat is restored.

8. All-or-None Law of the Heart

Increase the pressure on the AV ligature until the ventricle is completely quiet (no impulses are reaching the ventricle). Using electrodes, stimulate the ventricle to determine the threshold stimulus. Record the contraction height. Using a slow paper speed, stimulate the ventricle with increasing voltages (single stimuli), recording the contraction height at each voltage.

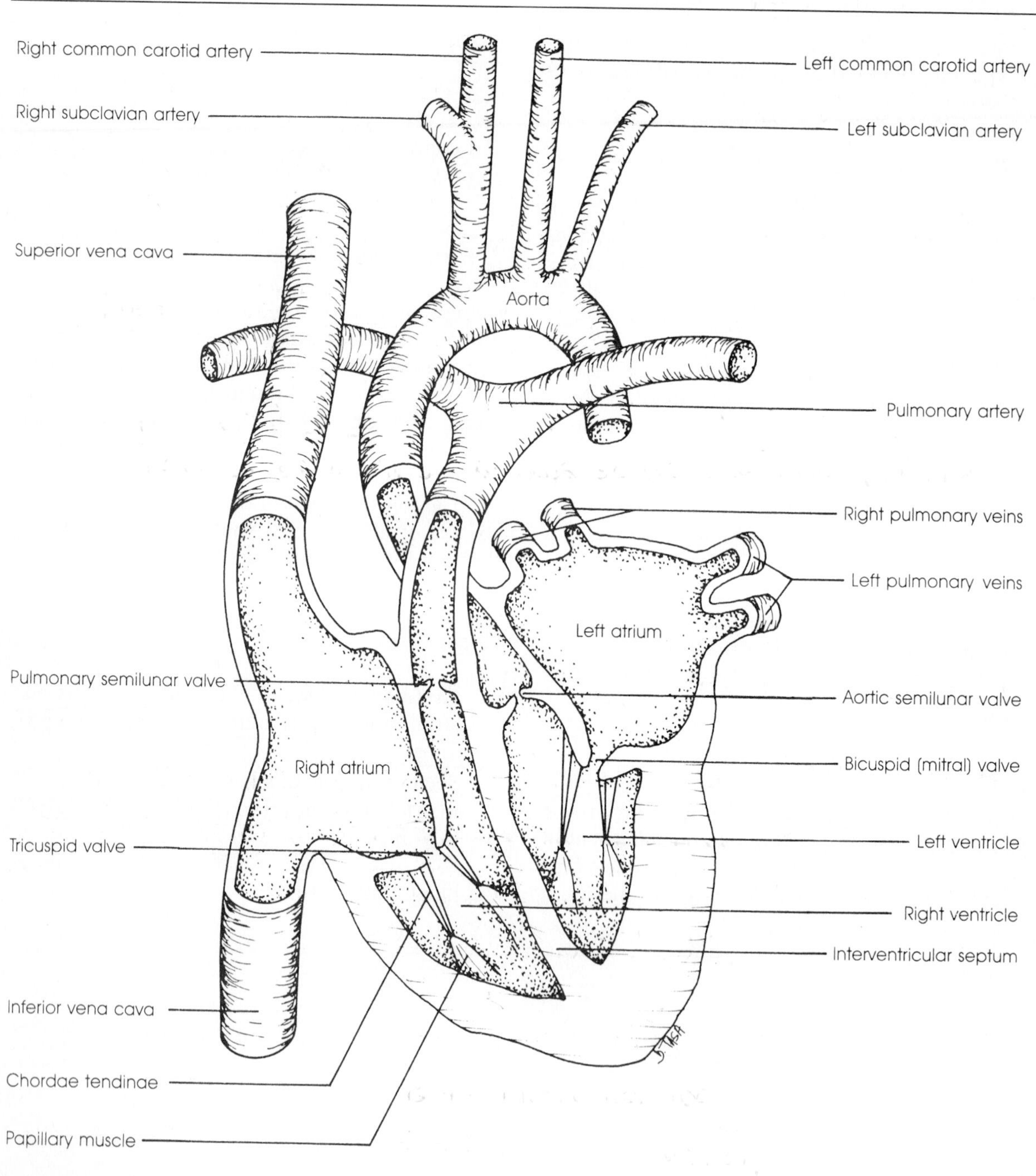

FIGURE 16.9. Mammalian heart.

LABORATORY REPORT

16. Cardiac Function

Name ______________________

Date ____________ Section ____________

Score/Grade ______________________

Physiology of Amphibian or Reptilian Heart

1. Normal Heartbeat

Place your record in the following space.

What is the heart rate at room temperature? 30-40 beats/min (2 sec. -1.5)

Duration of systole = 1/2 sec   Duration of diastole = 1.5-1 sec

What causes the delay between the beat of the atria and the ventricle?

Conduction fibers of A.V. node are difficult to excite, therefore, conduction thru A.V node is delayed

2. Refractory Period of Heart

Place recording of extra systole in the following space.

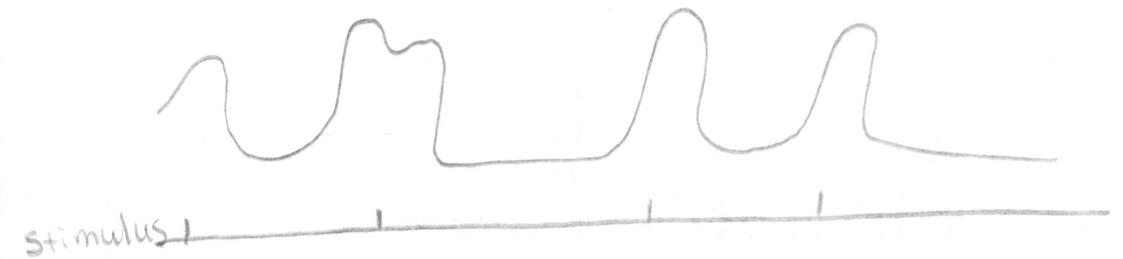

During which part of the ventricular cycle can extra systoles be obtained?

Diastolic

How does the duration of the ventricular refractory period compare with that for skeletal muscle? For a neuron?

Refractory Period

heart 300 m sec

Skeletal 10 m sec

nerve 1 m sec

Of what value is the length of the refractory period to the pumping action of the heart?

allows filling completely

What causes the compensatory pause following an extra systole?

Ventricle is in a refractory period of extra beat when the natural impulse from pace maker arrives & it loses one contraction

3. Tetanization of Heart

Place your record in the following space.

Explain the response obtained and its importance in the functioning of the heart as a pump.

the heart cannot be completely tetanized but does allow some relaxation due to long refractory period. Allows some pumping (reduced) even when another tetanizing stimulus is given

4. Starling's Law of the Heart

Place your record in the following space.

effect of increasing stretch

Under what conditions would Starling's law be of importance in the intact animal?

Allows the two sides of the heart to adjust stroke volume to the amount of blood returning to it

A person suffering from heart failure often has an enlarged heart (hypertrophy). How is this hypertrophy related to Starling's law?

It responds to increased load (fill) with increased contraction resulting in a need to increase muscle size.

Name ______________

Date ________ Section ________

Score/Grade ______________

5. Temperature Effects

Place your comparative records in the following space. 12 × 6

Cold Ringer's (HR = 20 /min) Warm Ringer's (HR = 150 /min)

How do you explain the changes seen when temperature is altered?

temperature alters protein activity and may alter ability of SA mode cells to depolarize

6. Drug Effects

DRUG	HEART RATE	CONTRACTILITY	TONE	EXPLANATION
Acetylcholine	↓	?	↓	Ach hyperpolarizes the SA node
Epinephrine	↑	↑	↑	Epi stimulates B receptors increasing rate + contraction
Pilocarpine				
Nicotine				
Atropine	↑	No change	↑	Blocks Ach receptors

Explain the mechanisms whereby acetylcholine and epinephrine alter the heart rate.

Acetylcholine: ↑ K^+ (Potassium) permeability allowing K^+ to exit creating a greater (+) charge outside + hyperpolarizes the cell results in a longer time to reach threshold membrane potential
Slows down heart rate

Epinephrine: ↑ Na^+ (sodium) permeability allows Na^+ to enter + shorten the time required to reach threshold membrane potential
increases heart rate

Explain the response of the heart to acetylcholine following application of atropine.

Ach blocks receptors on heart muscle therefore the effect above of Ach will be prevented + the effect of sympathetic (Epinepherine produced) nerves will dominate.

7. Heart Block

In the following space, place any interesting record that you obtained.

What types of heart block did you observe?

2:1, 3:1, 6:1, 4:1, 5:1

Exactly how does the ligature produce a heart block in this preparation?

Pressure on AV node prevents repolarization of conduction p
thus blocking impulses from SA node

How might a heart block be produced in pathological cases?

by an embolus
Hypoxia to tissue ion changes, drugs may
depress conduction thru the nodes

Explain how the excitation generated by the cardiac pacemaker spreads to the ventricular muscle fibers.

SA node - atrial muscle → AV Node → Bundle of His
→ R+L Bundle Branch → Purkinje fiber → ventricle
muscle cell A) Wenc back Bundle
B) thoral Bundle
C) Bachman Bundle

8. All-or-None Law of the Heart

Place your record in the following space.

Name ______________________

Date __________ Section __________

Score/Grade ______________________

What is the all-or-none law of the heart? when one of the cardiac muscles fibers is excited to threshold-all cardiac fibers contract to maximum

How does the anatomical structure of the heart make this law possible?

intercalated discs between cells allow intercellular communication & branching fibers allow the ventricle to act as 1 cell

How does your recording compare with that for skeletal muscle stimulated with increasing stimulus strengths? Why is there a difference in response?

there is no spatial summation (multiple motor unit) in cardiac muscle as seen in skeletal muscle

17 Human Cardiovascular Function

AUSCULTATION OF HEART SOUNDS

Auscultation of the heart means to listen to and study the various sounds arising from the heart as it pumps blood. These sounds are the result of vibrations produced when the heart valves close and blood rebounds against the ventricular walls or blood vessels. The heart sounds may be heard by placing the ear against the chest or by using a **stethoscope.** The vibrations producing the sounds can be visually displayed through the use of a heart sound microphone and physiological recorder to produce a **phonocardiogram.** There are four major heart sounds, but only the first two can be heard without use of special amplification.

First heart sound. Produced at the beginning of systole when the atrioventricular (AV) valves close and the semilunar (SL) valves open. This sound has a low pitched tone commonly termed the "lub" sound of the heartbeat.

Second heart sound. Occurs during the end of systole and is produced by the closure of the SL valves, the opening of the AV valves, and the resulting vibrations in the arteries and ventricles. Owing to the higher blood pressures in the arteries, the sound produced is higher pitched than the first heart sound. It is commonly referred to as the "dub" sound.

Third heart sound. Occurs during the rapid filling of the ventricles after the AV valves open and is probably produced by vibrations of the ventricular walls.

Fourth heart sound. Occurs at the time of atrial contraction and is probably due to the accelerated rush of blood into the ventricles.

Experimental Procedure

1. Using a stethoscope, listen to your partner's heart sounds, paying special attention to the four major **auscultatory areas** on the chest where the sounds from each valve can be heard most clearly (Figure 17.1).

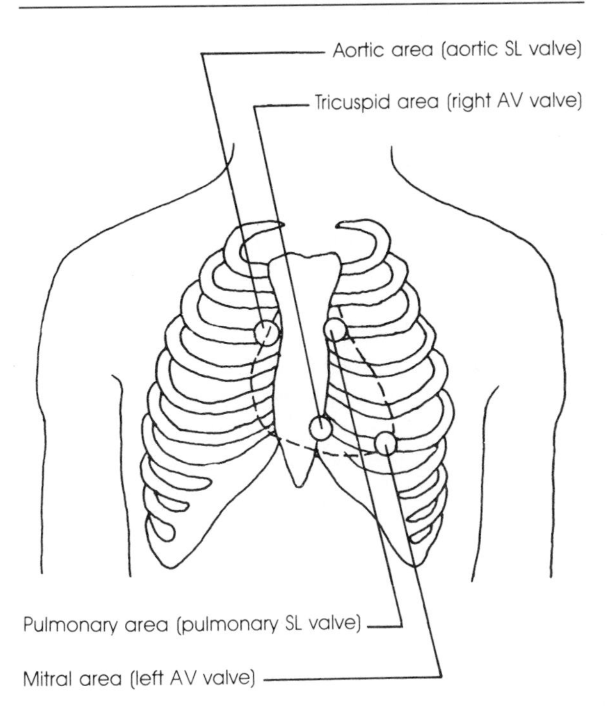

FIGURE 17.1. The four major auscultatory areas. SL, semilunar; AV, atrioventricular.

2. If equipment is available, make a recording of your partner's phonocardiogram, using a heart sound microphone and recorder. Obtain readings of the heart sounds at each of the auscultatory areas and compare them for differences in vibratory patterns.

MEASUREMENT OF BLOOD PRESSURE

The determination of an individual's blood pressure is one of the most useful clinical measurements that can be taken. By "blood pressure" we mean the pressure exerted by the blood against the vessel walls, the arterial blood pressure being the most useful, and hence the most frequently measured, pressure. You should become familiar with the following pressures used in cardiovascular physiology.

Systolic blood pressure. The highest pressure in the artery, produced in the heart's contraction (systolic) phase. The normal value for a 20-year-old man is 120 mm Hg.

Diastolic blood pressure. The lowest pressure in the artery, produced in the heart's relaxation (diastolic) phase. The normal value for a 20-year-old man is 80 mm Hg.

Pulse pressure. The difference between the systolic and diastolic pressures. The normal value is 40 mm Hg.

Mean blood pressure. Diastolic pressure plus one third of the pulse pressure. This is the average effective pressure forcing blood through the circulatory system. The normal value is 96 to 100 mm Hg.

The mean blood pressure is a function of two factors—cardiac output (CO) and total peripheral resistance (TPR). Peripheral resistance depends on the caliber (diameter) of the blood vessels and the viscosity of the blood.

$$\text{Mean BP} = \text{Cardiac output (ml/sec)} \times \text{Total peripheral resistance (TPR units)}$$

$$\text{Cardiac output (ml/min)} = \text{Heart rate/min} \times \text{Stroke volume (ml)}$$

Thus, the measurement of blood pressure provides us with information on the heart's pumping efficiency and the condition of the systemic blood vessels. In general, we say that the systolic blood pressure indicates the force of contraction of the heart, whereas the diastolic blood pressure indicates the condition of the systemic blood vessels (for instance, an increase in the diastolic blood pressure indicates a decrease in vessel elasticity).

Experimental Procedure

Blood pressure may be measured either directly or indirectly. In the **direct method,** a cannula is inserted into the artery and the direct head-on pressure of the blood is measured with a transducer or mercury manometer. In the **indirect method,** pressure is applied externally to the artery and the pressure is determined by listening to arterial sounds (using a stethoscope) below the point where the pressure is applied (Figure 17.2). This is called the **auscultatory method,** because the detection of the sounds is termed auscultation. An older and less accurate method is the **palpatory method,** in which one simply palpates, or feels, the pulse as pressure is applied to the artery. In either of these indirect methods, pressure is applied to the artery using an instrument called the **sphygmomanometer.** It consists of an inflatable rubber bag (cuff), a rubber bulb for introducing air into the cuff, and a mercury or anaeroid manometer for measuring the pressure in the cuff. Human blood pressure is most commonly measured in the brachial artery of the upper arm. In addition to being a convenient place for taking measurements, it has the added advantage of being at approximately the same level as the heart, so that the pressures obtained closely approximate the pressure in the aorta leaving the heart. This allows us to correlate blood pressure with heart activity.

1. Palpatory Method

Have the subject seated, with his or her arm resting on a table. Wrap the pressure cuff snugly around the bare upper arm, making certain that the inflatable bag within the cuff is placed over the inside of the arm where it can exert pressure on the brachial artery. Wrap the end of the cuff around the arm and tuck it into the last turn, or press the fasteners together to secure the cuff on the arm. Close the valve on the bulb by turning it clockwise.

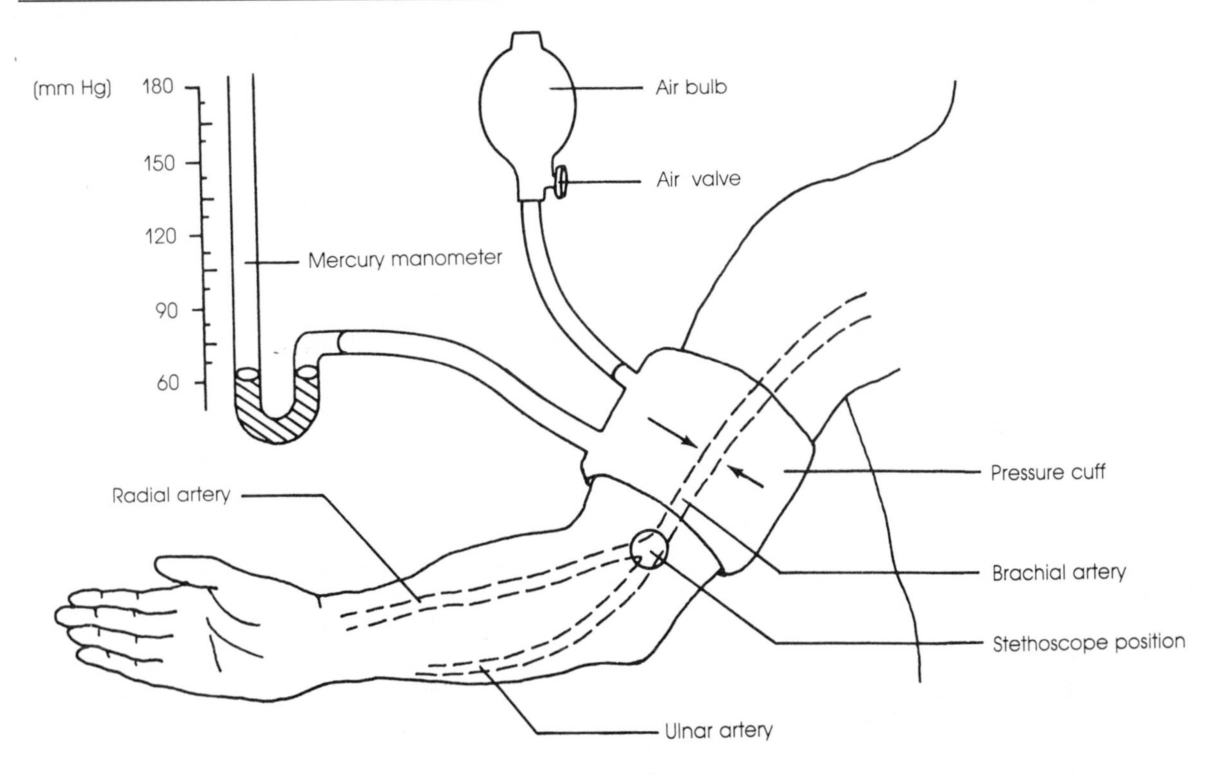

FIGURE 17.2. Apparatus for measuring blood pressure indirectly.

With one hand, palpate (feel) the radial pulse in the wrist. Slowly inflate the cuff by pumping the bulb with the other hand and note the pressure reading when the radial pulse is first lost. Then increase the pressure to around 20 mm Hg above this point. Slowly reduce the pressure in the cuff by turning the valve counterclockwise slightly to let air out of the bag. Note the pressure when the radial pulse first reappears. This is systolic blood pressure, the highest pressure in the systemic artery.

Let all the air out of the cuff, allow the subject to rest, and then run a second determination. Do *not* leave the cuff inflated for more than 2 minutes, because it is uncomfortable and will cause a sustained increase in blood pressure.

The systolic pressure recorded with the palpatory method is usually around 5 mm Hg lower than that obtained using the auscultatory method. A major disadvantage of the palpatory method is that it cannot be used to measure the diastolic pressure.

2. Auscultatory Method

In the auscultatory method, the pressure cuff is used as in the palpatory method, and a stethoscope is used to listen to change in sounds in the brachial artery. Place the bell of the stethoscope below the cuff and over the brachial artery where it branches into the radial and ulnar arteries (Figure 17.2). Use your fingers, rather than your thumb, to hold the stethoscope over the artery; otherwise you may be measuring the thumb arterial pressure rather than the brachial artery pressure. With no air in the cuff no sounds can be heard. Inflate the cuff so the pressure is above diastolic (80–90 mm Hg), and you will be able to hear the spurting of blood through the partially occluded artery. Increase the cuff pressure to around 160 mm Hg; this pressure should be above systolic pressure so that the artery is completely collapsed and no sounds are heard.

Now, open the valve and begin to slowly lower the pressure in the cuff. As the pressure decreases you will be able to hear four phases of sound changes; these were first reported by Korotkoff in 1905 and are called **Korotkoff sounds.**

Phase 1. Appearance of a fairly sharp thudding sound that increases in intensity during the next 10 mm Hg of drop in pressure. The pressure when the sound first appears is the **systolic pressure.**

Phase 2. The sounds become a softer murmur during the next 10 to 15 mm Hg of drop in pressure.

Phase 3. The sounds become louder again and have a sharper thudding quality during the next 10 to 15 mm Hg of drop in pressure.

Phase 4. The sounds suddenly become muffled and reduced in intensity. The pressure at this point is termed the **diastolic pressure.** This muffled sound continues for another drop in pressure of 5 mm Hg, after which all sound disappears. The point where the sound ceases completely is called the **end diastolic pressure.** It is sometimes recorded along with the systolic and diastolic pressures in this manner: 120/80/75.

The auscultatory method has been found to be fairly close to the direct method in the pressures recorded; usually the systolic pressure is about 3 to 4 mm Hg lower than that obtained with the direct method.

Blood pressure varies with a person's age, weight, and sex. Below the age of 35, a woman generally has a pressure 10 mm lower than that of a man. However, after 40 to 45 years of age, a woman's blood pressure increases faster than does a man's. The old rule of thumb of 100 plus your age is still a good estimate of what your systolic pressure should be at any given age. After the age of 50, however, the rule is invalid. The increase in blood pressure with age is caused largely by the overall loss of vessel elasticity with age, part of which is due to the increased deposit of cholesterol and other lipids in the blood vessel walls.

Practice taking blood pressure on your partner until you become adept at detecting the systolic and diastolic sounds. You will find this can be quite difficult in some people, especially those whose arteries are located deep in the body tissues.

3. Postural Effects on Blood Pressure

Measure your partner's blood pressure while she or he is lying down (supine), sitting, and standing. Record your results in the Laboratory Report and also briefly explain the changes in pressure that accompany these changes in body position.

4. Cold Pressor Test

This test is used to demonstrate the effect of a sensory stimulus (cold) on blood pressure. A normal reflex response to such a cold stimulus is an increase in blood pressure (both systolic and diastolic). In a normal individual the systolic pressure will rise no more than 10 mm Hg, but in a hypertensive individual the rise may be 30 to 40 mm Hg.

a. Have the subject sit down comfortably or lie supine.

b. Record the systolic and diastolic blood pressure every 5 minutes until a constant level is obtained.

c. Immerse the subject's free hand in ice water (approximately 5 °C) to a depth well above the wrist.

d. After a lapse of 10 to 15 seconds, obtain the blood pressure every 20 seconds for 1 or 2 minutes and record.

Does the blood pressure become normal after immersion in ice water, and if so, how long does it take? Explain the physiological mechanisms operating in this experiment.

ARTERIAL PULSE WAVE

The blood pressure within an artery varies during each cardiac cycle. The highest pressure (**systolic**) occurs when the ventricle contracts to force blood into the artery; the lowest pressure (**diastolic**) occurs when the heart is in its relaxation phase and no blood is flowing through the semilunar valves. The difference between the systolic and diastolic pressures is called the **pulse pressure.** A recording of these pressure changes in an artery during one cycle of the heart is called an arterial **pulse wave.** A normal pulse wave over the aorta is shown in Figure 17.3. The **dicrotic notch** results when the aortic semilunar valves close, causing the blood in the aorta to rebound against the arterial walls to produce a slight elevation in pressure.

The magnitude and contour of the arterial pulse wave are directly related to the stroke volume and inversely related to the compliance (elasticity) of the arterial vessels. As the vessels lose their compliance (as with age or in arteriosclerosis), the stroke volume increases and the height of the pulse wave increases (pulse pressure increases). Thus, an examination of the pulse wave can give valuable clues to the functioning of the arteries and heart, as is seen in the abnormal waves pictured in Figure 17.4.

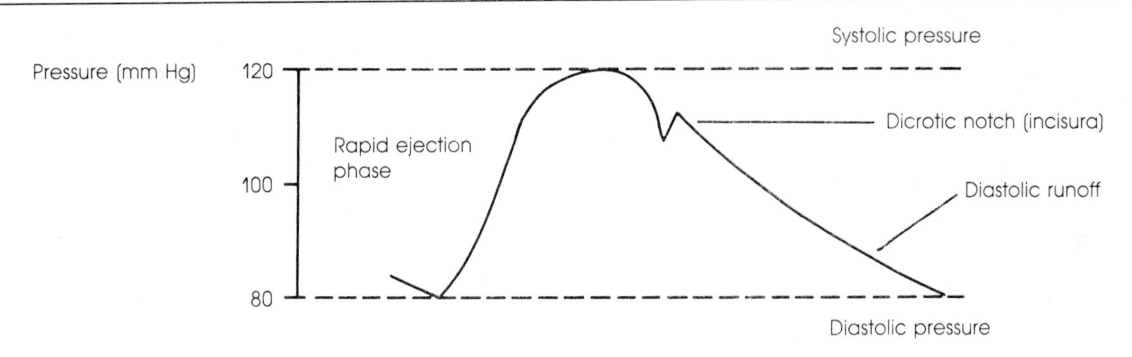

FIGURE 17.3. Typical arterial pulse wave.

The velocity of the pulse wave as it travels down the artery is also an important clinical measurement. The arterial pulse wave moves over the large arteries at a rate of 3 to 5 m/sec and over the small arteries at 14 to 15 m/sec. The difference in velocity is related to the compliance of the vessels—the less compliance a vessel has, the faster the pulse wave will move over it (as in the small arteries). Thus, a measurement of the velocity of the pulse wave can also provide useful information about changes in the vessel's elasticity (compliance). The change in vessel elasticity with age can be seen if you examine the velocity of the pulse wave over the aorta at various stages (Table 17.1).

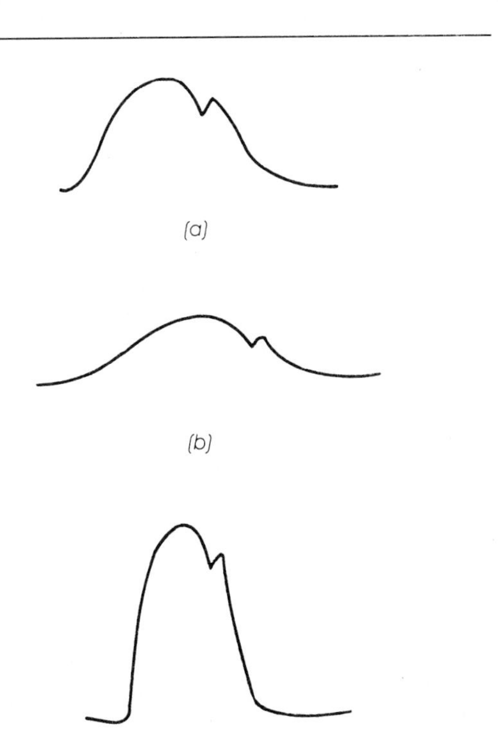

FIGURE 17.4. Pulse waves. (a) Normal. (b) In aortic stenosis. (c) In arteriosclerosis.

Recording the Peripheral Pulse

In this experiment, you will not record the pulse wave over an artery but from the tip of the finger; this is called a peripheral pulse. It is recorded using a photoelectric pulse transducer, which measures changes in blood volume (plethysmography). A light source in the transducer transilluminates the finger tip, and a photoconductor detects changes in light intensity within the finger caused by pulsatile variations in blood volume.

Experimental Procedures

1. With the subject seated, attach the transducer snugly to the palmar surface of the middle finger. Record the pulse for 10 seconds with the subject's arm resting on the lab table.

   Now have the subject raise the transducer above her head (arm extended) for 30 seconds and record the pulse during the last 10 seconds.

   Then have her lower the transducer (arm hanging at her side) for 30 seconds and record during the last 10 seconds.

TABLE 17.1. Change in Velocity of Pulse Wave with Age.

AGE (YR)	PULSE WAVE VELOCITY (M/SEC)
5	5.2
20	6.2
40	7.2
80	8.3

2. *Valsalva Maneuver*

   Note: a person with a history of cardiovascular disease should not be used as a subject for this experiment.

   With the subject's arm resting on the table, record the pulse for 10 seconds. Then have the subject inhale as deeply as possible, hold his breath and exert as much internal abdominal pressure as possible, while continuing to record the pulse. Compare the heart rate just prior to inhalation with the rate just prior to exhalation.

   Attempting to exhale forcefully against a closed glottis is called the Valsalva maneuver. It is commonly performed during forceful defecation or when lifting heavy weights. The contraction of the internal intercostal and abdominal muscles during this maneuver greatly increases the intrathoracic and intra-abdominal pressures, which impedes the venous return of blood to the heart.

3. After recovery from the Valsalva maneuver, wrap a sphygmomanometer cuff around the upper arm and record the peripheral pulse while the cuff is inflated to occlude the brachial artery. Continue recording the pulse as the cuff is slowly deflated at a rate of 5 mm per second.

VALVES IN THE VEINS

One of the classic experiments in the history of physiology was that performed by William Harvey on the venous valves. This simple demonstration helped Harvey formulate his theory that blood moves in a circular pathway around the body, rather than ebbing and flowing back and forth in the vessels as had been postulated by the physician Galen in ancient Greece.

Experimental Procedure

1. Tie a tourniquet around the arm above the elbow to obstruct venous return to the heart. Note the enlargement of the veins and the localized swellings that mark the position of the valves (Figure 17.5a).
2. Press one finger firmly down on a distal part of a vein. With another finger massage the blood out of the vein toward the heart by pushing up the vein past the next valve.

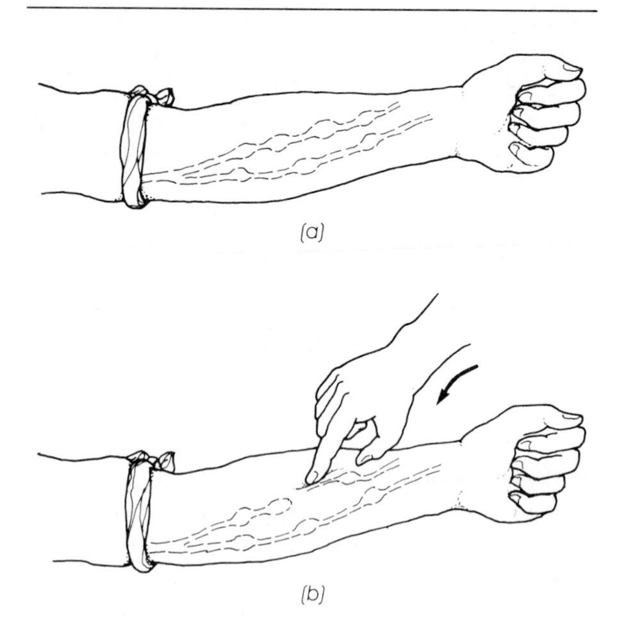

FIGURE 17.5. Harvey's experiment. (a) Location of venous valves. (b) Filling of vein only as far as valve.

3. Remove the second finger and note that the vein fills from above, but only as far as the valve (see Figure 17.5b). Press the blood from above toward the valve. What happens?
4. Remove the first finger and note the rapid filling of the vein. Starting near the elbow, push on the vein to move the blood toward the hand. What is the result?
5. Compare the filling time of the vein during rest and after exercise.

ELECTROCARDIOGRAM

Every living cardiac cell undergoes a regular sequence of electrical changes that initiate the contractile activity (**systole**) and the relaxation (**diastole**) of the cell. Thus, the contraction of the heart is associated with a compound action potential that is initiated at the sinus node and sweeps over the conduction path of the heart, preceding the mechanical contraction of the cardiac fibers. During this depolarization and repolarization of the myocardium, a potential difference is created between different regions on the surface of the heart. A separation of charge or potential difference is called a **dipole.** The electrical potential of the dipole is conducted through an electrolyte solution, such as the interstitial fluid and blood plasma, and eventually reaches the surface of the skin. By placing electrodes on the skin surface, we are able to detect and record the electrical activity over the heart surface prior to its contraction. By measuring the potential changes in various directions across the heart, it is possible to detect abnormalities.

The **electrocardiogram** (**ECG** or **EKG**) is a graphic record of the action potentials of the heart. It is recorded with an **electrocardiograph,** and the study of this cardiac electrical activity is called **electrocardiography.**

Electrocardiograph

The instrument that amplifies and records the heart's action potentials is actually a galvanometer, a device used by electricians to measure the passing of an electric current. Several types have been employed: the early string galvanometer used by Willem Einthoven, the electronic polygraph recorder, the cathode ray oscilloscope, and radiotelemetry recorders such as those used in space physiology research.

Electrodes

The limb electrodes generally used are slightly concave metal plates designed to fit snugly over the wrists and ankles. The chest electrode is a flattened disk. Because the skin has a high resistance, an electrolyte jelly (NaCl or KCl) in an abrasive base is first rubbed on the skin to remove the oil and dead cells and to form a conducting surface between the skin and the metal electrode, thus improving conduction of the impulse.

Standard Limb Leads

Various types of electrode positions can be used to record an ECG. The particular arrangement of the two recording electrodes is called a **lead.** The relative position of the two electrodes influences the direction and amplitude of the potentials recorded. To standardize the procedure so that results can be compared and evaluated from different laboratories, cardiologists have agreed on certain conventional requirements for the recording of the ECG.

The most common recording procedure is to use the standard limb leads. Electrodes are placed on the left arm, right arm, left leg, and right leg. The electrode on the right leg is a ground connection that prevents unwanted external potential fields from distorting the record. The other three electrodes are used in pairs to detect the cardiac potential. The electrocardiograph is calibrated so that 1 cm of vertical deflection represents 1 mV of potential difference, and the standard paper speed used is 25 mm/sec.

Einthoven's Triangle and Law

Einthoven, the father of electrocardiography, originated many of the conventions used in the recording of electrocardiograms. He visualized the three standard limb leads as enclosing the heart in a triangle, often referred to as Einthoven's triangle (Figure 17.6). Einthoven also found a relationship between the amplitude of the QRS complexes in each lead, such that lead I + lead III = lead II (Einthoven's law).

Lead I. Right arm to left arm

The right arm is connected to the negative terminal of the electrocardiograph, and the left arm to the positive terminal. When the right arm is negative to the left arm, the record shows an upward deflection. Thus, lead I measures the potential difference between the electrodes on the left and right arms, or across the base of the heart.

Lead II. Right arm to left leg

The right arm is connected to the negative terminal, and the left leg to the positive terminal.

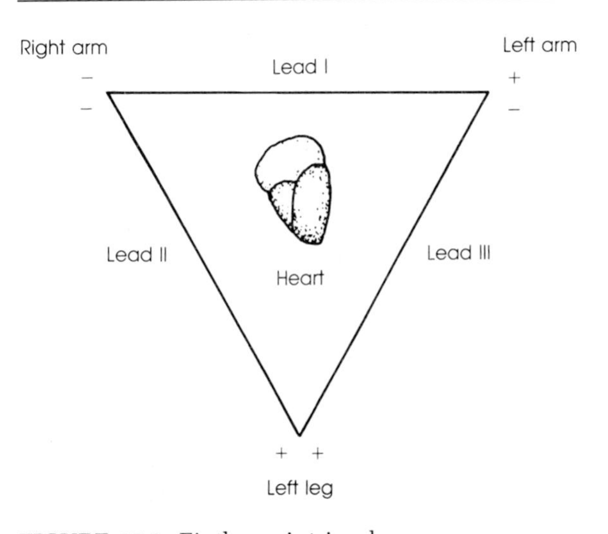

FIGURE 17.6. Einthoven's triangle.

Thus, lead II measures the potential difference between the left leg and the right arm, or along the long axis of the heart from base to apex.

Lead III. Left arm to left leg

The left arm is connected to the negative terminal, and the left leg to the positive terminal. This combination allows lead III to measure the potential difference between the left leg and left arm, or along the left side of the heart.

The sinoatrial (SA) node initiates the cardiac impulse (epicardium in this area becomes negative first), and this wave of negativity sweeps over the heart. Because the SA node is nearer the right arm, this area becomes negative while the left arm and left leg are still positive, and the deflection of the record is upward in those leads (I and II). The left arm is closer to the SA node, so in lead III the first deflection is also upward as the left arm becomes negative in reference to the left leg.

Components of Normal ECG Complex

The normal ECG is shown in Figure 17.7 for a single cardiac cycle.

P Wave. Represents the spread of electrical activity (wave of negativity) over the atria after the initial depolarization of the SA node.

QRS Complex. Represents the spread of the negativity wave (depolarization) through the ventricular musculature (Figure 17.7). A small amount of atrial repolarization also occurs at the same time.

PR Interval. Time from the beginning of the P wave to the beginning of the QRS complex; interval between activation of the SA node and the beginning of ventricular depolarization. Any abnormal lengthening of this interval suggests some interference with conduction of the impulse through the atria, atrioventricular (AV) node, bundle of His, and Purkinje fibers.

T Wave. Represents the repolarization of the ventricular musculature. It is of longer duration and lower amplitude than the depolarization wave (QRS complex), which indicates that the ventricular repolarization process is less synchronized and slower than the depolarization process.

QT Interval. Represents the time from the beginning of the QRS complex to the end of the T wave; that is, from the beginning of ventricular depolarization to the end of ventricular repolarization. The QT interval varies with the heart rate, becoming shorter as the heart rate increases.

PR Segment. From the end of the P wave to the beginning of the QRS complex. During this time the impulse is traveling through the AV node, AV bundle, and Purkinje fibers. These structures are within the heart myocardium; therefore, during this time there is no change in the negativity of the surface of the heart, and we say that the record is isoelectric (no change in potential is occurring).

ST Segment. From the end of the S wave to the beginning of the T wave. During this time the heart is completely depolarized,

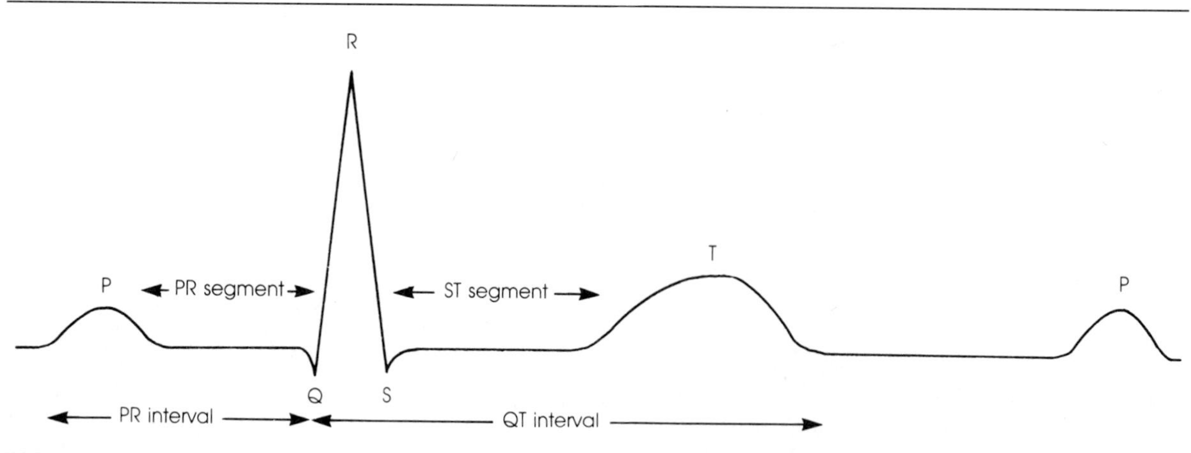

FIGURE 17.7. Normal electrocardiogram.

TABLE 17.2. Normal Values for Duration (and Voltage) of Different Phases of ECG Complex.

PHASE OF COMPLEX	DURATION (SEC) (VOLTAGE, MV)
P wave	0.1 (0.2)
QRS complex (lead II)	0.08–0.12 (1)
T wave	0.16–0.27 (0.2–0.3)
PR interval	0.13–0.16
QT interval	0.3–0.34
PR segment	0.03–0.06
ST segment	0.08

and therefore the record is again isoelectric. The position and shape of the ST segment are important in diagnosis.

Normal values for the duration, and in some cases voltage, of the different phases of the ECG complex are shown in Table 17.2.

Experimental Procedure

1. Check the electrocardiograph's calibration so that 1 cm = 1 mV of potential difference on the vertical axis. Run a length of record at the standard speed of 25 mm/sec. Check the record for artifacts.
2. Check the amplitudes of the QRS complexes in leads I, II, and III. Do they obey Einthoven's law?
3. Examine several inches of the record. Does the cycle length ever vary (arrhythmia)? Is there a change in cycle length (heart rate) with inspiration or expiration? Are any of the waves abnormal?
4. Make routine measurements of:

   PR interval
   P wave amplitude and duration
   QRS interval
   QT interval
   T wave amplitude and duration
   PR segment
   ST segment

   A PR interval greater than 0.2 second is abnormal and indicates first-degree heart block. In second-degree heart block there are P waves that are not followed by QRS waves; this may occur regularly or irregularly. Third-degree heart block is a complete AV dissociation in which P waves occur quite regularly but have no relation to R waves.

   The normal duration of the QRS complex is 0.08 to 0.12 sec. A duration of more than 0.12 sec indicates bundle branch block, or that the beat has arisen in one of the ventricles—a so-called ventricular beat or extra systole.

   Variations in the T wave are quite numerous and require an expert cardiologist for proper diagnosis. Inversion of the T wave is not abnormal, especially in lead III. Notching or splintering of the R wave (QRS) may be due to sudden changes in the electrical axis of the heart. Elevation of the ST segment by more than 2 mm is associated with acute injury or anoxia.
5. Have the subject exercise by doing several deep knee bends and record the ECG following the exercise. Are there any alterations in the patterns of the various waves or variations in the heart rate?
6. If the equipment is available, obtain a simultaneous recording of the ECG, phonocardiogram (heart sounds), and arterial pulse wave (using the finger or the radial artery). Compare the record obtained with that shown in Figure 17.8, a record of these three parameters during the cardiac cycle. The pressure pulse will be displaced to the right of the ventricular pressure curve shown because you will

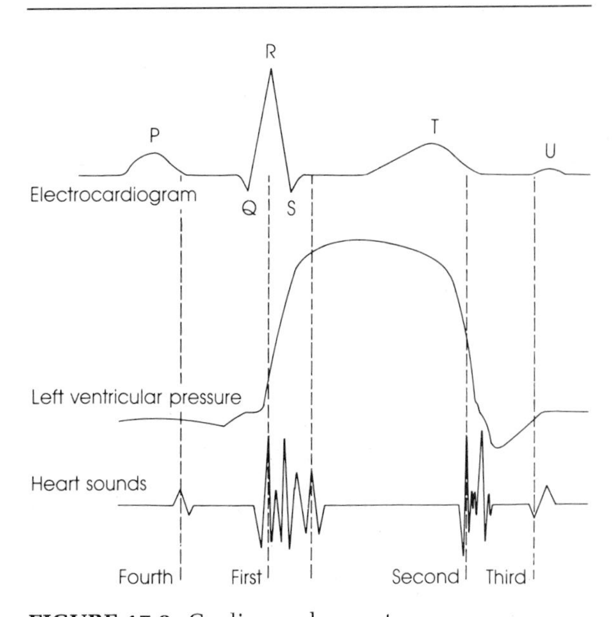

FIGURE 17.8. Cardiac cycle events.

be measuring the pulse farther from the heart. Can you explain the relationships between the various events in the cardiac cycle?

ELECTRICAL AXIS OF THE HEART

Using Einthoven's triangle and law, you can calculate the overall direction and magnitude of the electrical impulses conducted over the heart; this is called the **electrical axis** of the heart. The electrical axis is a valuable tool for clinical diagnosis because it indicates the approximate position the heart occupies in the thoracic cavity and possible hypertrophy of the heart chambers.

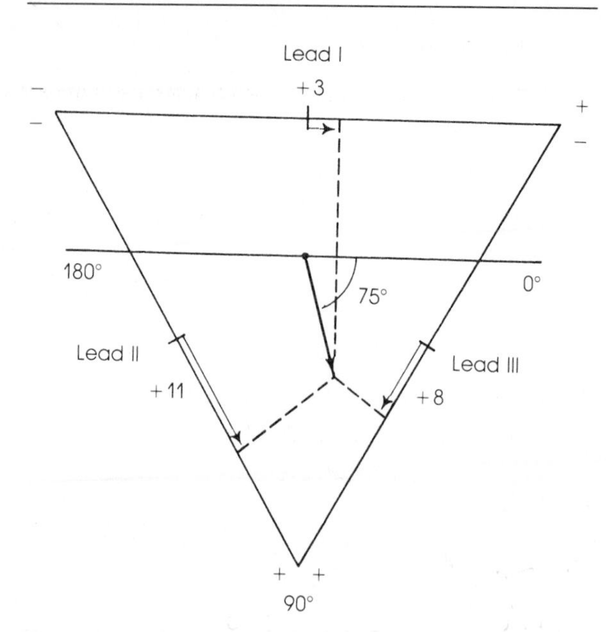

FIGURE 17.10. Electrical axis of the heart.

Experimental Procedure

1. Measure the height of the QRS deflections in each of the standard limb leads (I, II, and III). This height is usually measured in millimeters and converted to millivolts or microvolts. The height of the downward deflection is subtracted from the upward deflection height. Figure 17.9 shows examples.

2. The resultant deflections are plotted on the Einthoven triangle as vectors (Figure 17.10), appropriate units being used to represent the millimeter or millivolt height of each deflection. Each vector is drawn along the side of the triangle corresponding to the lead it represents. The vector is drawn from the center of each side and toward the (+) pole or electrode. If Einthoven's law is valid, perpendicular lines dropped from the point of each vector should meet at a single point inside the triangle. A line drawn from this intersection to the center of the triangle represents the electrical axis of the heart. The position of this axis is measured in degrees from the horizontal line drawn through the center of the triangle (75 degrees in Figure 17.10). The length of the axis vector is a measure of the overall electrical potential of the heart in the axis direction.

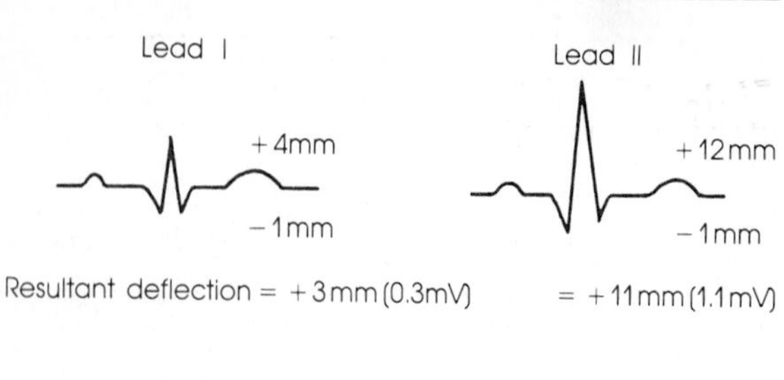

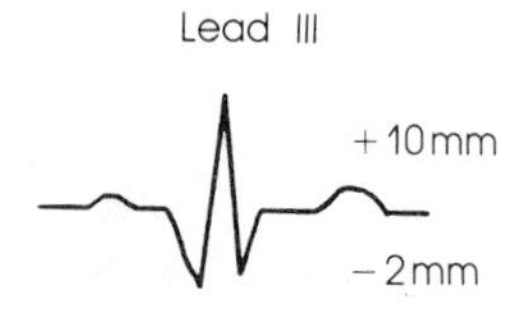

Resultant deflection = +8mm (0.8mV)

FIGURE 17.9. QRS deflections.

Normally the heart has an electrical axis of 59 degrees. If the axis is less than 0 degrees, it is termed a **left axis deviation.** If it is greater than 90 degrees, it is termed a **right axis deviation.** In some cases the heart itself may be pushed out of its normal position, producing a left or right shift in the axis deviation (as in pregnancy, excessive weight, or abdominal tumors). A left axis deviation may also be caused by hypertrophy of the left ventricle, as in systemic hypertension. A right axis deviation may be produced by right ventricle hypertrophy, as with pulmonary hypertension.

Calculate the electrical axis of the subject's heart and compare it with the normal expected deviation.

LABORATORY REPORT

17. Human Cardiovascular Function

Name ____________________

Date __________ Section __________

Score/Grade ____________________

Composite Class Data

NAME	HEART RATE	BLOOD PRESSURE				COLD PRESSOR TEST	HEART AXIS AND VECTOR
		PALPATORY	AUSCULTATORY				
			STANDING	SITTING	SUPINE		
Jani	78	126	186/98	130/80			
M. Anne	72	118	120/100				
Val	64	112	120/98				
Marion	60	100	100/68				
Alison	64	118	120/68				
Nat	80	96	130/86				
Tiffany	74	106	124/78				
Cyndi	76	110	138/82				
Lynn	80	105	110/64				
Deb	85	110	120/90				
Amy	70	100	105/70				
Mary	78	96	100/78				
Dan	86	124	150/90				
Melissa	82	106	182/76				
Tonya	92	106	120/78				
Class Average							

Auscultation of Heart Sounds

✗ 1. Why are the first and second heart sounds different in intensity or pitch?
first - AV valve closes at relative low pressure
second - SL valve closes at high pressure

2. What is a heart murmur? What causes it?
a soft rasping or blowing sound caused by leakage of blood through a damaged valve

Measurement of Blood Pressure

what is the first sound heard

* 1. What produces the systolic Korotkoff sound?
As the pressure in the cuff is reduced the blood will spurt through the partially constricted vessel at the time when the cuff pressure justs drops below the highest vessel pre[illegible]

2. Why is the muffling of the sound said to indicate the diastolic pressure?
when there is no longer a constriction on blood flow if the pressure in the vessel is above the pressure in the cuff the blood will begin to flow smoothly + lose the turbul[illegible] sound

3. What does systolic blood pressure represent? What does diastolic blood pressure represent?
systolic BP highest pressure in the artery during the cardiac cycle due to ventricular contraction
Diastolic B.P. lowest pressure in the artery during the cardic cycle due to elasticity of arteries

4. What is hypertension? What blood pressure indicates hypertension?
BP above normal usually above 140 systolic or 90 diastolic

5. If your TPR = 1, what is your CO in the sitting position? (Show your calculations.)
mean BP = 80 + 1/3 (40) =
if BP = 120/80   CO ml/sec = mean BP / TPR = 93.3 / 1 = 93.3 ml / sec × 60 sec = 5598 [illegible]

6. How do you explain the changes in blood pressure that occurred when body position was altered (supine, sitting, standing)?
As one moves to an erect position blood pools in lower parts resulting in failure of carotid bodies to be stimulated result. in increased sympathetic activity + blood vessel constriction resu[illegible] in higher pressure

7. Explain the physiological mechanisms operating in the cold pressor test. Why would the systolic pressure rise only 10 mm Hg in a normotensive individual?
Cold causes Vasoconstriction usually less than 10 mm Hg
if greater than 10mm Hg, it usually indicates a tendency for high B.P. In a normotensive a 10mm rise would result in reflex lowering of B.P. to compensate for the rise holding rise to 10 mm H[illegible]

Name ____________________

Date __________ Section __________

Score/Grade ____________________

Arterial Pulse Wave

1. Place your recordings in the following space.

Resting	Hand raised	Hand lowered
75	80	73

How do you explain the changes in the pulse wave when the arm is raised or lowered?

2. Place the pulse wave recording during the Valsalva maneuver in the following space and note the heart rate before inhalation and just prior to exhalation.

Explain the Valsalva effects on the peripheral pulse and heart rate. Why is it advisable to exhale when lifting weights?

3. In the following space, place your recording of the pulse during inflation and deflation of the sphygmomanometer cuff.

How do you account for the pulse wave changes seen as the cuff pressure was reduced?

Valves in the Veins

1. What is the importance of venous return to the pumping ability of the heart?

↑ blood flow from veins to heart stretches heart activiting starting mechanism → ↑ contraction during next beat → ↑ stroke volume & cardiac output

2. What factors aid venous return of blood to the heart?

a. constriction of veins
b. skeletal muscle contraction
c. Respiratory movements

Electrocardiogram

Lead I	Lead II	Lead III

1. Record your lead II ECG measurements for the following:

PR interval = 0.14 sec PR segment = 0.04 sec

QT interval = 0.32 sec ST segment = 0.06 sec

P wave amplitude = ____ mV P wave duration = 0.1 sec

QRS amplitude = ____ mV QRS duration = 0.1 sec

T wave amplitude = ____ mV T wave duration = 0.2 sec

Electrical Axis of Heart

Draw your axis and vector in the following triangle.

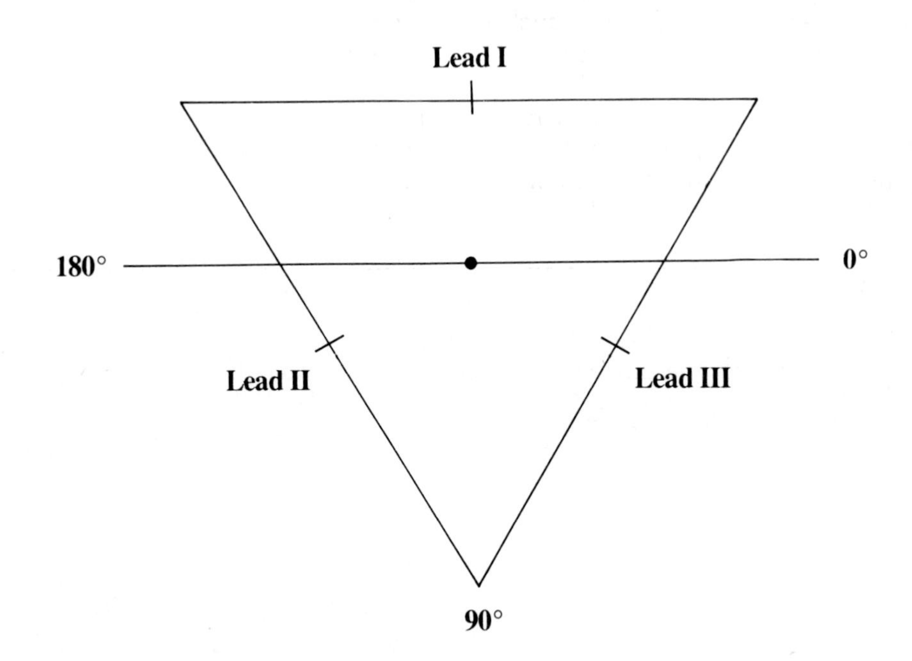

1. Mean electrical axis = ____ degrees. Heart vector length = ____ mm
2. How would the axis change during pregnancy? Why?

18 Respiratory Function

RESPIRATORY MOVEMENTS

Of the many processes occurring in our bodies each instant, those that function in the movement of oxygen to the tissues are among the most important. If tissues are deprived of oxygen for too long a time, they die; this time factor is especially critical for the cells of vital organs such as the heart and brain. Because of the importance of O_2 and CO_2, their concentration in the lungs and blood is finely regulated by a variety of receptors, reflexes, and feedback processes that control our respiratory patterns. You can gain insight into some of these control processes by observing a person's respiratory movements and the alteration of these movements caused by various factors.

Also important in oxygen delivery is the capacity of the lungs for air intake and the ability of the lungs to move air in and out quickly. You will analyze these functions when you study the various lung volumes and capacities and conduct the pulmonary function tests.

Respiratory movements are easily recorded by using a bellows pneumograph or impedance pneumograph around the subject's chest. The experimental setups for using these two types of pneumographs are shown in Figures 18.1 and 18.2. The subject should be seated close to the

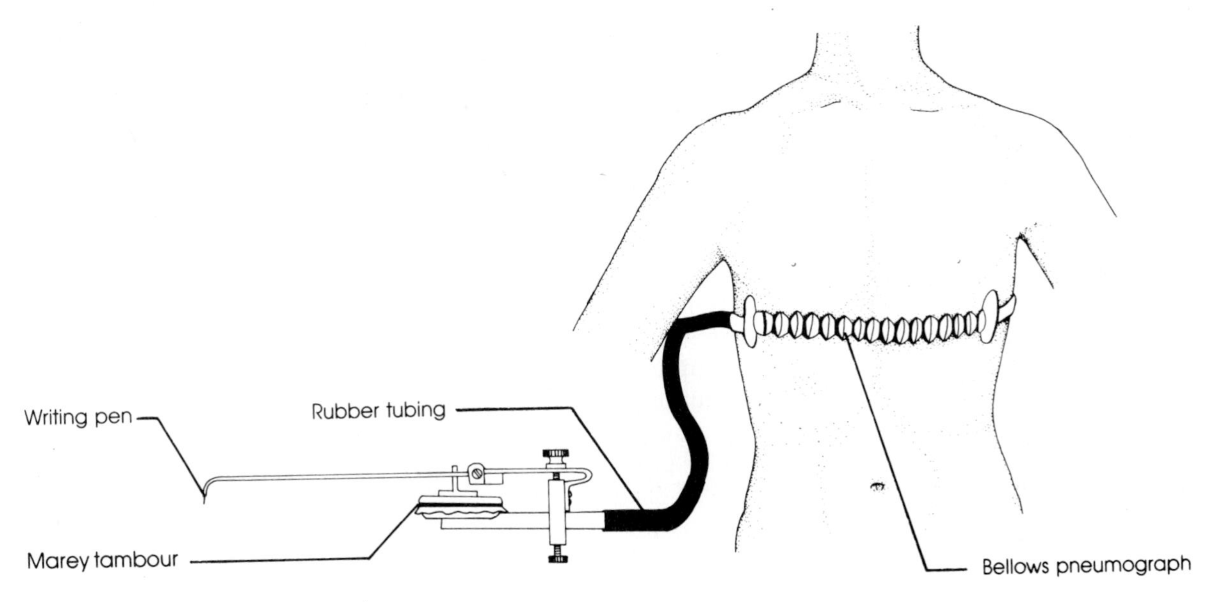

FIGURE 18.1. Attachment of bellows pneumograph.

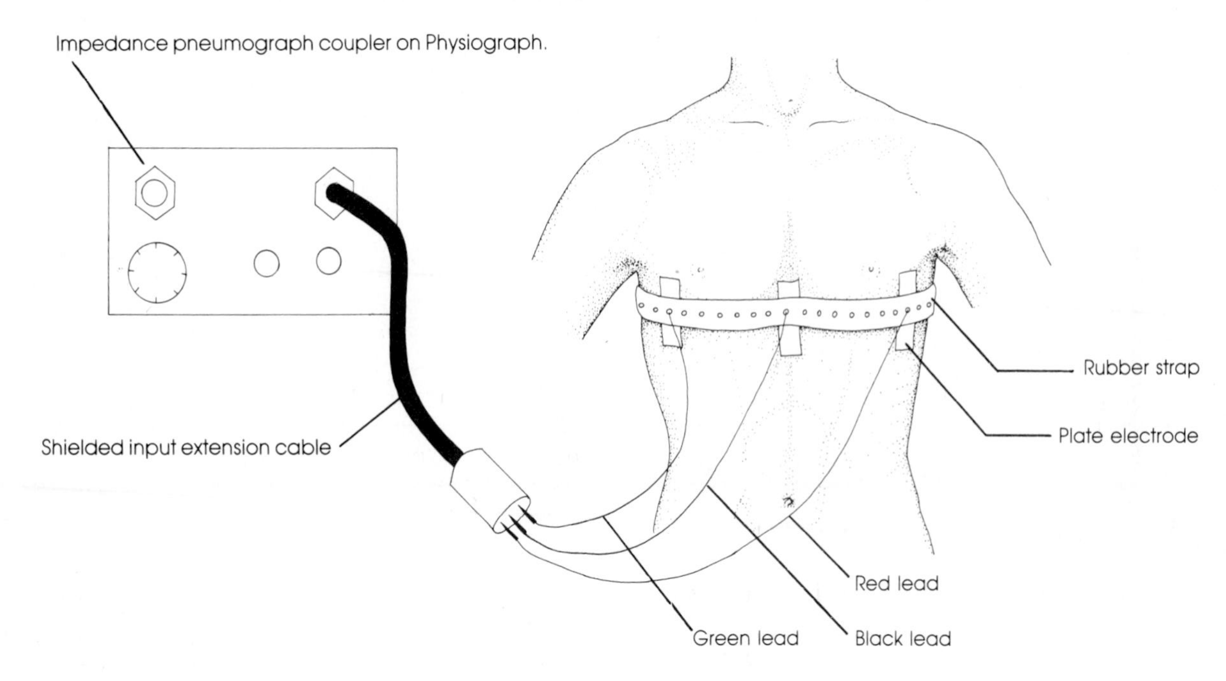

FIGURE 18.2. Attachment of impedance pneumograph

recorder when being tested but should not look at the record. A time line should be recorded so that respiratory rates can be determined.

Bellows Pneumograph (Kymograph System)

Attach the pneumograph snugly around the subject's chest at a level where the greatest thoracic movement during quiet breathing is observed. When the subject is halfway between the limits of normal inspiration and expiration, attach the pneumograph tube to a Marey tambour. Adjust the tambour recording pen so that respiratory movements are recorded in the middle of the pen excursion range.

Impedance Pneumograph (Physiograph System)

The impedance pneumograph measures the impedance between two plate electrodes applied to the thorax of the subject. A small current of 4 microamps is passed through the electrodes, and the voltage across the electrodes is directly proportional to the impedance between the plates. The impedance to the passage of current is reduced when air is inhaled into the lungs and increased when air is exhaled. The resulting voltage changes are amplified and used to drive the pen for recording respiratory movements.

Apply electrode gel to the plate electrodes and place them on opposite sides of the thoracic cavity. Fasten them in place with a long rubber strap. Attach an additional plate electrode on the center of the chest for grounding. Connect the red and green lead wires to the recording electrodes and the black wire to the ground electrode. Connect the lead wires to the impedance pneumograph coupler by means of a shielded input extension cable.

Respiratory Terms

You should become familiar with the following terms used in respiratory physiology:

Eupnea	Tachypnea
Apnea	Anoxia
Hyperpnea	Hypercapnia
Dyspnea	Asphyxia
Polypnea	Dead space

Experimental Procedure

1. Normal Respiratory Pattern

Record the subject's normal cyclic pattern of respiration for 1 to 2 minutes, first using a slow paper speed and then a faster speed. Note the amplitude of the inspiration and expiration cycles. What is the respiratory rate per minute?

2. Hyperventilation

Record normal ventilation for a few cycles at a slow paper speed. Then, at a given signal, stop the recording and have the subject breathe as fast and as deeply as possible for 30 seconds. At the end of this period, obtain a record of the aftereffects of the hyperventilation and report your observation in the Laboratory Report. The subject should allow his breathing to be as involuntary as possible during this posthyperventilation period. If the subject gets dizzy while hyperventilating, have him stop, but record the respiratory response.

What effect does hyperventilation have on the involuntary respiratory rate and amplitude of breathing? Does apnea or shallow breathing develop? What mechanism is responsible for these effects?

3. Hyperventilation in a Closed System

Repeat the hyperventilation experiment with the subject breathing in and out of a paper bag. Record the respiratory movements after hyperventilation. How does this record compare with the previous one? Is apnea as pronounced in this experiment? Explain.

4. Rebreathing

Record respiratory movements while the subject breathes in and out of a paper bag for several minutes. Do not record continuously, but only for 10 to 15 seconds during each minute. The bag may be held tightly around the nose and mouth or over the entire head. Be sure to avoid leakage of air from the bag. What happens to the rate and amplitude of breathing during the rebreathing of expired air? Explain the mechanism involved.

5. Effect of Mental Concentration

Record the respiratory pattern while the subject attempts to thread a needle or calculate a math problem. This experiment illustrates the effect of higher brain activity on the medullary respiratory centers. What changes in respiration are observed?

6. Effect of Speech on Respiration

Record respiratory movements while the subject reads audibly from a book. How is the pattern changed? Record the movements while she reads the same paragraph silently. Is there any effect on respiration in this case?

7. Breath Holding

Obtain a short record of normal respiration; then have the subject hold his breath as long as possible. Record respiratory movements after he reaches the breaking point and resumes breathing. Why is it impossible to hold the breath indefinitely?

8. Obstruction of Respiratory Passageways

Have the subject partially occlude her respiratory passageways by squeezing the nostrils. Record the respiratory patterns for a minute or so. What changes are seen in respiratory rate and amplitude? In what disorders is obstruction of airway passages a problem?

9. Effect of Exercise

Record respiratory movements after the subject has exercised by stepping up and down on a 12-in. stool 50 times or running in place for 200 steps.[1] What factors operate during exercise to increase the rate and amplitude of breathing?

RESPIRATORY VOLUMES

The total capacity of the lungs is divided into various volumes and capacities according to the function of these in the intake or exhalation of air. For a proper understanding of respiratory processes, it is necessary that you become familiar with these volumes and capacities.

As shown in Figure 18.3, the total amount of air one's lungs can possibly hold can be subdivided into four **volumes,** defined as follows:

Tidal volume (TV). The amount of air inspired or expired during normal, quiet respiration.

[1]This test causes cardiovascular stress; students who have any cardiovascular difficulties should not take part unless they have permission from their physician.

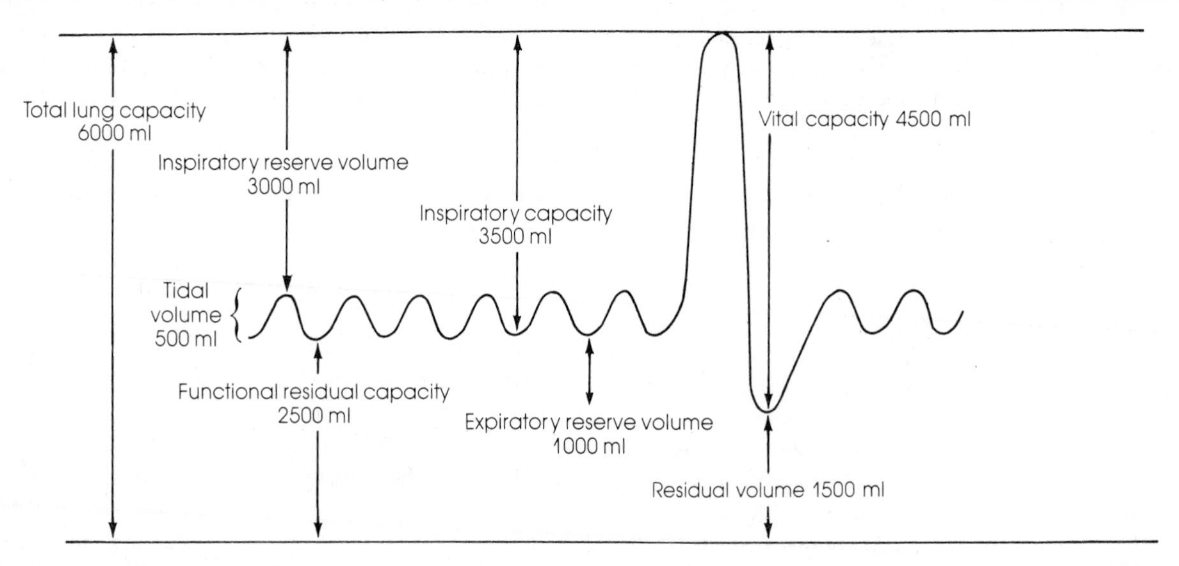

FIGURE 18.3. Lung volumes and capacities for normal adult male.

Inspiratory reserve volume (IRV). The amount of air that can be forcefully inspired above and beyond that taken in during a normal inspiration.

Expiratory reserve volume (ERV). The amount of air that can be forcefully expired following a normal expiration.

Residual volume (RV). The amount of air that remains trapped in the lungs after a maximal expiratory effort.

In addition to the four volumes, which do not overlap, there are four **capacities,** which are combinations of two or more volumes.

Total lung capacity (TLC). The total amount of air the lungs can contain—the sum of all four volumes.

Vital capacity (VC). The maximal amount of air that can be forcefully expired after a maximal inspiration.

Functional residual capacity (FRC). The amount of air remaining in the lungs after a normal expiration.

Inspiratory capacity (IC). The maximal amount of air that can be inspired after a normal expiration.

The milliliter values given for these volumes and capacities in Figure 18.3 are for a normal adult male. In the female they are all 20% to 25% smaller.

The respiratory volumes can be measured with a simple instrument called a **spirometer.** This consists of a lightweight metal bell inverted in a drum filled with water. A mouthpiece and hose allow the collection of air in the inverted bell. In this experiment you will use your own disposable mouthpiece. Record your results in the table in the Laboratory Report.

Experimental Procedure

1. Tidal Volume (TV)

Set the spirometer dial at zero. Take a normal inspiration, place your mouth over the mouthpiece, and exhale a normal expiration into the spirometer. You will have to make a conscious effort not to exceed your normal volume. Read the amount exhaled on the dial. Multiply your tidal volume by your respiratory rate per minute to give your resting **respiratory minute volume.** Have your lab partner count your respiratory cycles for 1 minute while you are seated at rest.

2. Expiratory Reserve Volume (ERV)

Set the spirometer dial at zero. After a normal expiration, place your mouth over the mouthpiece and forcefully exhale as much air as possible into the spirometer.

3. Vital Capacity (VC)

Set the spirometer dial at zero. Inhale as deeply as possible; place your mouth over the mouthpiece, hold your nose, and exhale into the spirometer with a maximal effort. Repeat the measurement three times and record the largest volume. Using nomograms C.1 and C.2 in Appendix C, determine your **predicted vital capacity** on the basis of your age, height, and sex. How does your predicted VC compare with your measured VC?

4. Inspiratory Reserve Volume (IRV) and Inspiratory Capacity (IC)

From the three previous volume measurements you can now calculate the IRV and the IC.

5. Heymer Test of Respiratory Reserve

In this test, take five deep breaths and then hold your breath as long as possible after the last inspiration. The breath-holding time gives an indication of your functional respiratory reserve and the efficiency of your respiratory system. Normal values are 50 to 70 seconds for men and 50 to 60 seconds for women. This test should be performed two or three times and the average taken. The Heymer test is often a better index of respiratory reserve than is the traditional vital capacity measurement.

The principal value of these pulmonary measurements lies in following volume changes caused either by disease or recovery from a disease. As an example, the VC is found to decrease in left heart disease. The decrease is due to blood congestion in the lung capillaries, which in turn leads to pulmonary edema and a decrease in VC. As the person recovers, his or her heart becomes stronger, pulmonary congestion and edema decrease, and the VC increases. The VC also decreases in paralytic polio, owing to partial paralysis of respiratory muscles, and in various other respiratory diseases.

PULMONARY FUNCTION TESTS

In most cases, measurements of pulmonary volumes are of limited value, because they are simply measurements of the anatomical size of lung compartments. They tell virtually nothing about the ability to move air in and out of the lungs, a critical factor in the delivery of oxygen to the blood. Consequently, in recent years we have seen increased use of **pulmonary function tests** as indexes of respiratory efficiency. Such tests are valuable because they give some measure of (1) lung compliance, or elasticity, (2) airway resistance, and (3) respiratory muscle strength. These three factors determine how much air a person can move into the lungs per unit of time, and this is what the pulmonary function tests measure.

Experimental Procedure

1. Maximal Breathing Capacity (MBC)

This is a measurement of the maximal volume of air that can be moved through the lungs in 1 minute. Clamp your nose and breathe as rapidly and deeply as possible for 15 seconds through a low-resistance respiratory valve and collect the expired air in a Douglas bag. The volume of air forced out in 15 seconds is determined by connecting the Douglas bag to a flowmeter and squeezing the bag to push all of the air collected through the flowmeter (Figure 18.4). Multiply the volume of air collected in 15 seconds by 4 to convert to liters (L) per minute. For men, the normal MBC is 60 to 170 L/min; for women it is 50 to 120 L/min. Tables and formulas are available for calculating a person's predicted MBC. These are based on age and body surface area. Two of the commonly used formulas are as follows:

Males: $\text{MBC} = [86.5 - (0.522 \times \text{Age})] \times \text{m}^2 \text{ body surface area}$

Females: $\text{MBC} = [71.3 - (0.474 \times \text{Age})] \times \text{m}^2 \text{ body surface area}$

The MBC is a good test of ability to move air rapidly but has the disadvantage of requiring a strenuous effort, so that considerable coaching and motivation are needed to obtain valid results. Also, many feeble or ill persons are unable to exert themselves for as long as 15 seconds. Because of these disadvantages, several single-breath tests have been devised that give almost the same information as the MBC test.

2. Forced Expiratory Volume (FEV) or Timed Vital Capacity (TVC)

This test measures the volume of air expired in 1, 2, or 3 seconds during a maximal exertion. This volume is then changed to percent of total vital

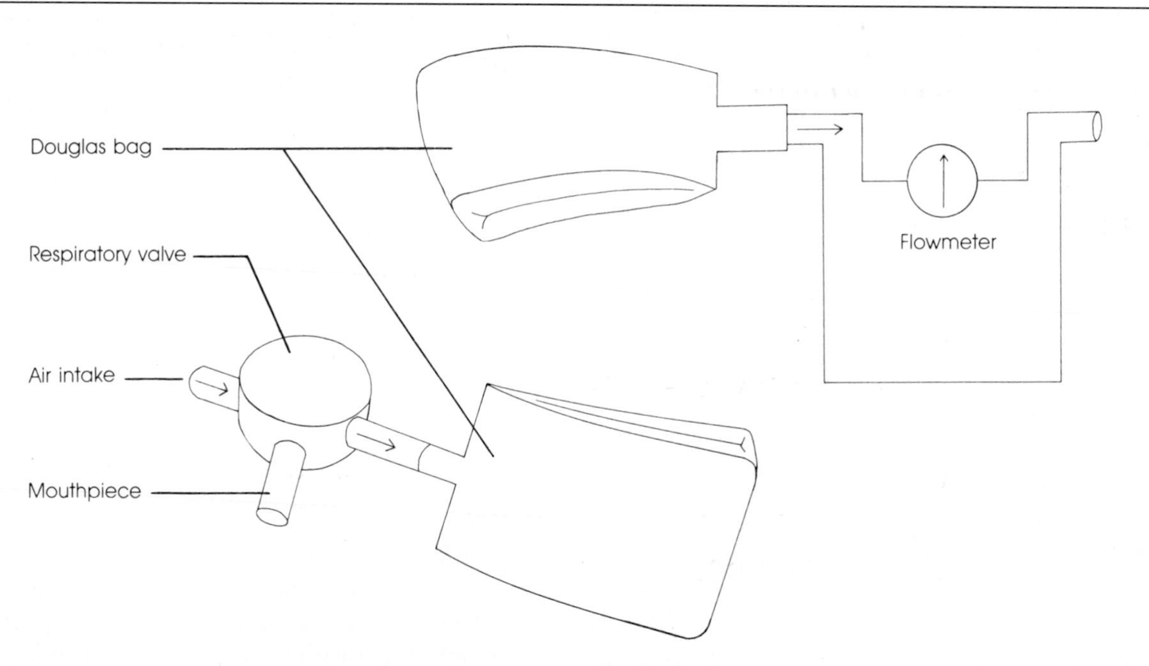

FIGURE 18.4. Douglas bag and flowmeter. Arrows denote air movement.

capacity expired during the entire expiratory period. A normal person should be able to forcefully exhale 83% of her VC in 1 second ($FEV_{1.0}$ = 83%), 94% of her VC in 2 seconds, and 97% of her VC in 3 seconds. The 1-second forced expiratory volume ($FEV_{1.0}$) has been found to correlate best with MBC measurements. The correlation coefficients between them are .88 to .92.

The FEV is measured with a **timed vitalometer** or a **recording vitalometer.** Simply take a deep breath and then exhale as forcefully and fully as possible into the vitalometer. The timed vitalometer records, on a dial, the total VC and the volume of air forced out in either 0.5, 0.75, 1, 2, or 3 seconds. The recording vitalometer makes a graph of the forced exhalation, and the FEV can then be calculated from the graph. As in the other sets, the predicted FEV may be calculated from equations and tables based on age and height (nomograms C.1 and C.2 in Appendix C). Is your predicted FEV close to your measured FEV?

LABORATORY REPORT

Name ______________

Date ________ Section ________

18. Respiratory Function

Score/Grade ______________

Respiratory Movements

1. What happens to the respiratory rhythm following hyperventilation?

rhythm becomes erratic, shallow, + periods of apnea may be interspersed

2. What causes the apnea that sometimes occurs after hyperventilation?

Excessive removal of CO_2 from lungs + blood. CO_2 at increased concentrations is a stimulant for rapid breathing, its loss will cause decreased breathing

3. Why does a person often get light-headed and dizzy after blowing up a balloon or blowing a horn vigorously? What mechanism causes these sensations?

Blowing causes the removal of CO_2 which leads to a reduction of Carbonic Acid + both may be normal stimulants for vasodilation of cerebral blood vessels; When lacking CO_2 the vessels contract reducing blood flow to the brain, result dizzy.

4. Explain the difference in a person's respiratory movements when hyperventilation takes place in and out of a paper bag.

The person's respirations increase as a result of the accumulation of CO_2 in the paper bag.

5. How does reading or concentration alter respiratory movements? What is the teleological explanation for such alterations?

When one reads aloud, conscious centers control respiratory centers, allowing the body to be controlled in motion or allowing for controlled exhalation or inhalation.

6. What mechanisms are responsible for the changes in the respiratory rhythm seen with rebreathing, breath holding, or obstruction of airway passages?

Chromoreceptors activate inspiratory + expiratory centers which result from an increase of CO_2 concentration

7. What factors operate during exercise to increase the rate and depth of breathing?

increase of CO_2, H^+
decrease of O_2

Respiratory Volumes

1. Vital capacity can be used to evaluate recovery from a myocardial infarction (MI). Explain why the measurement of vital capacity can be used as an index of recovery from a heart attack.
   After being weakened by M.I., the heart may be unable to pump all of the blood returned to it, resulting in a backup of blood into the lungs, causing pulmonary edema & loss of vital capacity

2. Using your own measurements of respiratory rate and tidal volume, calculate your alveolar minute volume if you had a normal dead space volume of 150 ml or if a respiratory disorder developed that elevated your dead space volume to 225 ml.
   Respiratory Rate
   500 − 150 = 350
   350 × 15 = 5250
   500 − 225 = 275
   275 × 15 = 4125

   Alveolar minute volume: with 150 ml dead space 5250 ml/min
   with 225 ml dead space 4125 ml/min

   Which respiratory disorders produce an increase in dead space volume?
   emphysema, pneumonia, fibrosis

   How would this increase in dead space affect cellular respiration and metabolism?
   Reduction in ventilation / perfusion ratio causing decrease in areobic metabolism & energy capture

3. Explain the pathophysiology of the following respiratory disorders:

   Pneumonia:
   Inflamation of lungs due to bacteria, virus, etc. resulting in fluid in Alveolar Spaces

   Asthma:
   harsh breathing caused by restriction of bronchioles as a result of an allergic reaction

   Atelectasis:
   Partial or total lung collapse caused by chest penetration

   Pleurisy:
   infection which leads to inflamation of the pleural membrane

   Emphysema:
   air trapped in alveoli, due to airway resistance caused by infectious agents or foreign products as dust causing stretching of chest & loss of elasticity; barrel shaped chest.

Name ____________________

Date __________ Section __________

Score/Grade ________________

NAME	TV	RESP. MIN. VOL.	ERV	VC	PREDICTED VC	CALCULATED		HEYMER TEST	MBC	PREDICTED MBC	$FEV_{1.0}$	PREDICTED $FEV_{1.0}$
						IRV	IC					
Class Average												

19 Regulation of Circulation and Respiration

In this experiment, a large rabbit (8–9 lb) will be anesthetized and observed for its cardiovascular and respiratory responses to various manipulations and drug injections. Direct blood pressure will be recorded via a cannula inserted into the carotid artery and connected to a transducer. Respiratory movements and heart rate will be recorded via needle electrodes inserted under the skin in the thoracic region. Drugs will be injected into the marginal ear vein via a catheter.

INSTRUMENTATION

Blood Pressure Apparatus

The blood pressure transducer, arterial cannula, stopcocks, and syringe are set up as shown in Figure 19.1. It is extremely important that you study the stopcock positions carefully so that you understand the direction of fluid movement when the stopcock is set in each position.

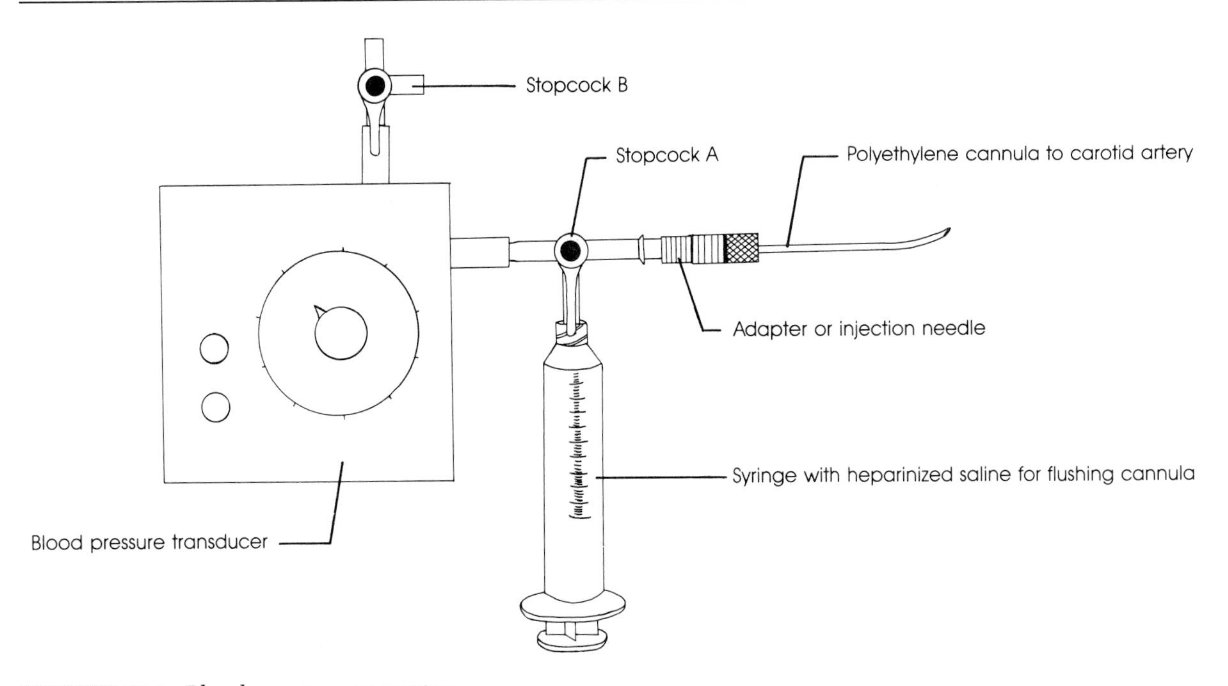

FIGURE 19.1. Blood pressure apparatus.

Open the stopcocks so that heparinized saline can be pushed in to fill the transducer. Tap the transducer to ensure removal of air bubbles. To calibrate the transducer, open stockcock B and position stopcock A so that all three openings are closed (stopcock lever between the cannula and syringe openings). Calibrate so that 100 mm Hg of pressure equals 2.5 cm, using either the internal calibration on the transducer or a reservoir bottle pressurizing assembly (the instructor will demonstrate this assembly). Close both stopcocks to the transducer and push heparinized saline into the arterial cannula. All air bubbles must be removed from the cannula and transducer or trapped air will compress and produce erroneous pressure readings. During the experiment you will need to flush the cannula from time to time to reduce blood clots at the cannula tip that dampen the pulse pressure. Close stopcock A to the syringe and you are ready to cannulate the artery.

Respiratory Assembly

The respirator is used to artificially ventilate the animal if breathing stops or to maintain respiration in the open-chest preparation when the heart is manipulated. When needed, the respirator is connected to the tracheal cannula by means of a tygon tube fitted with a three-way connector to provide an escape valve for pressure release (Figure 19.2). When the assembly is first attached to the tracheal cannula, the escape valve should be wide open. Then, to increase the pressure and lung inflation, the valve is slowly closed to restrict the air escaping through the valve.

Caution! *Overinflation of the lungs will produce alveolar hemorrhage and death.*

ANESTHETICS FOR RABBITS

Urethane (Ethyl Carbamate)

This is the anesthetic of choice for rabbits because of its wide range between anesthetic and lethal doses. Inject intraperitoneal (IP) 8.5 ml/kg of a 20% urethane solution to induce anesthesia. Supplementary doses of urethane (2-3 ml/kg) may be given IP to maintain anesthesia, or ether anesthesia may be used with caution.

Caution! *Recent research indicates that urethane is potentially carcinogenic and teratogenic. Warning labels now carry this statement: "Avoid contact and inhalation. Wear protective clothing, gloves, and mask. Avoid exposure to women of childbearing age. Wash thoroughly after handling."*

Pentobarbital (Nembutal)

This is somewhat tricky to use because different rabbit breeds respond differently to the same dose of pentobarbital. The best procedure is to give the drug in sequential doses until a surgical plane of anesthesia is reached. Inject IP 10 mg/kg of pentobarbital initially, then supplement with 5 mg/kg injections at 2-minute intervals until anesthesia is obtained.

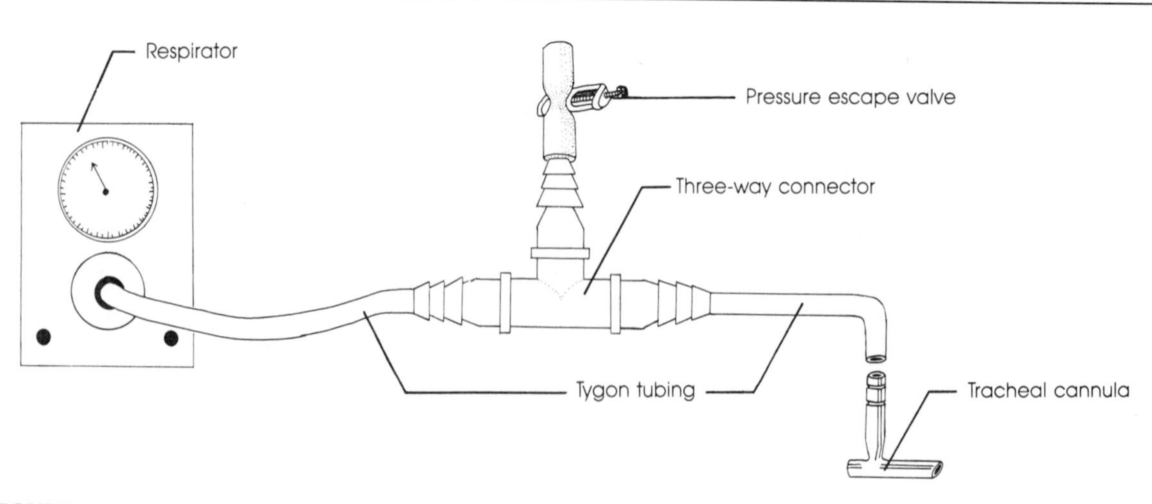

FIGURE 19.2. Respirator assembly.

Magnesium Sulfate

Inject IP 4 ml/kg of a 25% $MgSO_4$ solution. If anesthesia is not obtained in 10 minutes, inject another 2 ml/kg. Respiratory depression can occur with magnesium sulfate; this can be overcome by injecting intravenously 8 ml of 3% calcium chloride solution.

Surgical Procedure

1. Weigh the rabbit and anesthetize it with an IP injection of one of the anesthetics just described. Throughout the experiment the animal should be watched closely to see that anesthesia is maintained.
2. Using scissors or animal clippers, remove the fur from the neck and from under the front legs and the dorsal surface of both ears.
3. Insert a minicath infusion set (23-gauge needle) into the marginal ear vein and connect the end to a three-way stopcock fitted with two syringes (Figure 19.3). These syringes will be used for injecting drugs and flushing the infusion tubing with heparinized saline following the injections.
4. Tape the infusion set to the ear so that the needle is not easily pulled out of the vein. Using heavy cord, tie the rabbit to the operating board, ventral side up.
5. Insert needle electrodes under the skin just below the front legs for recording respiratory movements and ECG. These electrodes are connected to the impedance pneumograph coupler of the Physiograph and this coupler in turn is connected by a cable to the hi-gain coupler (amplifier) for recording the ECG. Other recording systems may use a pneumograph around the chest for recording respiration. You may have to experiment with different positions for the needle electrodes to record a good respiratory pattern.
6. Make a midline, longitudinal incision through the skin of the neck from the jaw to the sternum. Using blunt dissection (no scalpel or scissors) push aside the neck muscles to expose the trachea. Free the connective tissue from the trachea and pass a heavy cord under the trachea to aid in manipulations. Using a scalpel, cut halfway through the trachea between the rings of cartilage, insert a tracheal cannula, and tie it in place. This cannula will allow the animal to breathe freely (the nasal airway passageways often become obstructed with mucus during anesthesia).
7. Using blunt dissection, expose the left and right carotid arteries lying deeper in the neck on either side of the trachea. Carefully free the carotid artery and adjacent vagus nerve from the connective tissue surrounding them. Isolate as long a segment of the carotid as possible so that carotid cannulation can be performed more easily. Place loose ligatures under each vagus nerve so they can be lifted up for stimulation later.
8. Pass two threads under the right carotid. Use one thread to tie off the carotid as

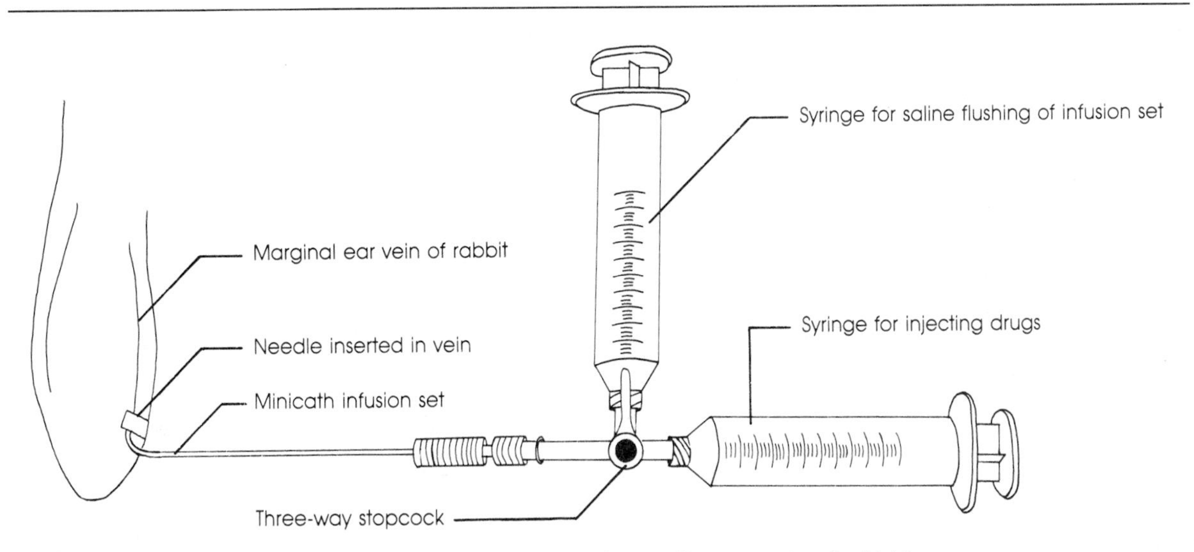

FIGURE 19.3. Minicath infusion set inserted by way of a needle into vein of rabbit's ear.

far anteriorly (toward the head) as possible. Clamp off the posterior end of the carotid with a bulldog clamp to prevent blood loss when the carotid is cut. Place the flat handle of a scalpel under the carotid and, using another scalpel, cut approximately halfway through the artery—*Careful!* Using a small probe to locate the incision, guide the arterial cannula into the carotid and tie it securely in place with the thread previously placed around the carotid. Release the bulldog clamp and you are ready to begin recording blood pressure. Details of the cannulation are shown in Figure 19.4. For a rabbit this size, a cannula with an outside diameter of 0.062 in. works well.

Experimental Procedure

Perform each experimental procedure shown in the table in the Laboratory Report, record the results, and explain the physiological responses observed. Allow time between experiments for the parameters measured to return to normal. Drugs are injected into the ear vein and flushed out of the injection cannula using heparinized saline.

Near the end of the lab period, connect the trachea to the respirator and open the abdominal cavity to expose the contracting diaphragm. Then cut through the diaphragm and the rib cage to expose the heart and lungs. Use bone cutters or heavy scissors to cut through the ribs. Remove the pericardial sac around the heart. Examine the inflation and deflation of the lungs. Place your hand around the heart and feel the contractile power of the ventricle during systole.

Try to induce fibrillation of the ventricles by stroking them with a stimulating electrode (tetanizing frequency and high voltage). Note the effect of fibrillation on the blood pressure and ECG. Attempt to restore a normal sinus rhythm by shocking the heart with a defibrillator. The heart may escape from fibrillation on its own if it is healthy. Feel the ventricle when it is fibrillating. How does its contractile power compare with the power felt during synchronized contraction? Does it feel like a bag of worms? Finally, inject 10 ml of concentrated KCl to induce fibrillation chemically and terminate the animal.

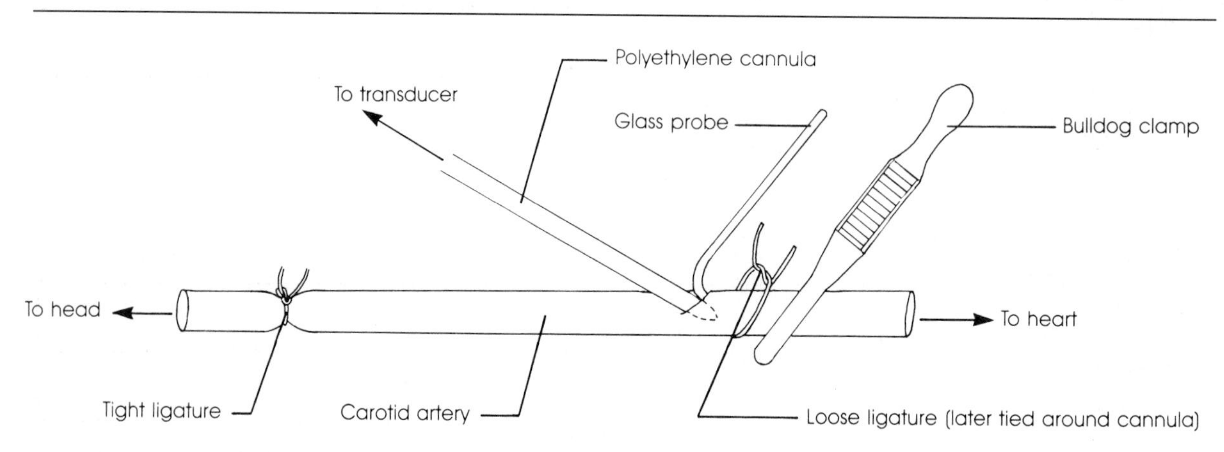

FIGURE 19.4. Cannulation of carotid artery.

LABORATORY REPORT

Name ____________

Date ________ Section ________

Score/Grade ____________

19. Regulation of Circulation and Respiration

EXPERIMENTAL PROCEDURE	HEART RATE	BLOOD PRESSURE	RESPIRATION RATE AND DEPTH	EXPLANATION
Normal, under anesthesia				
Recovery following asphyxia for 1 min				
Recovery after hyperventilation for 2 min				
Elevation of hindquarters				
Clamp of left carotid artery				
Release of carotid clamp				
Injection of 3% NaH_2PO_4 (0.3 ml)				
Injection of 0.2% KCN (0.5 ml)				
Right vagus stimulation (30 V at 50/sec)				
Left vagus stimulation				
Cutting of both vagus nerves				
Stimulation of central end of right vagus				
Stimulation of peripheral end of right vagus				
Epinephrine, 0.1 ml of 1:1000				
Histamine, 3 ml of 1:50,000				
Norepinephrine, 1 μg/kg				
Isoproterenol, 0.5 μg/kg				
Vasopressin, 0.1 unit/kg				
$CaCl_2$, 20 mg/kg				
Nitroglycerin, 0.3 mg/kg				
Ouabain, 75 μg/kg				
Ventricular fibrillation				

1. Which of the experimental procedures demonstrated major factors that control the respiratory rhythm? Briefly outline the relationship between the respiratory centers and the receptors that alter these centers.

2. The following drugs are injected into an animal following the infusion of propranolol, which blocks the beta receptors. Explain the effect of each drug on heart rate, blood pressure, and respiration and the physiological mechanisms through which each exerts its effect.

   Vasopressin:

   Norepinephrine:

   Ouabain:

   Isoproterenol:

3. How does injection of concentrated KCl produce fibrillation? Explain the mechanism.

20 Blood Physiology I: Erythrocyte Functions

Caution! *Parts of this lab may involve working with human blood. You should handle only your own blood. Dispose of all supplies (cotton, gauze, lancets, etc.) that come in contact with blood in properly marked containers. ALL BODY FLUIDS AND SUPPLIES MUST BE TREATED AS POTENTIALLY INFECTIOUS. Please read Appendix A, Precautions for Handling Blood.*

FUNCTIONS OF BLOOD

Blood serves the cells of complex organisms in the same way that the aquatic environment serves unicellular organisms. That is, it provides a medium for the maintenance of homeostasis in the cells' environment. To do this in complex organisms, blood must function as a transportation system, bringing nutrients and oxygen to the cells and removing wastes and carbon dioxide from the interstitial fluid around the cells. This transportation system also serves to link the various organs of the body together, integrating them through the action of hormones. Blood also performs other functions that are not as obvious, such as providing buffers for acid-base balance, destroying foreign organisms through phagocytosis and antibody action, distributing and conserving body heat, and preventing its own loss through hemostatic (coagulation) mechanisms.

In the following set of experiments, you will examine the important characteristics of the red blood cells (erythrocytes), which transport oxygen from the lungs to the tissues. A decreased ability to transport oxygen produces the condition called *anemia,* which can be caused by a decrease in the number or size of the red cells, or the amount of hemoglobin in the blood. To accurately diagnose the cause of anemia, the complete status of the erythrocytes must be examined—hematocrit, blood hemoglobin concentration, RBC count, RBC size, and percent hemoglobin per cell. These parameters will be measured in this lab using either your own blood or blood samples from another animal (cow, pig, rat, frog, etc.). You will also examine the movement of erythrocytes in the microcirculation (arterioles, capillaries, and venules) where oxygen is delivered to the tissues.

BLOOD HEMATOCRIT

The **hematocrit** (Hct) is the percent volume of whole blood that is occupied by red blood cells (erythrocytes). It is determined by centrifuging the blood in special hematocrit capillary tubes. The percent of whole blood composed of cells is determined by the height of the red cells in the tube compared with the height of the total column of blood. The average normal hematocrits and their ranges for males and females are as follows:

	AVERAGE	RANGE
Males	46%	43%–49%
Females	41%	36%–45%

The hematocrit may fall to as low as 15% in severe anemia or rise to as high as 70% in **polycythemia.**

Experimental Procedure

1. Puncture your finger using a sterile lancet to obtain a drop of blood. Wipe off the drop that forms (why?) and allow a second drop to accumulate.
2. Touch the red-circled end of a heparinized capillary tube to the drop. Hold the tube in a horizontal position and allow the blood to enter until the tube is one-half to three-fourths full.
3. Seal one end of the tube by pushing it into a tablet of sealing compound and rotating it to form a plug.
4. Place the capillary tube in a microhematocrit centrifuge with the plug end to the outside, and centrifuge for 4 minutes.
5. At the end of 4 minutes, measure in millimeters the height of the red cell column and the height of the cells plus the plasma. Calculate the hematocrit using the following formula and record it in the Laboratory Report.

$$\text{Hct}(\%) = \frac{\text{Height of red cells (mm)}}{\text{Height of red cells and plasma (mm)}} \times 100$$

Some labs employ a hematocrit "reader" that reads the hematocrit value directly on a scale. Where are the white blood cells in the hematocrit tube after the tube is centrifuged?

HEMOGLOBIN DETERMINATION

In clinical practice, the blood hemoglobin (Hb) is usually measured by a colorimetric method such as the cyanmethemoglobin method described in this section. Other simpler tests are often performed in the laboratory to give an approximate Hb value. The values obtained by using the simpler Tallquist or Sahli methods should be checked against those yielded by the more precise cyanmethemoglobin method.

Experimental Procedure

1. Tallquist Method

This test uses a book of special Tallquist blotting papers and a color comparison chart having different intensities of red. These intensities correspond to different concentrations of Hb found in human blood.

Obtain a drop of blood and place it on a piece of blotting paper. Before the blood becomes dry or coagulated, match its color with the closest color on the comparison chart. The number by each color represents the percent of Hb in the blood. This number is multiplied by the Tallquist standard of 16.5 g to give you grams of Hb per 100 ml of blood.

How does this reading compare with the value obtained through the colorimetric method? Is your Hb within the normal range for your sex?

2. Sahli Method

In this method the blood hemoglobin is converted to a brownish hematin compound by the action of hydrochloric acid. The higher the hemoglobin concentration, the more intense the hematin color will be.

a. Place 5 drops of 0.1 N hydrochloric acid (HCl) in the bottom of a graduated Sahli tube. This amount should fill the tube to around the 10% mark on the scale.

b. Lance your finger to obtain a drop of blood. Place the tip of the Sahli pipette[1] in the drop and gently suck a *solid column of blood* into the pipette up to the 20-mm mark (0.02 ml). When sucking, use the mouthpiece and rubber tubing attached to the pipette. If you draw in too much blood, touch the pipette tip to a filter paper or tissue to draw the excess blood out. Do not allow air to enter the pipette column or you will invalidate your results.

Note: You cannot spend too much time in filling the Sahli pipette, or the blood will coagulate in the pipette and block the bore. To clean it out, flush the pipette repeatedly in the following solutions in this order: distilled water—alcohol—ether or acetone.

If the Sahli pipette is difficult to clean, use hydrogen peroxide to clean it. Use cau-

[1] A new, disposable Sahli capillary pipette is available and can also be used.

tion, because hydrogen peroxide is a strong oxidizing agent. A fine wire also may have to be used to ream out the bore of the pipette. Coagulation of blood in the pipette can be delayed by drawing a heparin solution (1:1000) in and out of the pipette prior to filling it with blood.

c. Insert the tip of the pipette beneath the surface of the HCl in the Sahli tube and gently blow out the blood. Rinse the pipette of any blood by drawing the solution in and out of the pipette two times.

d. Mix the blood and HCl by stirring with a glass rod, and then let the tube stand for 10 minutes.

e. Place the tube in the comparator block and hold it up to a strong light. Add distilled water drop by drop to the hematin solution (stir after each addition) until its color matches the color of the standard color on the comparator.

f. Read the scale on the Sahli tube to obtain the percent of Hb and grams of Hb per 100 ml of blood. Note that the Hb standard used in calibration may vary from tube to tube. The standard (g Hb) used is imprinted on each tube.

3. Cyanmethemoglobin Method

more precise

The amount of oxygen that blood can carry is closely related to the concentration of hemoglobin in the blood. Each gram of hemoglobin is capable of carrying 1.34 ml of oxygen when the hemoglobin is completely saturated. Normally, each 100 ml of blood contains approximately 15 g of hemoglobin, which is distributed among the 500 billion red cells in each 100 ml of whole blood. The red blood cell is essentially a bag of hemoglobin, because as much as 34% of the red blood cell by weight is hemoglobin. The average normal concentrations of hemoglobin and their ranges for males and females are as follows:

	AVERAGE	RANGE
Males	15.4 g/100 ml of blood	13.6–17.2 g
Females	13.3 g/100 ml of blood	11.5–15.0 g

A concentration of less than 10 g/100 ml of blood is usually considered anemia, but major health difficulties seldom develop until a level of 7.5 g/100 ml of blood is reached.

In this experiment you will employ a very accurate colorimetric method for the determination of hemoglobin concentration in your own blood. (The colorimeter is shown in Figure 20.1.)

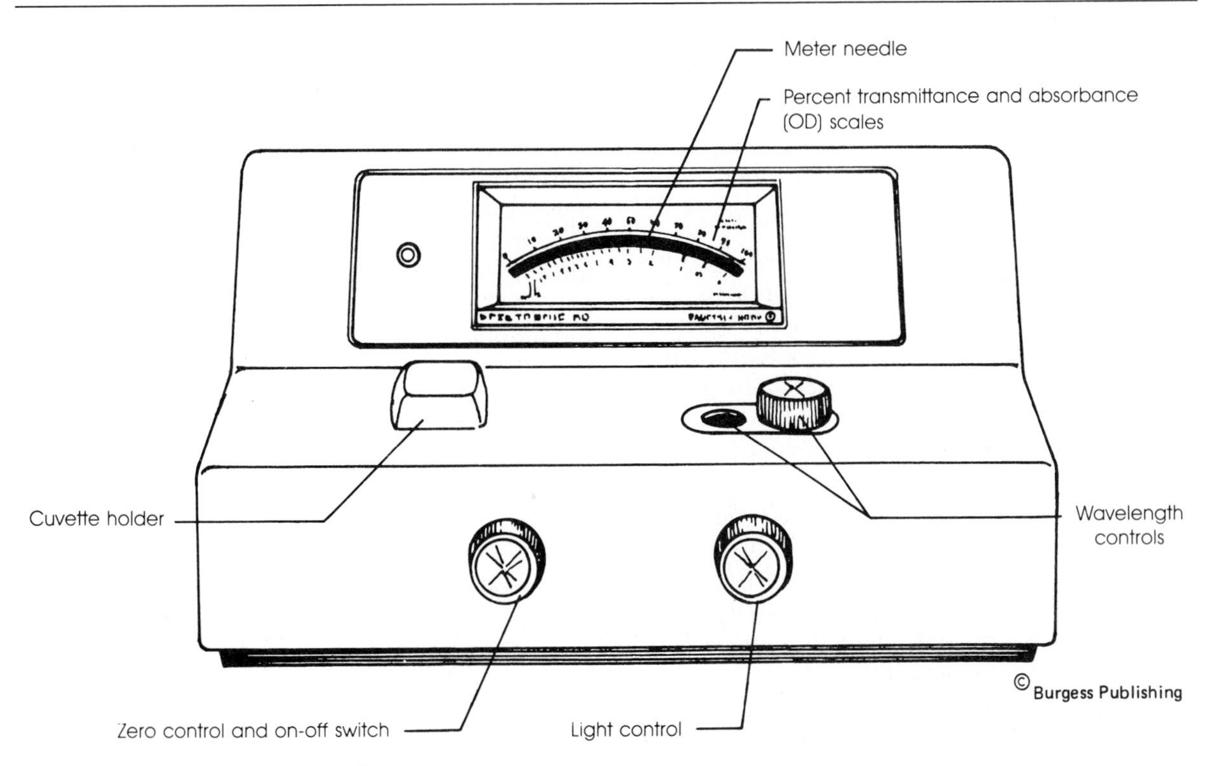

FIGURE 20.1. Spectronic 20 colorimeter.

The cyanmethemoglobin test is based on the reaction of hemoglobin with *reagent* solution containing potassium cyanide (KCN) to form cyanmethemoglobin, a colored compound. The concentration of cyanmethemoglobin is determined by comparing on a **standard curve** the amount of light the compound can absorb with the amount of light absorbed by compounds with known concentrations of hemoglobin.

Operation of the Colorimeter

a. Turn on the instrument and let it warm for several minutes.

b. Set the wavelength to the desired position.

c. Adjust the meter needle to 0% transmittance (absorbance at infinity) by using the zero control.

d. Place the *blank* (0%) in the light path (in the cuvette holder).

e. Adjust the meter needle to 100% transmittance by using the light control.

f. Remove the blank and place the unknown in the light path.

g. Read and record the percent transmittance and absorption.

h. Repeat steps d through g for your other unknowns.

For a more accurate conversion between optical density and percent transmittance, use Table C.1 in Appendix C.

a. Setting Up the Standard Curve

The standard curve is set up by increasing dilutions of cyanmethemoglobin *standard solution;* 5.0 ml of the commercially obtained and undiluted standard corresponds to 20.0 g% hemoglobin. Dilutions must be made with the cyanmethemoglobin *reagent solution*—never with water. The dilutions are prepared using the volumes of standard and reagent shown in Table 20.1.

Transfer the dilutions to well-matched cuvettes. Set the wavelength of the colorimeter to 540 nanometers (nm). Adjust the instrument so the blank tube (0%) has zero absorbance (optical density) or 100% transmittance. Take the readings for the standard and plot percent transmission versus concentration on semilog graph paper (or absorbance [optical density] versus concentration on standard graph paper).

TABLE 20.1. Volumes of Standard Solution and Reagent Used in Setting up Standard Curve.

DILUTION (ML)		Hemoglobin (g%)
Standard	Reagent	
5	0	20*
4	1	16
3	2	12
2	3	8
1	4	4
0	5	0

*Standard solution.

b. Determining an Unknown Sample

Place 5.0 ml of cyanmethemoglobin *reagent* in a test tube. Using a Sahli pipette, add exactly 0.02 ml of your blood. Mix the contents by inverting the test tube several times. Transfer the contents to a cuvette and read the percent transmittance as compared with the reagent blank. Transfer the reading to the standard curve and obtain the hemoglobin concentration in grams percent.

BLOOD CELL COUNTING

The blood contains three specialized classes of cells, or formed elements: (1) red blood cells (RBCs), or erythrocytes, which transport oxygen and carbon dioxide, (2) white blood cells (WBCs), or leukocytes, which combat infections and invading organisms, and (3) platelets, or thrombocytes, which prevent loss of blood. For these cells to carry out their functions properly they must be present in sufficient numbers, but not in excess. Thus, the counting of blood cells is an important technique, in that it helps establish the blood's capacity for performing these functions. The following are normal blood cell values (M = million).

Red blood cells

Males 5.4 ± 0.8 M/mm^3
Females 4.8 ± 0.6 M/mm^3

White blood cells

Males 7000–9000/mm^3
Females 5000–7000/mm^3

Platelets

150,000–400,000/mm^3
Average = 300,000/mm^3

Erythrocytes and platelets are not true "cells" as we have come to define the term. Both lack nuclei and are unable to undergo mitosis to form daughter cells. Actually, they are nothing more than "bags" to carry specific chemicals: hemoglobin in the RBC and platelet factor 3 in the platelet. If a gram of hemoglobin in the red cells is maximally saturated with oxygen, it can carry about 1.34 ml of O_2. In each 100 ml of blood there is roughly 15 g of Hb; hence around 20 ml of oxygen can be carried in every 100 ml of blood.

Anemia often results from an abnormal decrease in the number of erythrocytes, so that insufficient oxygen is carried to the tissues and they become oxygen starved. Other factors may also cause anemia, such as decreased hemoglobin in each cell, decreased cell size, and hemorrhage.

Hemocytometer Counting Chamber and Dilution Pipettes

Although many clinics are now using automatic devices such as the Coulter counter to make their cell counts, the standard techniques are still based on the use of the hemocytometer counting chamber (Figure 20.2).

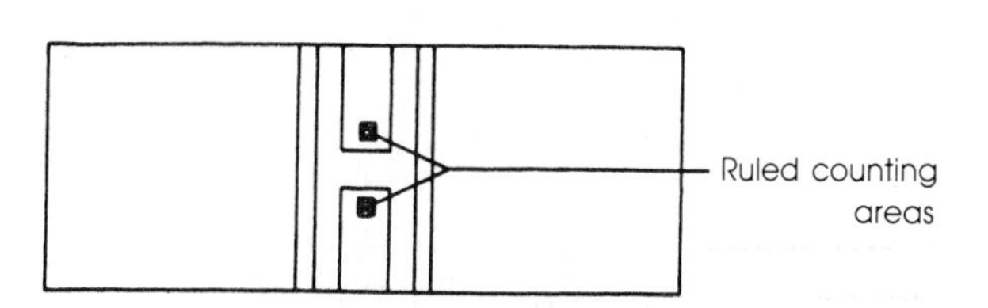

FIGURE 20.2. Hemocytometer.

Each of the two counting chambers is 9 mm^2 and is divided into nine squares, each measuring 1 mm^2 (Figure 20.3). The four corner squares are used for counting leukocytes and are divided into 16 smaller squares to make the counting process easier. The center 1-mm^2 square is divided into 25 small squares (1/25 mm^2), and each of these is further subdivided into 16 smaller squares for counting ease. The 1/25-mm^2 squares are bounded by double lines and are used for counting red blood cells.

The blood cells are so numerous that they must first be diluted before they are placed in the hemocytometer for counting. Special Thoma pipettes (Figure 20.4) are usually used in making these dilutions.

Once the dilutions are made, the hemocytometer is charged by touching the tip of the

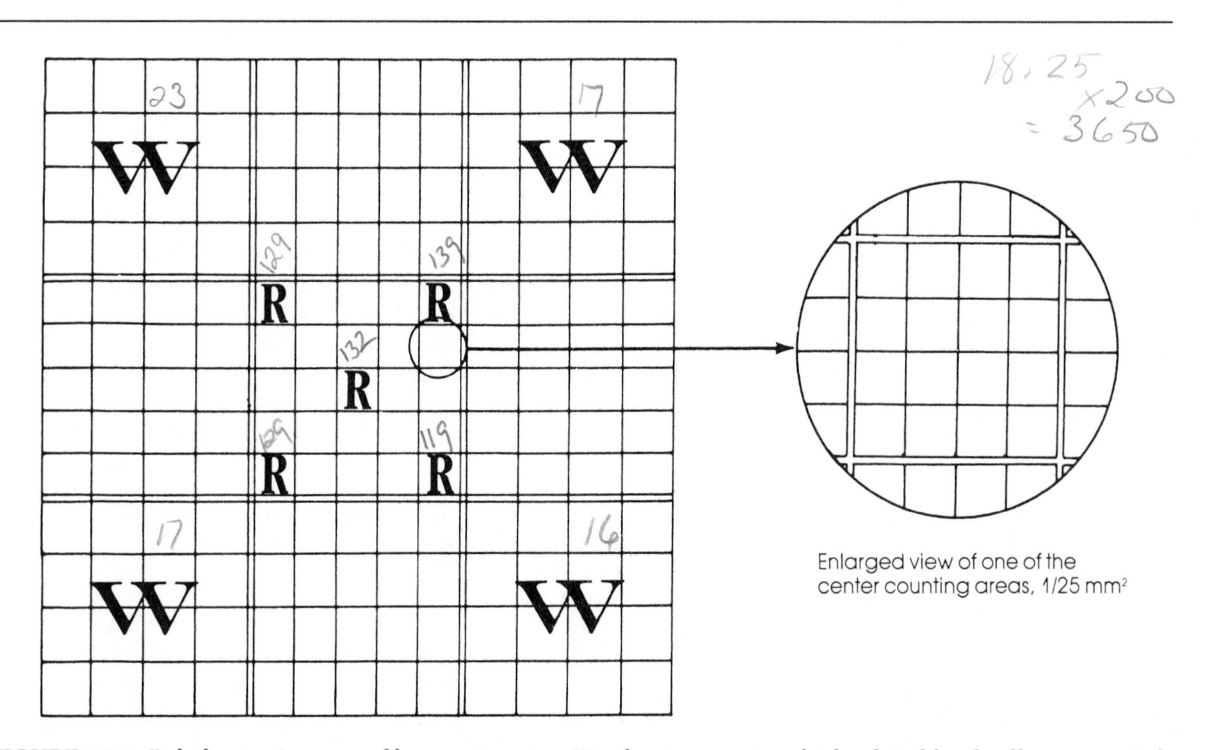

FIGURE 20.3. Ruled counting area of hemocytometer. W refers to areas in which white blood cells are counted, R refers to areas in which red blood cells are counted.

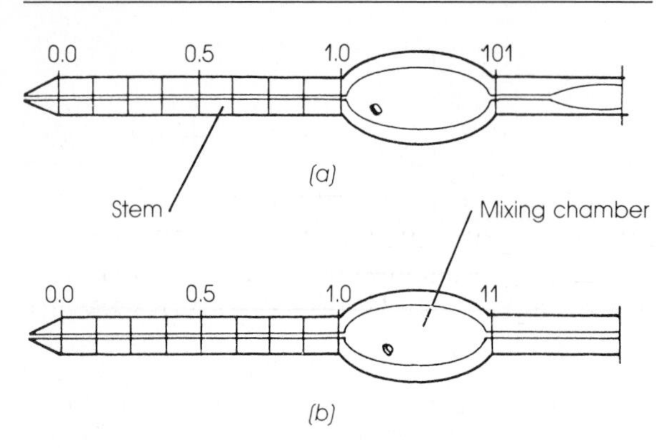

FIGURE 20.4. Thoma blood dilution pipettes. (a) Red cell pipette. (b) White cell pipette.

pipette to the junction of the hemocytometer and its coverslip (Figure 20.5). The diluted blood will flow in by capillary attraction to charge the counting chambers. After 2 to 3 minutes the cells will have settled to the bottom of the chamber, and you will be ready to begin counting.

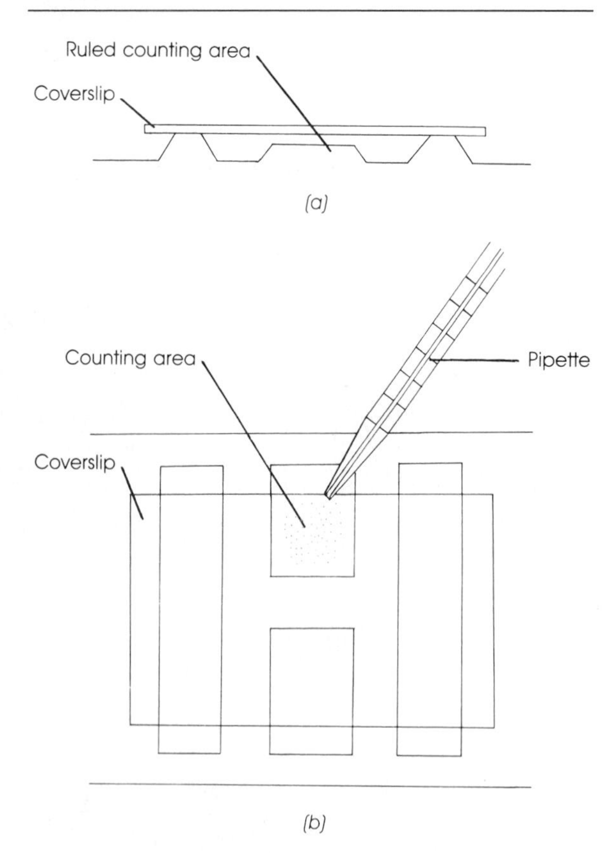

FIGURE 20.5. Loading the hemocytometer counting chamber. (a) Side view of counting area. (b) Procedure for loading the chamber.

Experimental Procedure

1. Red Blood Cell Counting

a. Obtain an RBC pipette and hemocytometer. Place the hemocytometer on the microscope stage and examine it so you are able to identify the counting areas. Use the low power to find the center 1-mm^2 square and high power to focus on the smaller 1/25-mm^2 squares.

b. Clean and dry the pipette *before* and *immediately after* using it, with the following solutions in the following order: distilled water (hydrogen peroxide for coagulated blood)—alcohol (95%)—ether or acetone.

Allow the ether to drain out by gravity from the upper end of the pipette. Then draw air through the pipette, using an aspirator. Never blow air through the pipette—this leaves water droplets in the pipette, which may alter the count.

Note: Coagulation of blood in the pipettes can be greatly reduced by drawing a heparin solution (1 mg of heparin in 500 ml of Ringer's) into the pipette and then blowing it out before drawing blood.

c. Clean the hemocytometer with distilled water and dry it with a soft tissue. *Do not* use an organic solvent such as alcohol, acetone, or ether because it will damage the hemocytometer counting area.

d. Place a few milliliters of Hayem's or Gower's solution in a clean watch glass. These solutions preserve the corpuscles and prevent coagulation.

e. Clean the tip of your finger with 70% alcohol, let it dry, and then lance the finger to obtain a drop of blood. Wipe off the first drop with a tissue and allow a second drop to collect.

f. Attach a rubber suction tube to the upper end of the RBC pipette and place the mouthpiece between your lips. Keeping the pipette in a horizontal position, insert the tip into the drop of blood. *Suck gently* on the rubber tube to draw blood up to *exactly* the 0.5 mark on the pipette. Be sure that the pipette tip is in the blood drop at all times so that a *solid column* of blood fills the pipette. If air bubbles enter the pipette, blow them out, clean the pipette, and try again. Hold your tongue against the mouthpiece to hold the blood column intact.

Wipe the excess blood from the pipette tip with a tissue. If excess blood is obtained, touch the pipette tip to a tissue to draw the blood back to the 0.5 mark.

g. Place the pipette tip in the diluting fluid and draw the fluid to exactly the 101 mark. Remove the rubber tubing, cover the two ends of the pipette with your thumb and forefinger, and move the pipette in a circular figure-eight motion for 2 minutes.

h. After diluting the cells, blow out about 3 to 5 drops to remove the diluting fluid from the stem of the pipette. Place a coverslip over the counting area of the hemocytometer. Touch the tip of the pipette to the junction of the coverslip and the hemocytometer. The diluted cells will flow in by capillary attraction to charge the chamber. Allow 2 minutes for the cells to settle before beginning your count.

i. Focus on the center square of the counting area using high power. Count the number of red cells in five of the 1/25-mm^2 squares and take their average. Usually the four outer squares and the middle one are counted. In your counting, you will find that some cells touch the boundary lines around the squares. Count the cells that touch on two sides of the square and omit those that touch on the other two sides.

j. Calculate the number of RBCs per cubic millimeter of blood by taking into account the following *multiplication factors:*

   The blood was diluted 200 times in the pipette.
   Therefore, you must multiply the average number per square by 200 (× 200).
   The depth of the counting chamber is 0.1 mm.
   Therefore, you must multiply the number of cells by 10 (× 10).
   The square counted was only 1/25 the area of the center square (1 mm^2).
   Therefore, you must multiply by 25 (× 25).

Multiplication factor = 200 × 10 × 25 = 50,000

For example, if you count an average of 120 RBCs per square, your RBC count is 120 × 50,000 = 6,000,000 RBC/mm^3.

Record your results in the Laboratory Report. Is your red cell count within the normal range for your sex?

Calculation of Total Oxygen-Carrying Capacity

Once the hemoglobin concentration of blood has been determined, it is possible to make an estimate of the total milliliters of oxygen a person's blood can carry. This estimate is based on the assumption that each gram of hemoglobin is carrying a maximum of 1.34 ml of oxygen. Estimate your blood volume as follows (1 kg = 2.2 lb):

Males: 79 ml blood/kg body weight ± 10%

Females: 65 ml blood/kg body weight ± 10%

$$\frac{\text{Total grams}}{\text{Hb in blood}} = \frac{\text{Blood volume}}{\text{(in 100 ml)}} \times \begin{array}{c}\text{Hemoglobin}\\ \text{concentration}\\ \text{(g/100 ml of blood)}\end{array}$$

$$\frac{\text{Total } O_2\text{-carrying}}{\text{capacity}} = \frac{\text{Total grams}}{\text{Hb in blood}} \times 1.34 \text{ ml } O_2/\text{g Hb}$$

Mean Corpuscular Hemoglobin Concentration (MCHC)

Using your hematocrit and hemoglobin values, calculate the mean corpuscular hemoglobin concentration (MCHC) for your red blood cells. The normal value is 34% ± 2%.

$$\text{MCHC (\%)} = \frac{\text{Hemoglobin (g/100 ml blood)}}{\text{Hematocrit (\%)}} \times 100$$

Calculation of Mean Corpuscular Volume (MCV)

Use your RBC count and hematocrit to calculate the average volume of your RBCs.

$$\text{MCV } (\mu m^3) = \frac{\text{Hematocrit (\% RBC)} \times 10}{\text{RBC count (millions/mm}^3)}$$

$$\text{Normal} = 87 \pm 2 \mu m^3$$

Anemia may be caused by several factors, such as a plastic bone marrow, RBC fragility, maturation deficiency, and hemorrhage. The determination of hematocrit, hemoglobin, RBC count, MCV, and mean corpuscular hemoglobin concentration (MCHC) allows one to classify the type of anemia more precisely.

MICROCIRCULATION

The real business of the circulatory system takes place in the exchange of substances between the interstitial fluid and the small blood vessels. The collection of vessels through which this exchange occurs is often referred to as the **microcirculation** and consists of capillaries, metarterioles, arterioles, and venules. Our current conception of this microcirculatory unit in the systemic circulation is depicted in Figure 20.6. It is often called the **Chambers-Zweifach capillary unit** after the investigators who first described its major features.

The **arterioles** contain a layer of smooth muscle in their walls that is under neural control by the autonomic nervous system (ANS). Contraction of this smooth muscle causes the lumen of the arteriole to become narrower (constrict) and thereby decrease the blood flow to that region. Relaxation of the smooth muscle causes the vessel to dilate. In this way, the ANS can divert the blood in the body from one area to another in response to the overall needs of the body areas for blood. The **venules** also have a smooth muscle layer, but it is not as extensive as that around the arterioles. The **metarterioles** are direct channels that are always open (patent) between the arterioles and the venules. They contain scattered amounts of smooth muscle and hence can constrict or dilate actively.

The actual exchange of substances occurs between the **true capillaries** and the interstitial fluid. The true capillaries are offshoots of either the arterioles or the metarterioles and consist of a single layer of endothelial cells and no smooth muscle. Their constriction or dilation is a purely passive process, depending on whether blood is flowing through them or not. At the entrance to each true capillary is a ring of smooth muscle called a **precapillary sphincter.** Its constriction or dilation controls the entrance of blood into the true capillaries. The precapillary sphincters are influenced largely by local factors resulting from tissue metabolism. Factors such as H^+ (low pH), pCO_2, temperature, histamine, and various other metabolites cause the sphincter to dilate, thus allowing blood to enter the capillaries. When the tissues are inactive and these products are not produced, the sphincters constrict and thereby shut off the blood flow through the capillaries. In this way the tissues control their own blood flow locally—a phenomenon called **autoregulation.**

The microcirculation may be examined using several preparations, for example, the tongue, mesentery, or foot web of the frog. Because it is a simple preparation, you will use the web of the hind foot in this experiment.

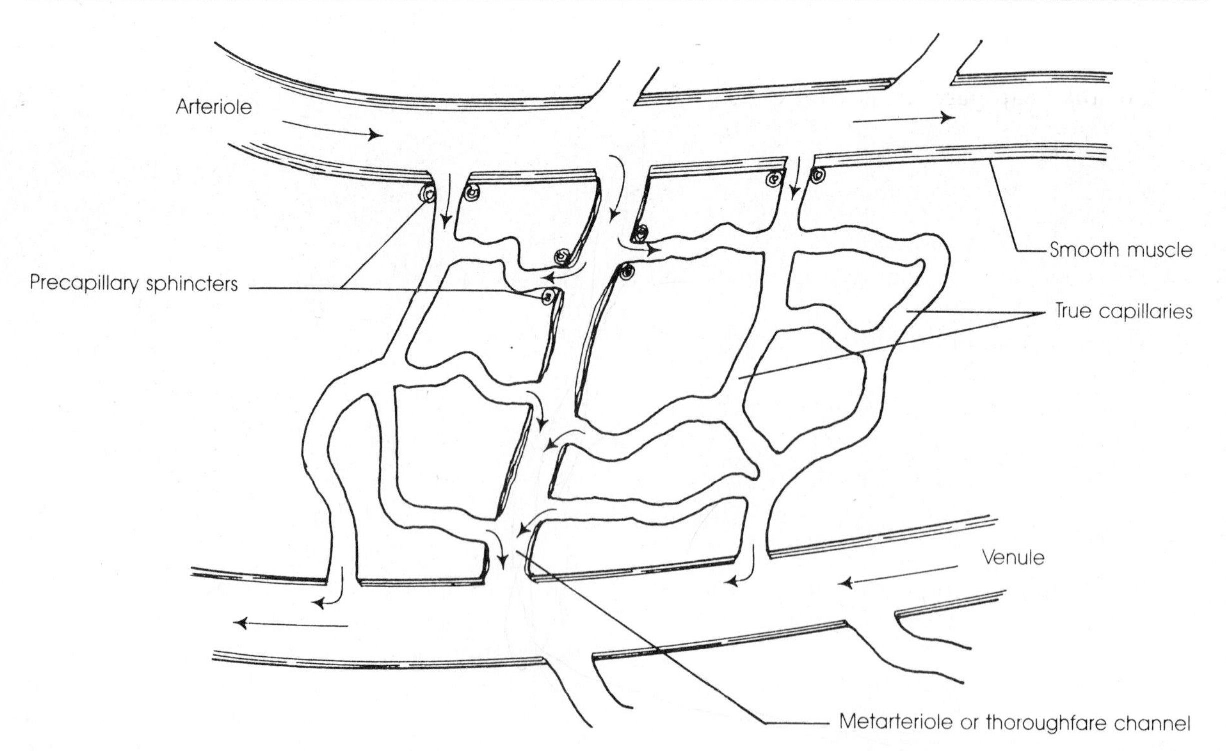

FIGURE 20.6. Chambers-Zweifach capillary unit.

Experimental Procedure

1. Anesthetize the frog by double pithing, or by injecting 1.0 ml of a 10% urethane solution under the skin on the back of the frog.

   Caution! *Urethane is potentially carcinogenic and teratogenic. Warning labels now carry this statement: "Avoid contact and inhalation. Wear protective clothing, gloves and mask. Avoid exposure to women of childbearing age. Wash thoroughly after handling."*

2. After anesthetizing, place the frog ventral side down on a frog board having a hole at one end. Spread the toes of one hind foot over the hole and fasten them in place loosely with pins. Do not stretch the foot too tightly or you will shut off the circulation. Keep the web moist with frog Ringer's and cover the frog with a wet towel to aid its respiration. Place the board on the microscope stage.

3. Examine the field under low and high power. How can you distinguish between arterioles, venules, and capillaries? Compare the rate of blood flow in each vessel. Does it flow in a pulsating manner or in a smooth flow? Estimate the diameter of each type of vessel using the red cell size (7 to 8 μm in diameter) as a measuring device. Notice the pliability of the red cells as they move through the small capillaries. Look for an alteration of dilation and constriction of the capillaries over a period of several minutes. This process is called **vasomotion.** What causes vasomotion? Can you locate any leukocytes? Where are they usually found?

1. Vasoactive Agents

Apply 4 to 5 drops of each of the solutions in the following list to the frog web, washing off the web with Ringer's between different solutions. The application of the solutions or sciatic nerve stimulation will disrupt your view of the web temporarily, so you will have to remember how the normal flow looks and quickly compare it with the postexperimental flow. Record your results in the Laboratory Report.

To isolate the sciatic nerve, make a dorsal incision through the skin over the thigh muscle. Cut the fascia over the muscles and use a glass probe to separate the muscles and reach in to pull out the nerve (see Figure 15.5, Experiment 15). Place a thread under the nerve so that the nerve can be pulled out for stimulation.

Warm Ringer's (40 °C)

Cool Ringer's (10 °C)

Histamine, 1:10,000[2]

Vasopressin or epinephrine, 1:1000[2]

Lactic acid or acetic acid, 0.5%[2]

Light sciatic nerve stimulation

Strong sciatic nerve stimulation

Mechanical stimulation (draw a pin across the web)

[2]Most effective if the mesentery preparation is used instead of the web of the foot.

LABORATORY REPORT

Name ______________________

Date __________ Section __________

Score/Grade ______________________

20. Blood Physiology I: Erythrocyte Functions

NAME	HEMATOCRIT (%)	HEMOGLOBIN (G/100 ML)	RBC COUNT (M/MM^3)	O_2-CARRYING CAPACITY (ML)	MCHC (%)	MCV (μM^3)
Lyn	40%	13.4	6,480,000	17.9	33.5	61.7
Janie	45%	17.0	4690,000	22.8	37.7	95.9
Cyndi	46.2%	14	6,860,000	18.8	30.3	67.3
Cam	47.4%	14	6,860,000	18.8	29.5	69.1
Natilie	50%	14	6,860,000	18.8	28	72.9
Tiffany	45.5%	14	6860,000	22.1	30.8	66.3
Melissa	44%	16.5	7,609,000	22.1	37.5	58
Tonya	44	16.5	7608,000	22.1	37.5	58
Mary	50	16.5	7,608,000	22.1	33	66
Dan	50	16.5	6,480,088	22.1	33.5	61.7
Class Average — Female						
Class Average — Male						

1. What is anemia? Which blood measurements provide information on a possible anemic condition? anemia is when there is a decreased ability for the red blood cells to transport oxygen due to the decrease in number or size of the cells, or the amount of hemoglobin in the b

hematocrit, blood hemoglobin concentration, RBC count, RBC size & % of hemoglobin to cell MCV + MCHC

2. Briefly explain the function of the following in erythrogenesis:

Vitamin B_{12} essential for formation of RBC; prevents anemia to normoblasts

Erythropoietin response of kidney cells to low oxygen stimulates conversion o stem cell to hemocytoblast

Iron essential for hemoglobin formation; bind oxygen

Intrinsic factor allows for the absorption of vitamin B12 promotes absorption in intestine

3. Polycythemia (excess number of red cells) occurs in patients with chronic emphysema. Explain the mechanism responsible for this response.
emphysema results in decreased oxygen transfusion therefore low blood O_2 (causes erythropoietn release & increased RBC production)

4. How does hemoglobin carry both oxygen and carbon dioxide in the blood?
by the interstitial fluid around the cells + true capillaries
O_2 binds to Fe^{+2} in heme (blood); O_2 binds to amino acids in globin
Bonding of CO_2 to globin weakens attraction Fe for O_2 + oxygen is released to tissue

5. Why is the inhalation of automobile exhaust fumes life threatening? Explain the physiology involved. carbon monoxide is a gas that is found in the exhaust fumes which interferes with the bloods ability to transport oxygen to the cells
Co binds Fe^{+2} in blood competing with O_2 but has an attraction of 210x that of O_2 ∴ prevents tissues from gitting oxygen

6. Why are hematocrits, hemoglobin concentrations, and erythrocyte counts generally lower for females than for males? because of the amount of body mass females & smaller in stature

blood volume + body weight
1. males are larger
2. males generally more active physically
3. male hormone stimulates RBC formation more than female hormones
4. female decrease during menstral loss

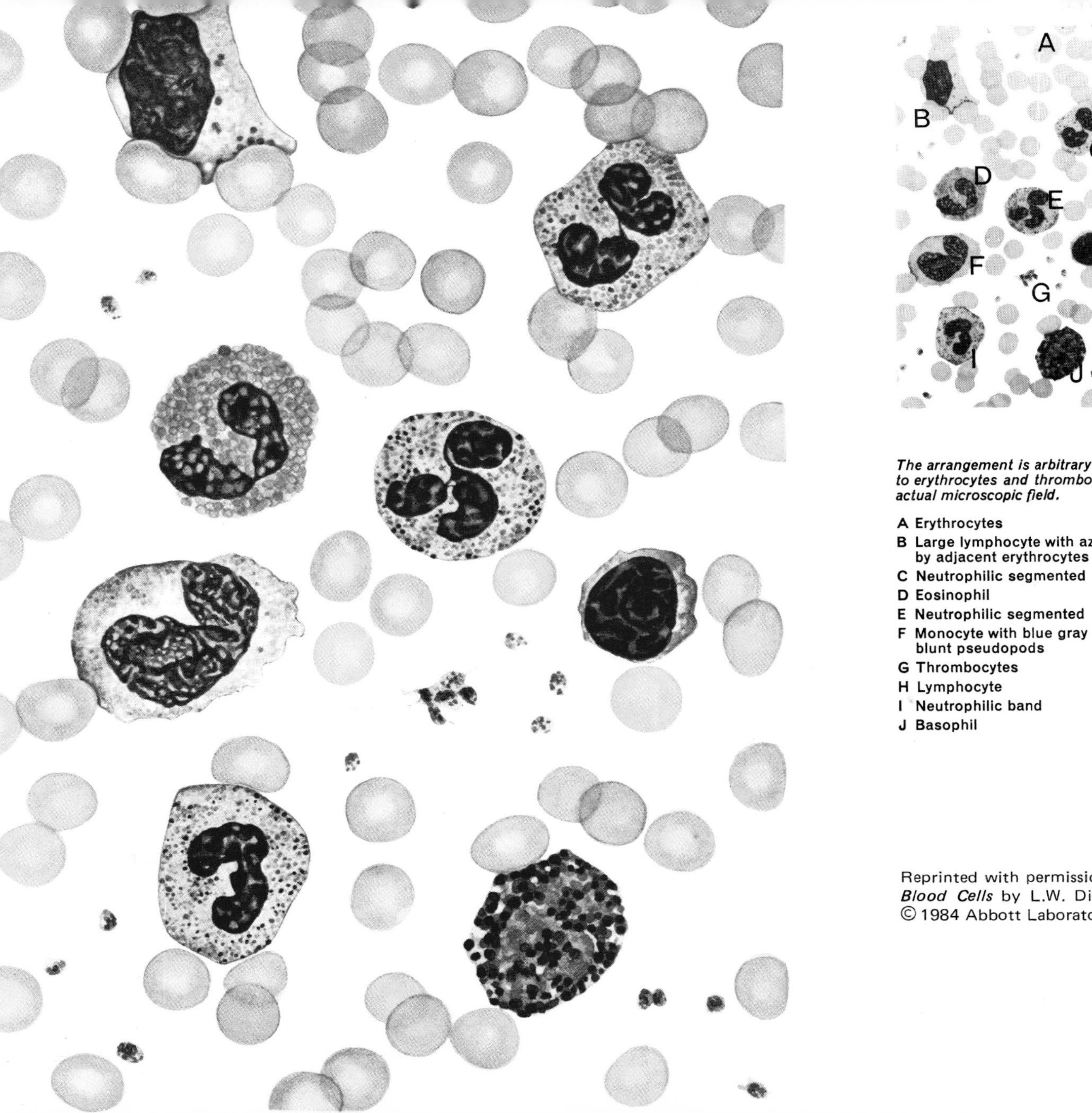

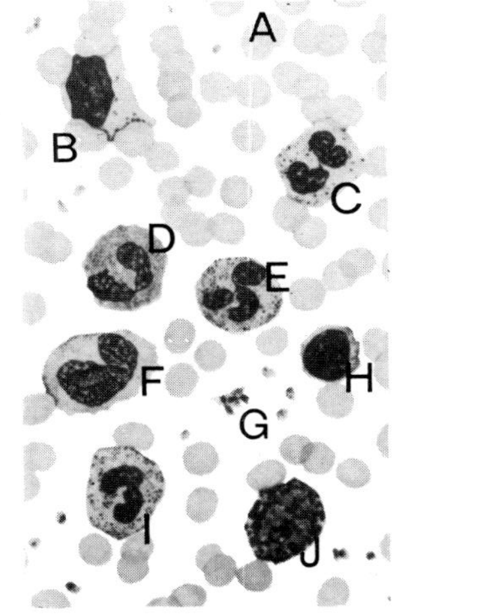

The arrangement is arbitrary and the number of leukocytes in relation to erythrocytes and thrombocytes is greater than would occur in an actual microscopic field.

A Erythrocytes
B Large lymphocyte with azurophilic granules and deeply indented by adjacent erythrocytes
C Neutrophilic segmented
D Eosinophil
E Neutrophilic segmented
F Monocyte with blue gray cytoplasm, coarse linear chromatin and blunt pseudopods
G Thrombocytes
H Lymphocyte
I Neutrophilic band
J Basophil

Name ______________________

Date ____________ Section ____________

Score/Grade ______________________

Microcirculation

1. Record your observations after each experiment and explain the physiological action of each solution or procedure.

EXPERIMENTAL PROCEDURE	OBSERVATIONS AND EXPLANATIONS
Warm Ringer's (40 °C)	dilation + increased blood flow in the web
Cool Ringer's (10 °C)	constriction + decreased flow
Histamine, 1:10,000	dilation of arterioles may cause venule constriction cause increased flow + tissue fluid formation
Epinephrine or vasopressin, 1:1000	Constriction of blood vessels pre cap sphincter shuts down flow
Acetic acid or lactic acid, 0.5%	dilation + increased flow
Light sciatic nerve stimulation	activates vasodilator by decreasing sympathetic tone ↑ flow
Strong sciatic nerve stimulation	vaso constriction due to pain response
Mechanical stimulation	trauma to blood vessel causes smooth muscle contraction + reduces blood flow

2. What factors are responsible for the large increase in blood flow through the skeletal muscles during exercise? dilation of arterioles + cappillaries supplying the muscles caused by autoregulation 1) O_2 deficiency 2) waste produced production Lactic Acid, Histamine, Carbonic Acid, Adenosine 3) higher temperature

3. How is blood flow through the capillary unit controlled by central integrative centers in the medulla of the brain? Vasometer centers control Vaso constriction or dialation ↑ sympathetic activity releases norepinephrine constriction arterioles ↓ sympathetic activity allows arterioles to passively dilate but there is some tonic activity due to sympathetic nerve stimuli

4. Outline the forces that cause movement of fluid and solutes in and out of tissue capillaries.

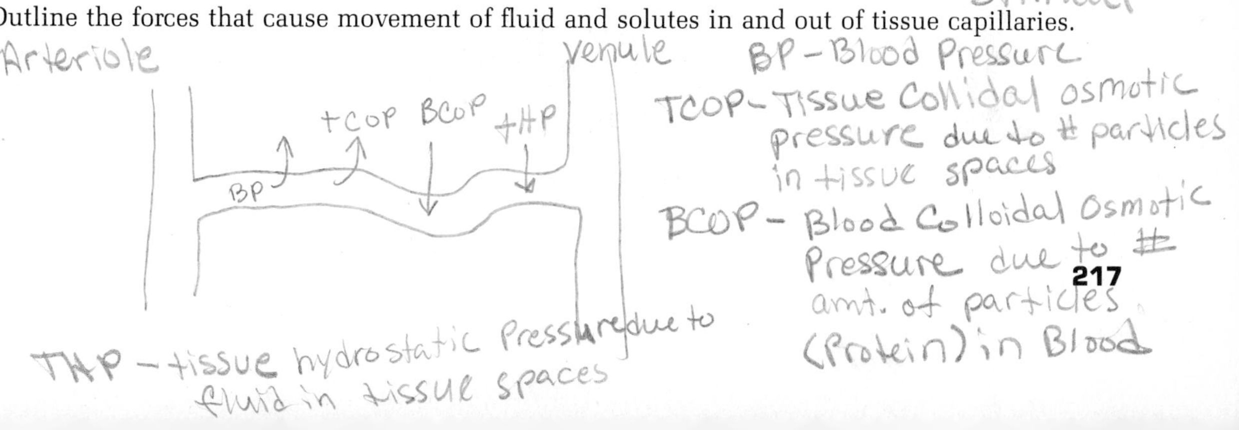

21 Blood Physiology II: Leukocytes, Blood Types, Hemostasis

Caution! *Parts of this lab may involve working with human blood. You should handle only your own blood. Dispose of all supplies (cotton, gauze, lancets, etc.) that come in contact with blood in properly marked containers. ALL BODY FLUIDS AND SUPPLIES MUST BE TREATED AS POTENTIALLY INFECTIOUS. Please read Appendix A, Precautions For Handling Blood.*

IDENTIFICATION OF WHITE BLOOD CELLS

In contrast to the red blood cells, the white blood cells (leukocytes) are nucleated and exist in several distinct types. They perform a variety of functions related to defense of the body against invading organisms.

Use the microscope to examine the prepared blood slides. Learn to identify each type of white blood cell (WBC) by its characteristic size, nuclear arrangement, or cytoplasmic granulation. Six types of WBCs are recognizable. The color plate following page 216 will help you identify the leukocytes.

Granulocytes (Polymorphonuclear Leukocytes)

Neutrophils

65% of total WBCs.

10- to 12-μm diameter.

Three-lobed nucleus.

Small pink cytoplasmic granules, purple nucleus.

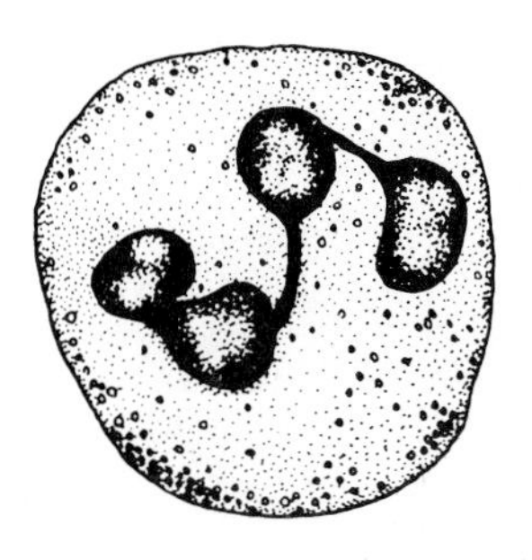

Eosinophils

2%–4% of total WBCs.

13-μm diameter.

Bilobed nucleus.

Coarse red-orange cytoplasmic granules, blue-purple nucleus.

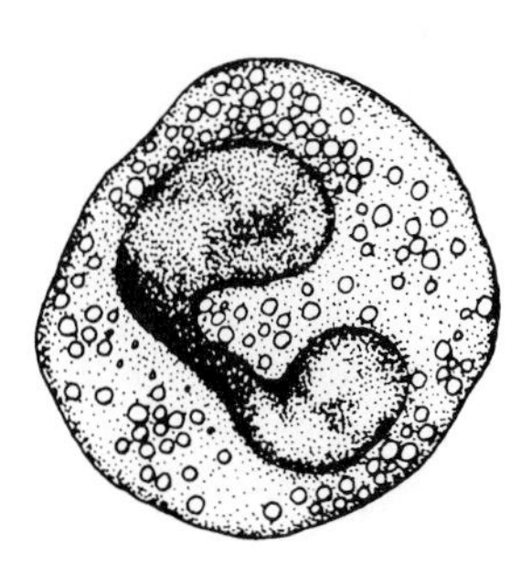

Basophils

0.5% of total WBCs.

7-μm diameter.

Bilobed nucleus.

Large deep blue or reddish purple cytoplasmic granules, blue-black nucleus.

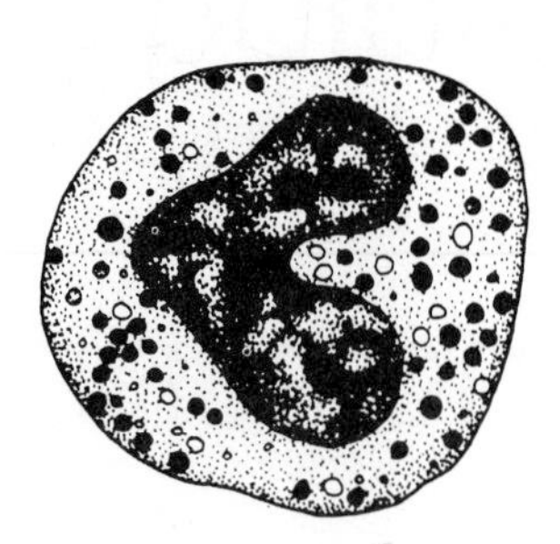

Agranulocytes (Mononuclear Leukocytes)

Small Lymphocytes

25% of total WBCs.

7-μm diameter.

Very large, spherical nucleus surrounded by thin cytoplasm.

Light blue cytoplasm (nongranular), deep blue or purple nucleus.

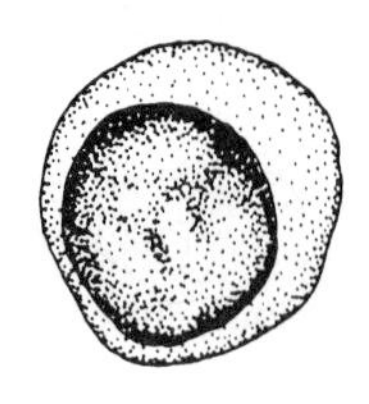

Large Lymphocytes

3% of total WBCs.

10-μm diameter.

Large oval, indented nucleus.

Light blue cytoplasm (nongranular), dark purple nucleus.

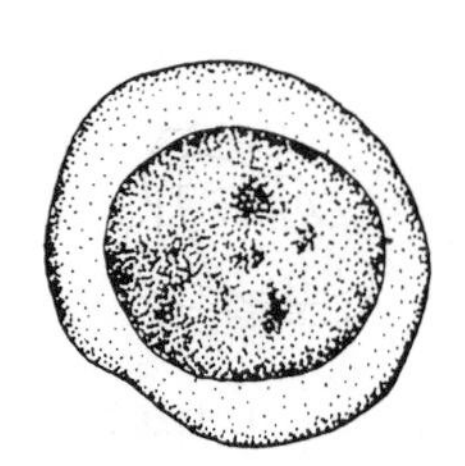

Monocytes

3%–7% of total WBCs.

15-μm diameter.

Large blue-gray cytoplasm (nongranular), blue or purple nucleus.

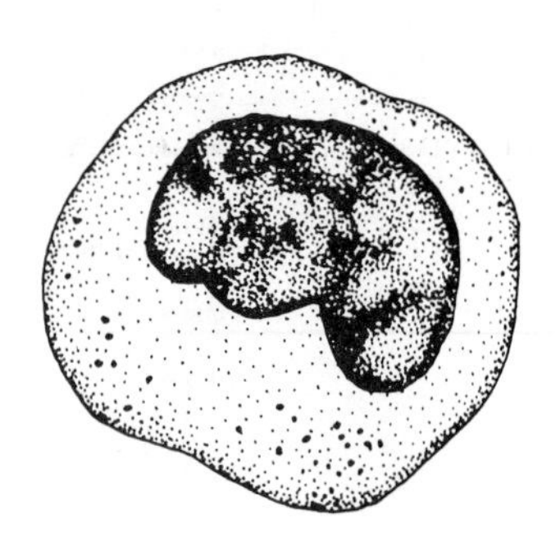

White Blood Cell Counting

This technique is similar to that used for counting erythrocytes (see Blood Physiology I), with the following exceptions:

The white cell counting pipette that dilutes the cells 20 times is used.

The blood is diluted with *Turk's* solution, which consists of 1 ml of glacial acetic acid, 1 ml of aqueous 1% gentian violet solution, and 100 ml of distilled water. This acid solution lyses the red cell membranes and converts the hemoglobin to hematin. The gentian violet stains the WBC to make identification easier.

The number of white blood cells is counted in each of the four large 1-mm^2 squares in the corners of the ruled area, and the average count determined. Use the low power for these counts.

The following *multiplication factors* are used to calculate the number of WBC per cubic millimeter:

The cells were diluted 20 times ($\times$ 20).
The counting chamber is 0.1 mm deep ($\times$ 10).
The average number of cells in 1 mm^2 were counted ($\times$ 1).

Multiplication factor $= 20 \times 10 \times 1 = 200$

How does your leukocyte count compare with the normal range for your sex? What terms do we use for a deficient number of leukocytes? For an abnormally high number of leukocytes?

Differential Leukocyte Count

A determination of the total leukocyte count (7500/mm^3 average) is an important clinical measurement, but a more accurate diagnosis is obtained by making a differential WBC count. In

a differential count, the percentage of each type of leukocyte in the total leukocyte population is determined. Each type of leukocyte performs a different function in the battle against infection, and each disease causes different responses by the WBCs. A few examples of alterations in the leukocyte population for various diseases are given in Table 21.1.

Blood Smear Staining Procedure

1. Obtain a drop of blood by finger puncture. Place a small blood drop on one end of a clean glass slide (Figure 21.1).
2. Hold a second slide (the spreader) at a 45-degree angle to the first slide and move it toward the drop of blood. Allow the blood to spread along the edge of the spreader slide; then move the spreader in a smooth, fast motion to the other end of the first slide. This motion will deposit a thin, evenly spread film of blood across the slide. Allow the slide to air dry.
3. Using a medicine dropper, cover the slide with Wright's stain. Count the number of drops used. Allow the stain to stand 2 minutes.
4. After 2 minutes, add an equal number of drops of distilled water to the slide to dilute the stain (better results will be obtained if a buffer solution is used in place of distilled water—1.63 g of KH_2PO_4 and 3.2 g of $NaHPO_4$ in 1000 ml of distilled water). Blow gently on the slide to mix the buffer and stain. Let the slide stand for 4 minutes.
5. After 4 minutes, flush the slide gently with tap water. Dry the bottom of the slide and allow the top to air dry.

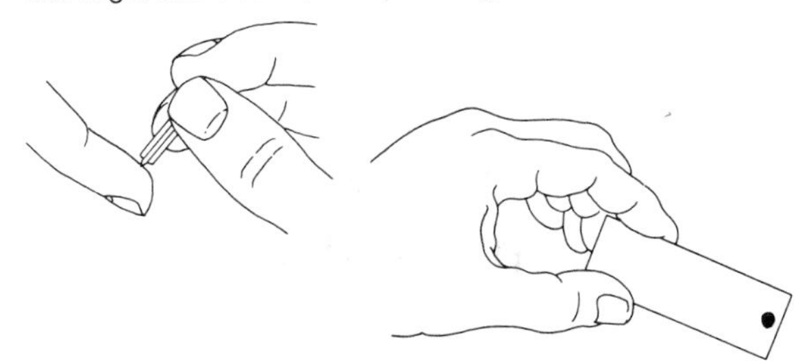

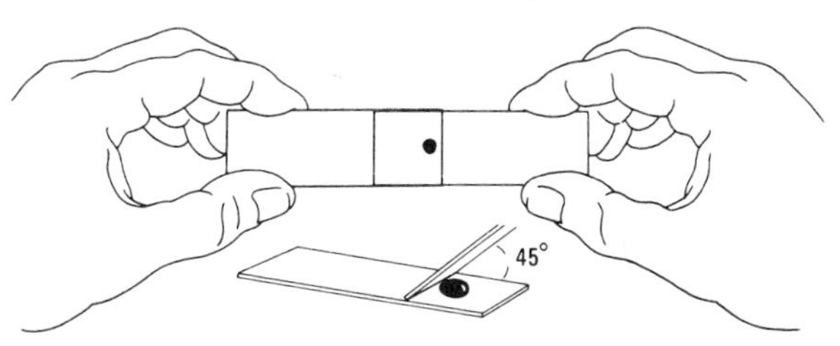

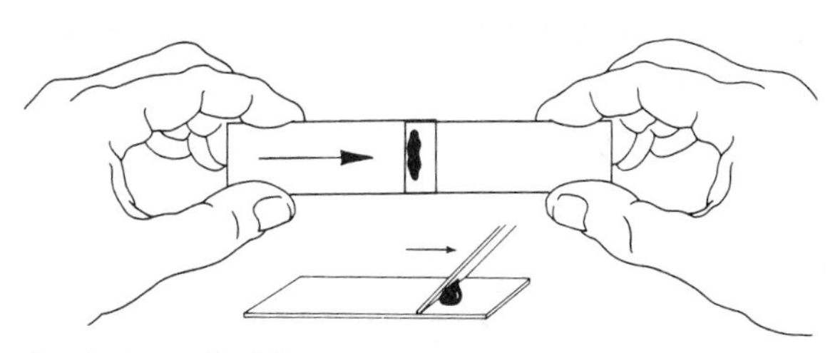

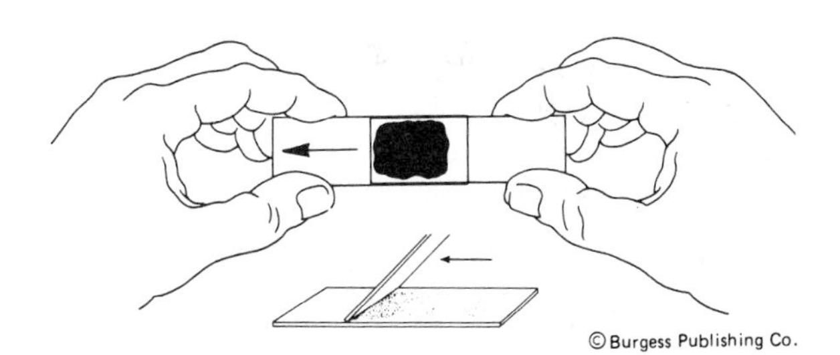

FIGURE 21.1. Preparation of a blood smear.

TABLE 21.1. Leukocyte Alterations Occurring with Various Diseases or Conditions.

DISEASES OR CONDITIONS	SYMPTOMS
Protozoan infections, malnutrition, aplastic anemia	Neutrophilic leukopenia
Strenuous exercise, rheumatic fever, severe burns	Neutrophilic leukocytosis
Mumps, German measles, whooping cough	Lymphocytosis
Scarlet fever, parasitic infections, allergic reactions	Eosinophilia
Chronic diseases, such as tuberculosis and leukemia	Monocytosis
Administration of glucocorticoid drugs	Lymphocytopenia

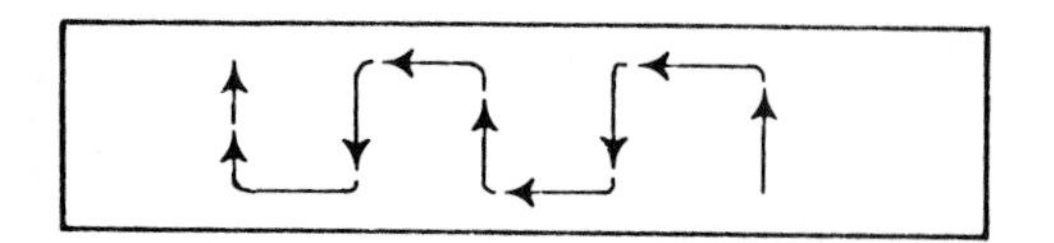

FIGURE 21.2. Scanning procedure for counting WBCs.

6. Examine the smear, first under low power and then under oil immersion, to identify the various leukocytes. If your smear is not satisfactory, prepare another one. The color plate will help you identify the leukocytes.
7. Count the number of each type of WBC on the slide, recording each on a tally sheet as you identify it. Count and identify 100 WBCs and record your results in the Laboratory Report. Use the scanning procedure shown in Figure 21.2 when examining the slide.
8. After counting 100 cells, express the results in percentages. How do your percentages compare with the normal percentages?

BLOOD TYPING

Many clinical conditions require the transfusion of whole blood. Transfusions cannot be performed indiscriminately between persons, however, because of the possibility of antigen/antibody reactions producing **agglutination** of red cells. Agglutination refers to a clumping of red cells together. Why would agglutination be dangerous?

The human red cell has around 30 commonly occurring **antigens** on its membrane. In blood typing terminology, these are called **agglutinogens.** These agglutinogens may react with complementary **antibodies,** or **agglutinins,** in the donor's or recipient's plasma to cause agglutination of red cells.

$$\text{Agglutinogens} + \text{Agglutinins} \rightarrow \text{Agglutination}$$

Although any of the 30 antigen-antibody combinations can cause agglutination, in actual practice most agglutinations in transfusion are caused by two antigen-antibody systems—the ABO and Rh systems.

ABO System

A person may have A-, B-, or O-type antigens on the red cells, or any two of these together. O antigens are very weak, as are the anti-O antibodies; hence they rarely cause any agglutination. For this reason, a person who has O-type blood is usually regarded as having *no* antigens on the red cells. Only the A and B antigens are regarded as having strong antigenicity.

Antigens are genetically determined. It should be pointed out that the ABO system is the only one in which the person's plasma *automatically* contains the noncomplementary antibodies to the red cell antigens. These antibodies are also determined genetically. All other antibodies found in the plasma must be formed through the entrance of an antigen into the body to stimulate antibody production.

The antigens and antibodies for each blood type are summarized in Table 21.2, along with the percentage of each type found in various races.

The differences among the races in percentage of each ABO type indicate the role of genetic determination for these blood groups.

Agglutination results from the reaction of an antigen with its complementary antibody. For example,

$$A + \alpha \rightarrow \text{Agglutination}$$

$$B + \beta \rightarrow \text{Agglutination}$$

A person with type O blood is referred to as the **universal donor,** and a person with type AB

TABLE 21.2. Antigens and Antibodies Found in Each Blood Type and Percent Distribution of Blood Types in Various Races.

BLOOD TYPE	AGGLUTINOGEN (ANTIGEN)	AGGLUTININ (ANTIBODY)	PERCENT FOUND IN: CAUCASIAN	BLACK	ARABIC
A	A	β (beta or anti-B)	43	22	5
B	B	α (alpha or anti-A)	7	29	0
AB	AB	None	3	4	0
O	None	α and β	47	45	95

as the **universal recipient.** Explain these designations in the Laboratory Report. What are antibodies? Where are they produced in the body? What is the current theory of the mechanism of antibody production?

Experimental Procedure

1. Obtain a clean microscope slide. Using a glass-marking pencil, mark one end A and the other end B.
2. Lance your finger to obtain blood. Place 1 drop of blood on each end of the marked slide.
3. Add 1 drop of anti-A serum to the A side. Add 1 drop of anti-B serum to the B side. Mix the antiserum and blood on each side with a toothpick, using a different toothpick for each side. Spread each mixture over an area of about 3/4 in. in diameter. Make certain you do not mix the anti-A and anti-B antisera.
4. Observe the slide for any agglutination of red cells. If agglutination occurs on side A only, you have the blood type A. If it occurs on side B only, you have type B. If a reaction occurs on both sides, you have type AB. If no reaction occurs on either side, you have type O. Explain the antigen-antibody basis for these reactions. The strength of the agglutination reaction is not the same for every person; in some cases it may be necessary to observe the cells under the microscope to ascertain if agglutination has actually taken place.

Use Figure 21.3 to aid in your identification of blood type.

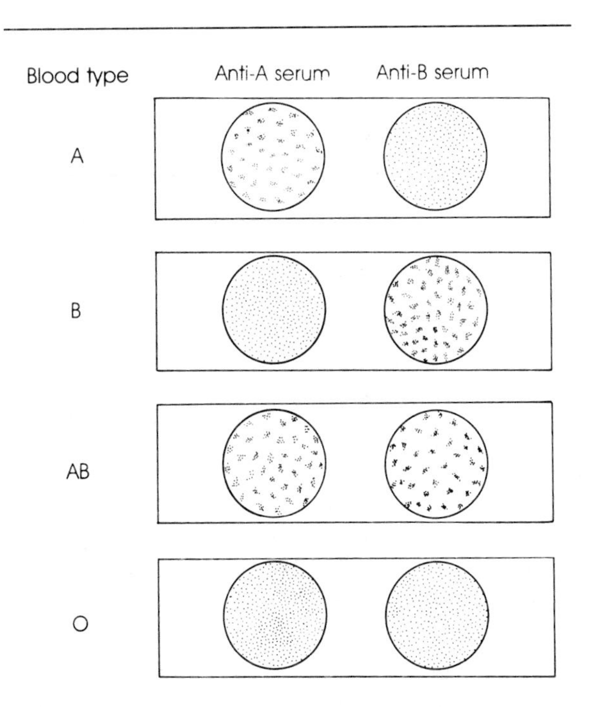

FIGURE 21.3. Antigen-antibody reactions, ABO blood type system.

Rh System

In 1940, Landsteiner and Wiener discovered a system of antigens in the cells of the Rhesus monkey that is different from the ABO system. After producing an antiserum (antibody) against the "Rh" factor, they tested it with human RBCs and found that 85% of the human population also has this Rh factor (are Rh positive). The other 15% of the population does not have this factor (Rh negative). In contrast to antigens in the ABO system, the Rh factor is found in all body cells, not just on the erythrocytes. Actually, there are eight different types of Rh agglutinogens. However, the four strongest types react with anti-Rho (anti-D) antiserum. Hence, if your blood agglutinates with anti-Rho antiserum, we say you are Rh+. If it does not, you are Rh−.

If a person who has Rh− blood receives a transfusion of Rh+ blood, there is usually no adverse reaction the first time, but the entrance of the Rh factor stimulates an accumulation of anti-Rh antibodies in the recipient's blood. If the same Rh− recipient receives a second transfusion of Rh+ blood, the antibodies are ready and will cause an agglutination reaction. The best known agglutination reaction of the Rh factor is **erythroblastosis fetalis,** a destruction of red cells in the newborn baby (hemolytic disease of the newborn).

Experimental Procedure

1. Prewarm a clean microscope slide on a slide warming box.
2. Mix 2 drops of your blood with 1 drop of anti-Rho (anti-D) antiserum on the slide.
3. Place the slide back on the warming box and tilt the slide occasionally to aid the mixing.
4. Observe the agglutination within the first 2 minutes after mixing. Check under the microscope if you are in doubt. The Rh factor is usually weaker than the AB antigens, and the agglutination reaction is not as strong or as easy to detect.

Obtain the ABO and Rh blood types of all members of the class. How do the percentages of persons having each type compare with the national percentages?

BLOOD COAGULATION (HEMOSTASIS)

The blood contains its own chemical system to **coagulate** it and thereby to prevent blood loss from the body. Coagulation is a fantastically complex process that begins as soon as blood platelets are ruptured or body tissues are damaged. In the following experiment you will examine some of the simpler processes in the coagulation mechanism.

Experimental Procedure

1. Bleeding Time

Clean the tip of your finger with 70% alcohol, and then dry it with a piece of cotton. Puncture the finger with a lancet and record the time. At 15-second intervals, wipe the blood drop away completely with a filter paper (do not touch your finger when wiping the blood away). Continue this procedure until no more blood stains appear on the filter paper. Record this time. Calculate the bleeding time. Is it close to the normal bleeding time of 1 to 3 minutes?

2. Clotting Time

Lance your finger to obtain a large drop of blood. Note the time when the drop appears. Rapidly draw blood into a nonheparinized capillary tube by holding the tube in the drop of blood in a horizontal position.

At 30-second intervals, break off a small piece of the capillary tube (0.5 cm) and see if clotting has occurred. Clotting has occurred when a thread of coagulated blood is visible between the two pieces of tubing.

How does your clotting time compare with the normal time of 5 to 8 minutes?

3. Observation of Fibrin Strand Formation

Place a drop of freshly drawn blood on a clean glass slide, cover it with a coverslip, and focus on it with the microscope. Touch a drop of methyl violet solution to the edge of the coverslip so that the solution will run under and stain the blood. In the next few minutes, watch for the formation of fibrin strands. Probe these strands to test their elasticity.

LABORATORY REPORT

Name ____________________

Date __________ Section __________

Score/Grade ____________________

21. Blood Physiology II: Leukocytes, Blood Types, Hemostasis

NAME		WBC COUNT	DIFFERENTIAL LEUKOCYTE COUNT (%)				
			NEUTROPHILS	EOSINOPHILS	BASOPHILS	LYMPHOCYTES	MONOCYTES
Normal Values:	Female	$6000/mm^3$	50% to 65%	1% to 4%	0% to 1%	25% to 33%	3% to 7%
	Male	$8000/mm^3$					
Janie		7550	54	2	2	40	2
Cam		9400					
Lynn		3650	54	2	2	40	2
Class Average	Female						
	Male						

1. List the major functions of the leukocytes:

Neutrophils responsible for increasing phagocytes in acute infections

Eosinophils fight diseases + parasitic infestations

Basophils Convert to mast cells; release histamine + heparine

Lymphocytes hemoral + cellular immunity

Monocytes phagocytic in chronic infections become macrophages

2. Define the following terms:

$10^6 mm^3$ Leukemia disease where there is rapid & unrestrained growth of white blood cells

above $10^4 mm^3$ Leukocytosis ↑ number of WBC caused by an infection

below $5000/mm^3$ Leukopenia abnormal decrease of WBC (Bone marrow failure or drugs)

Mononucleosis ↑ number of mononuclear WBC in blood

Thrombocytopenia ↓ in number of blood platelets

Hemophilia the inability of the blood to clot (hereditary)

3. Why is a person with type O blood called the "universal donor"?

A person with Type O blood has no antigens on the red cells therefore produces weak antibodies. Therefore it can be given to anyone without risk of agglutination

blood will find nothing to react with

4. Why is a person with type AB blood called the "universal recipient"?

AB recipient has no A or B antibodies, therefore recieving A or B antigen will have no possibility for reaction unless a large amount of plasma accompanies the donation

5. Under what conditions is erythroblastosis fetalis possible? Why is the condition given this name?

When a woman carries an Rh+ fetus whose blood may enter the mother — then is stimulated to produce antibodies which cross over to the child destroying RBC. The child attempts to replace destroyed RBC with immature RBC (erythroblasts)

6. What treatment can be given for erythroblastosis fetalis? intrauterine transfusion or giving the infant a blood transfusion immediately following birth

give potential mothers RhoGam (immune globulin)

7. What is the difference between active and passive immunity?

Active - immunity which results from a person having a disease or by innoculation of a vaccine, the person produces own antibodies against the antigen.

Passive - immunity acquired in utero from antibodies that pass from the mother to the fetus

or by the newborn ingesting the mother's milk

or by receiving an antibody to prevent a prevelant infection

8. Many of the early heart transplant patients died not from rejection of the transplanted heart but from bacterial infections. How can this be explained? to prevent rejection the immune system is suppressed by immunosuppressive drug ie. cyclosporin, This weakens immune response to bacteria

9. Record your values for:

   Bleeding time ______________ Clotting time ______________

   Why is bleeding time shorter than clotting time? Damaged blood vessels constrict shutting off blood flow,
   The clotting process requires many chemical reactions

10. What role do the blood platelets play in coagulation?
    serotonin - vascular spasms
    PF3 - catalyses coagulation
    ATP - provide energy
    ADP - increases platelet stickiness → vascular plug

11. Outline briefly the intrinsic and extrinsic pathways for blood coagulation.

Extrinsic Pathway
tissue damage
↓
thromboplastin → Stuart Prower factor
↓
Proacceleration
↓
Prothrombin → thrombin
↓
Fibrogen
↓
Fibrin
↓
Clot

Intrinsic Pathway
Collagen contact
↓
Hageman factor
↓
Clotting factors → (Stuart Prower factor)

12. What is the difference between a thrombus and an embolus? Why are they potentially dangerous?

    thrombus - stationary clot
    embolus - circulatory clot
    may shut off blood vessels to vital areas, resulting in death of cell beyond blockage

22 Physical Fitness

All of us are aware of the concept of "physical fitness" and the decrease of fitness in recent generations owing to the increased mechanization of our age. Yet the term "fitness" is an elusive one that means different things to different people. For our study, fitness will be defined as the capacity to meet the physical stresses encountered in life. Major components of this fitness include **muscular strength, flexibility,** and **cardiorespiratory endurance (aerobic fitness).** We will also include **obesity** as a negative aspect, because it profoundly reduces the other components of physical fitness. In this exercise you will be introduced to some simple measurements of these fitness components. It is hoped that this will spark your interest to undertake a more thorough examination of your total fitness, using more complete and exact methods of testing. For this lab you should wear athletic apparel (T-shirt, shorts, tennis shoes), for you will be doing moderate exercise and measuring skin folds at various sites on the body.

MUSCULAR STRENGTH AND ENDURANCE

Strength is one of the first things most people think of when fitness is mentioned. Muscle tone does play an important role in good posture, helping prevent lower back problems, giving us better performance in sports, and improving our figures, which gives us a definite psychological boost. An increase in muscular strength results when fast twitch muscle fibers develop more myofilaments (actin and myosin), which provides more cross-bridges to produce tension. Isometric exercises like weight lifting are especially good for producing muscle hypertrophy and increased muscle strength.

Muscular endurance is the ability to contract muscles repeatedly or to sustain a single contraction. Endurance is important for many of our daily work activities and athletic endeavors. Endurance is a property of the slow-twitch muscle fibers, which increase their concentration of oxidative enzymes and capillaries with isotonic training. This allows these fibers to contract repeatedly with greatly prolonged fatigue time. It is interesting to note that increases in muscular endurance can occur with little increase in hypertrophy of the muscle fibers. In this exercise we will examine one aspect of muscular strength and endurance by testing hand grip strength.

Experimental Procedure

1. Hold the grip tester in the center of your right palm with your arm extended at your side. Squeeze the tester as hard as possible and record the grip strength. Do this three times and use the average as your maximum grip strength. Repeat for the left hand.
2. After 1 minute of rest, perform 20 consecutive grip contractions at a rate of one every 2 seconds, trying to make each a maximal contraction. Repeat for the other hand.

Record the total kilograms of force exerted for the 20 contractions and the average force of these contractions.

3. Record, on the chalkboard, the maximum grip strength, the total force for 20 contractions, and the average force of the 20 contractions for all members of the class. Calculate the mean values and ranges for the class. Compare your grip strength measurements with those of other class members of the same sex.

FLEXIBILITY

Flexibility is the ability to move the limbs through their normal range of motion. Movement is limited by the connective tissue that covers the muscles and by the tendons that link the muscles to bone. With increasing age or inactivity, these tissues lose their elasticity, the range of motion decreases, and we become more susceptible to muscle and joint injuries. Static stretching exercises help maintain flexibility if they are performed on a daily basis, especially before and after an exercise bout. The proper technique for stretching involves a slow movement until the limit of the motion range is reached, holding of the position for 10 seconds, and then relaxing. The stretching should not be done in a jerky, rhythmic fashion, as this may damage tissues. Athletes who train their muscles (e.g., by running or power lifting) without doing stretching exercises may find their range of limb motion so limited they appear to be "muscle bound." In this exercise we will test trunk flexibility, a very important factor in preventing lower back problems.

Experimental Procedure

1. While sitting on the floor, extend your legs so your heels touch the foot stop of the flexibility tester.

2. Place your fingers against the sliding mechanism and with a slow, steady motion push the slider as far as possible to the point you can reach and hold for 3 seconds without bending your knees. The scale reading at the slider position is your flexibility score. How does your score compare with those of the other class members?

BODY COMPOSITION

Obesity has become a critical detriment to fitness in many countries such as the United States, where approximately 80 million people are classed as obese. Obesity is an excess accumulation of fat beyond what is considered normal for the person's age and sex. Overweight does not always mean obesity if the excess weight is muscle rather than fat.

How much fat classifies a person as obese? Rough guidelines are over 20% for men and over 30% for women. The average fat percentage varies with age and sex, as shown in Table 22.1.

So what's the problem with a little excess fat? Fat is a good storage form of energy, and everyone knows that fat people are happier than the rest of the population—right? One reason fat is such a villain is that it reduces our muscle strength, flexibility, and cardiorespiratory endurance—the three fitness components we are measuring in this exercise. Equally important is that excess fat contributes to the development of three of our most serious health problems: cardiovascular disease, hypertension, and cerebral vascular accident (stroke). As you can see, there are many good reasons we should shed the excess fat we are carrying around. The first step, of course, is to determine if we are actually obese or just overweight with excess muscle mass.

How can we measure the relative fat and lean weight in the body? The most accurate method is

TABLE 22.1. Body Fat in Relation to Age and Sex.

AGE RANGE	AVERAGE FAT PERCENTAGE	
	MEN	WOMEN
15	12.0	21.2
18–22	12.5	25.7
23–29	14.0	29.0
30–40	16.5	30.0
40–50	21.0	32.0
Minimum	2–5	7–11

Modified from B. Sharkey, *Physiology of Fitness,* 2nd ed. (Champaign, Ill.: Human Kinetics Publishers, 1984). Used by permission.

underwater weighing, which compares weight in air to weight underwater. This technique requires trained technicians and subjects who are comfortable underwater. A simpler, less expensive technique uses skin fold calipers to measure the thickness of skin and fat at representative sites around the body. These measurements are used in equations or nomograms to give an estimate of body density and percent fat. Because approximately 50% of the body fat is subcutaneous, skin fold measurements are able to give us a fairly good estimate of fat weight. A variety of calipers are commercially available for measuring skin folds, including the inexpensive Fat-O-Meter used in this exercise (available from Carolina Biological Supply Co., Burlington, N.C.).

Experimental Procedure

1. Acquaint yourself with the proper technique for using the calipers: Hold the skin fold between the thumb and middle finger of your left hand and the caliper in your right hand, with the scale facing you. Slide the caliper open, place it around the skin fold, and slowly close around the fold. Measure the thickness in millimeters for three trials and record the average of the two closest readings. Measurements are taken with the subject in a standing position, and usually on the right side of the body.
2. Measure the appropriate skin fold sites needed for calculating the density using the equations given in Table 22.2 for each sex and age group. Consult Figure 22.1 for the proper location of each site.
3. Calculate your percent fat from the density, using this equation:

$$\%\ \text{fat} = \frac{457}{\text{Density}} - 414.2$$

4. An alternate method for determining percent fat involves use of the nomogram in Figure 22.2. Determine the sum of three skin fold thicknesses as follows:

   Males: Mid axilla, umbilicus, thigh
   Females: Triceps, suprailiac, thigh

   How do the nomogram and equation methods compare in the percent fat obtained?

TABLE 22.2. Calculation of Body Density by Means of Skin Fold Measurements.*

AGE GROUP	DENSITY FORMULA
Women	
5–9	1.0936 – .0073(S1) + .0037(S4) + .0014(S5) – .0029(S7) – .0015(S8)
10–14	1.0740 + .0079(S2) + .0012(S4) – .0054(S5) – .0028(S8) – .0042(S10)
15–19	1.0802 – .0002(S2) – .0007(S6) – .0003(S7) – .0006(S8) – .0040(S9)
20–24	1.0861 – .0005(S2) – .0011(S5) – .0015(S6) – .0005(S9) + .0006(S10)
25–29	1.0757 – .0017(S2) – .0008(S3) – .0003(S5) – .0004(S6) – .0005(S8)
30–39	1.0827 – .0010(S2) + .0012(S3) – .0007(S5) – .0007(S6) – .0012(S7)
40–49	1.0726 – .0005(S1) + .0002(S5) + .0007(S6) – .0003(S8) – .0013(S9)
50+	1.0674 – .0003(S1) – .0008(S3) – .0007(S5) – .0002(S7) – .0009(S8)
Men	
5–9	1.0642 + .0025(S1) – .0025(S5) – .0015(S9)
10–14	1.1201 – .0028(S2) – .0080(S6) + .0069(S10)
15–19	1.0832 + .0006(S1) – .0009(S2) + .0004(S5) + .0021(S6) – .0023(S7)
20–24	1.0971 + .0005(S1) – .0023(S8) – .0005(S4) – .0005(S5) + .0005(S10)
25–29	1.0862 – .0016(S2) + .0016(S3) – .0005(S4) + .0025(S6) – .0009(S9)
30–39	1.0926 + .0008(S3) – .0008(S6) – .0009(S7) – .0010(S9)
40–49	1.0737 – .0012(S2) + .0005(S3) – .0004(S4) – .0015(S6) – .0010(S10)
50+	1.0639 + .0004(S1) + .0003(S2) – .0003(S5) – .0013(S7) – .0002(S10)

*Refer to Figure 22.1 for skin fold measurement sites. Table courtesy of Health and Education Services, Addison, Ill.

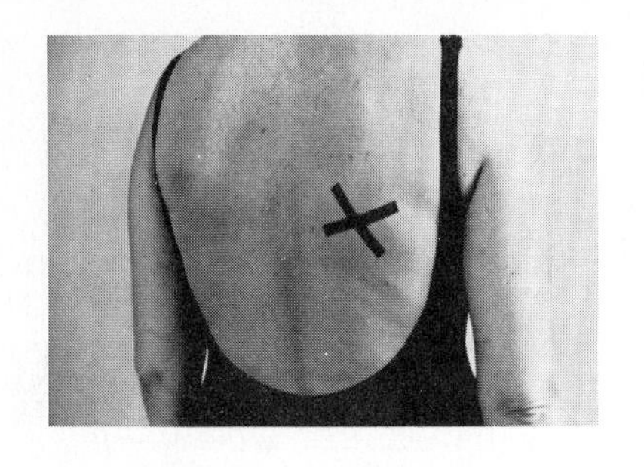

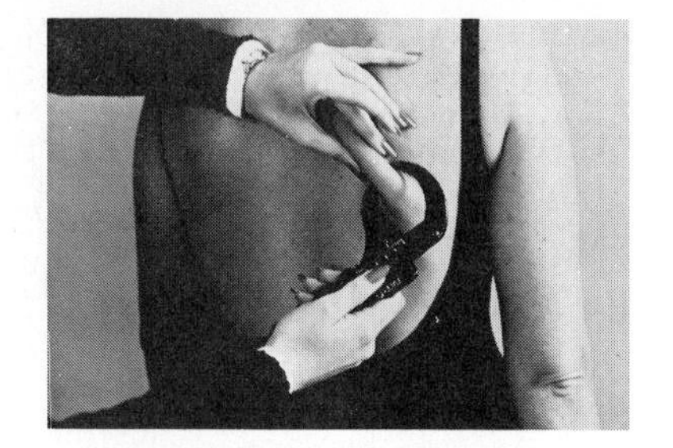

SITE 1 (S1)
SUBSCAPULAR
Inferior angle of the scapula, following the natural fold of the skin, about 1 cm below the angle.

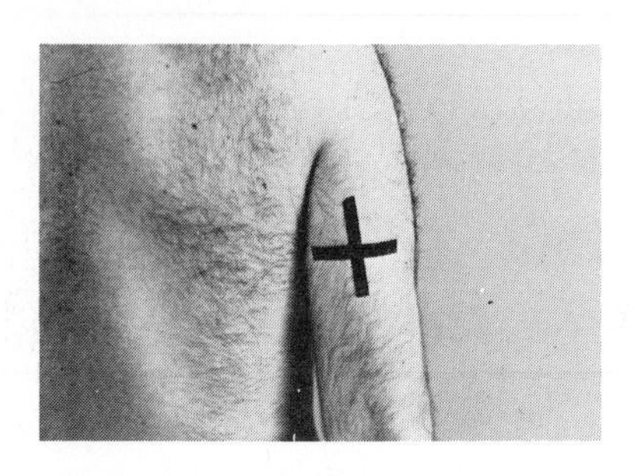

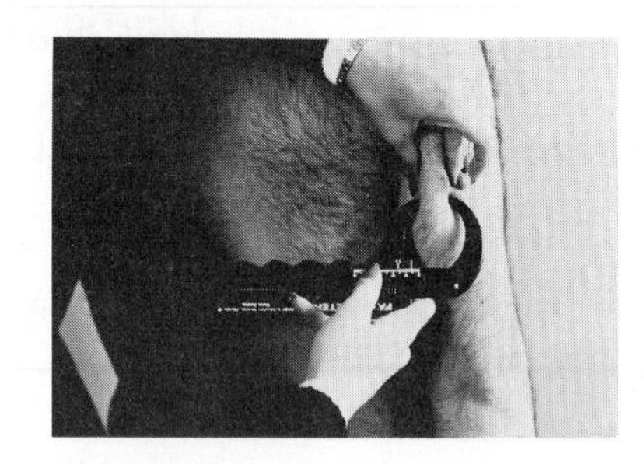

SITE 2 (S2)
TRICEPS
Halfway between the acromian process of the scapula and olecreanon process of the unlna on the dorsum (back) of the arm.

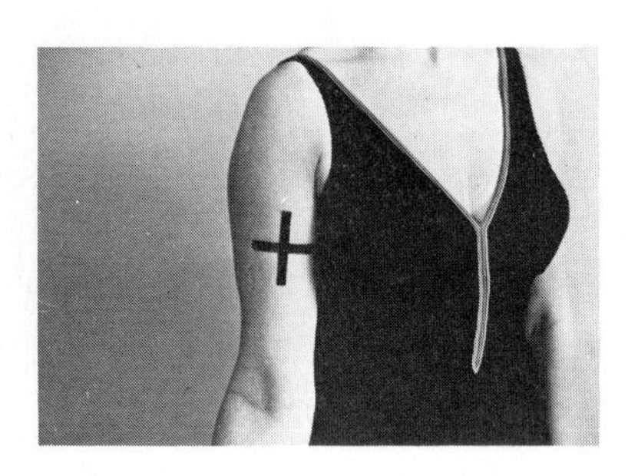

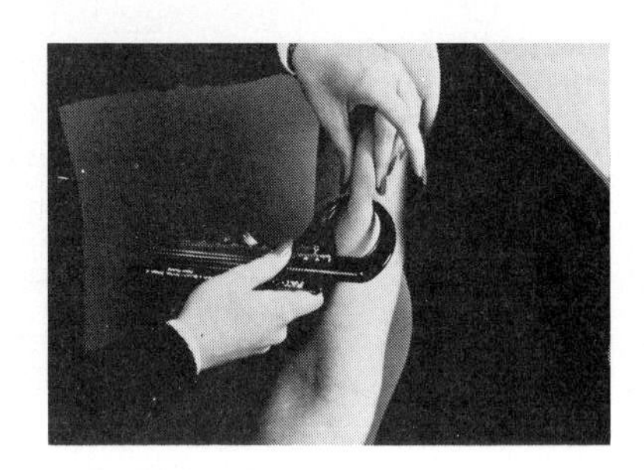

SITE 3 (S3)
BICEPS
Anterior of the arm, halfway between the greater tubercule of the humerus and the coronoid fossa.

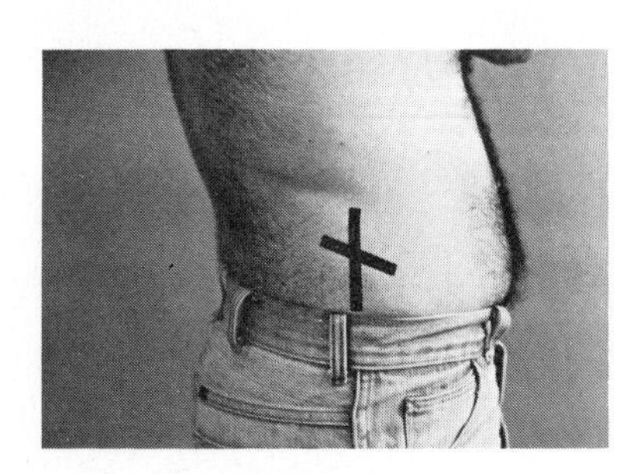

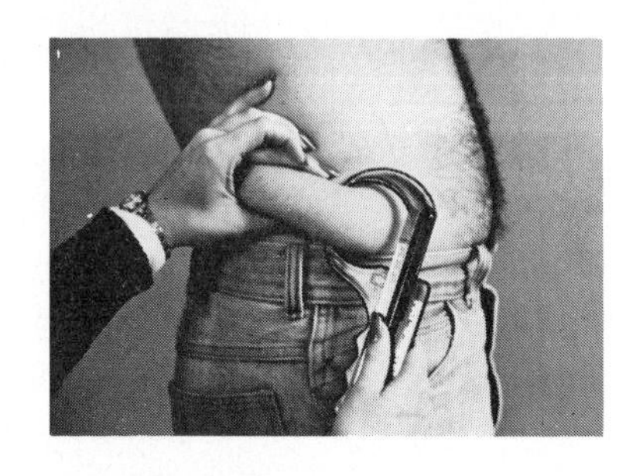

SITE 4 (S4)
SUPRAILIAC
A diagonal fold immediately above the iliac crest following the natural fold of the skin.

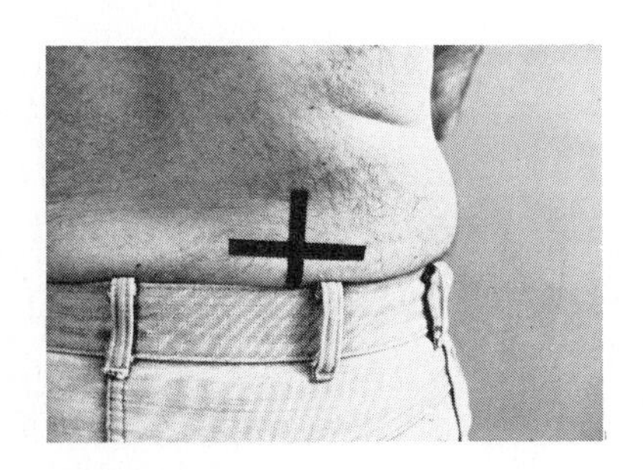

SITE 5 (S5)
POST SUPRAILIAC
5 cm to the right of the first lumbar spine.

FIGURE 22.1. Sites for measuring skin fold thickness. The S numbers in parentheses show where measurements are inserted in formulas in Table 22.2. Photos courtesy of Health and Education Services, Addison, Ill.

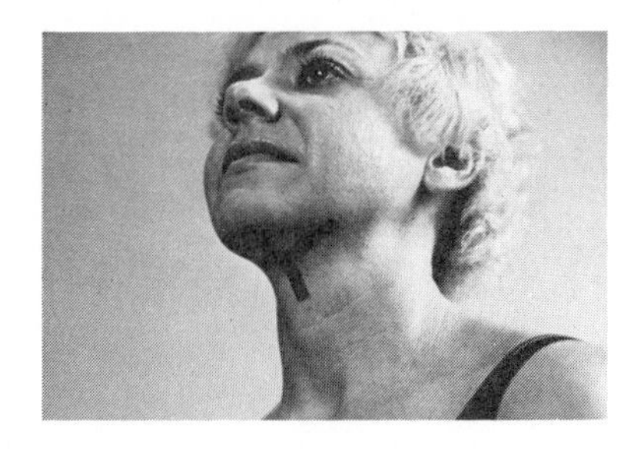
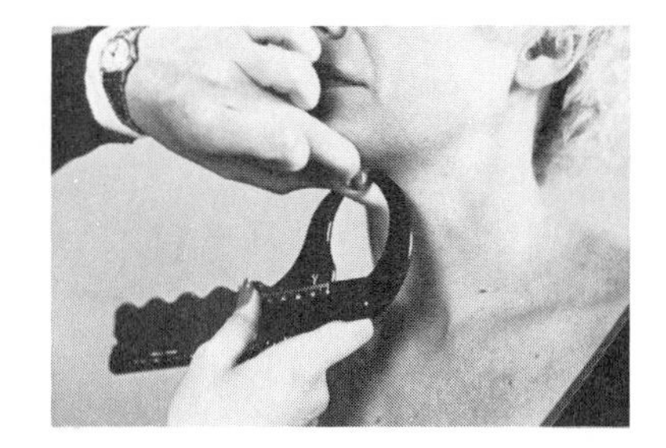

SITE 6 (S6)
CHIN
Under the chin above the hyoid bone.

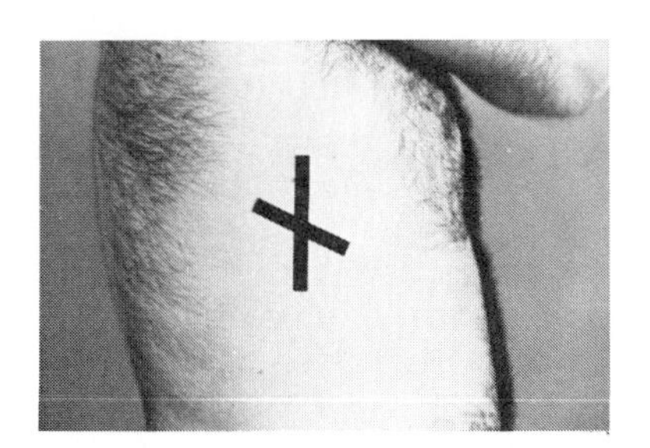
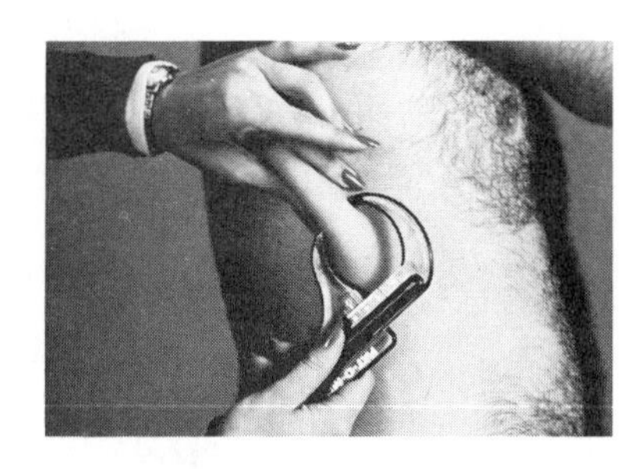

SITE 7 (S7)
MID AXILLARY
Anterior diagonal fold in the mid axilla at the level of the 5th rib.

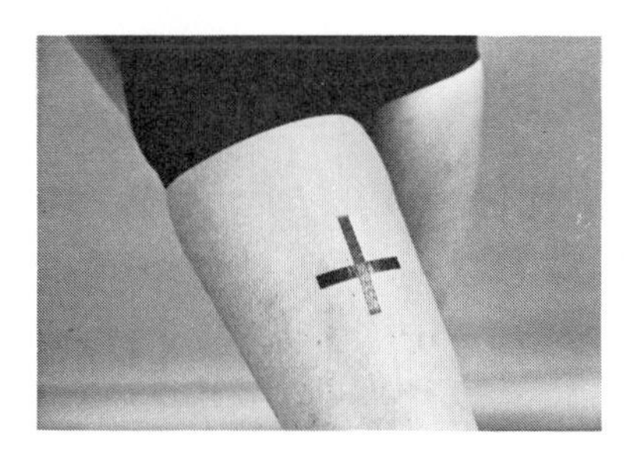
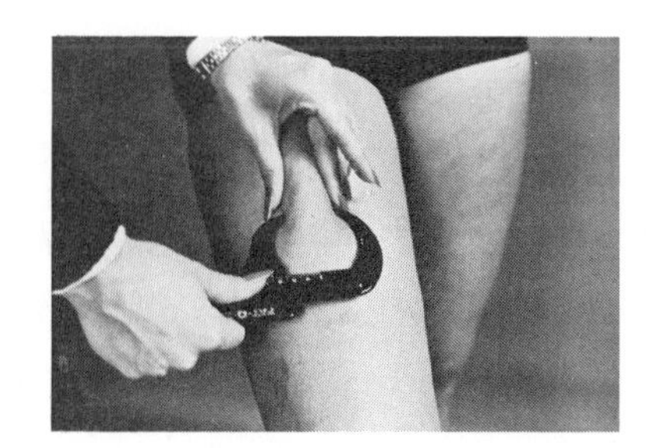

SITE 8 (S8)
THIGH
Vertical fold on front of the thigh midway between the greater trochanter of the femur and the top of the patella.

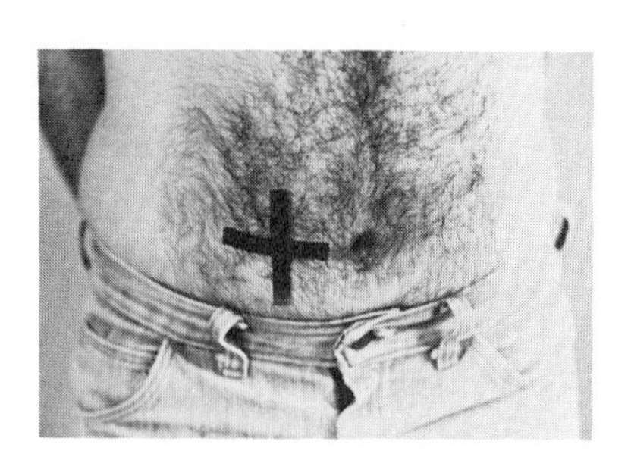
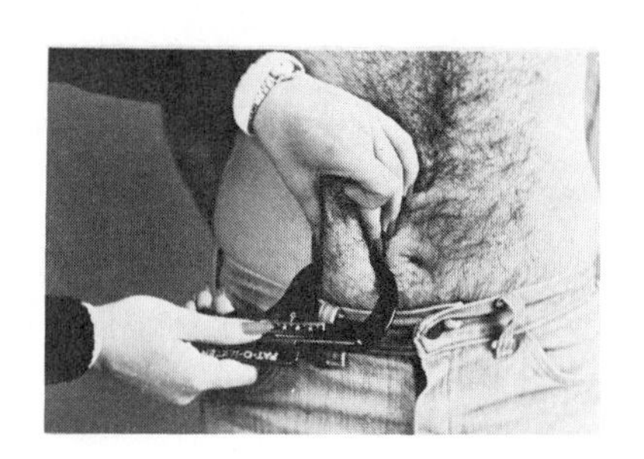

SITE 9 (S9)
UMBILICUS
A vertical fold to the side of the umbilicus.

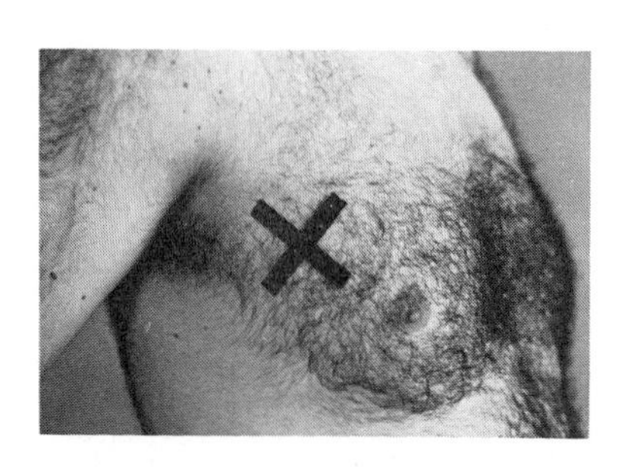
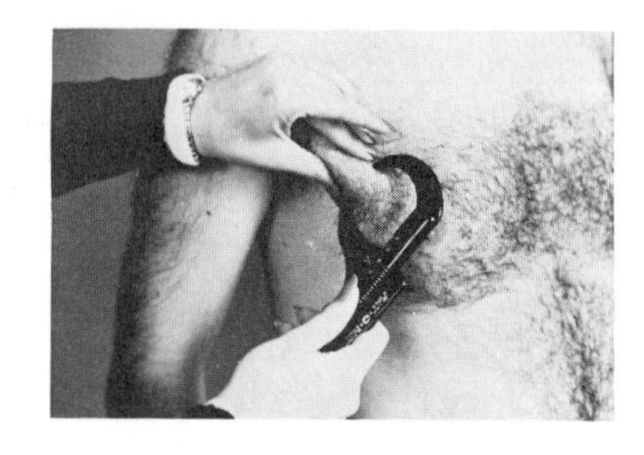

SITE 10 (S10)
PECTORAL
Midway between the axillary fold and the nipple in a fold parallel to the muscle tendon.

FIGURE 22.1. *(continued)*

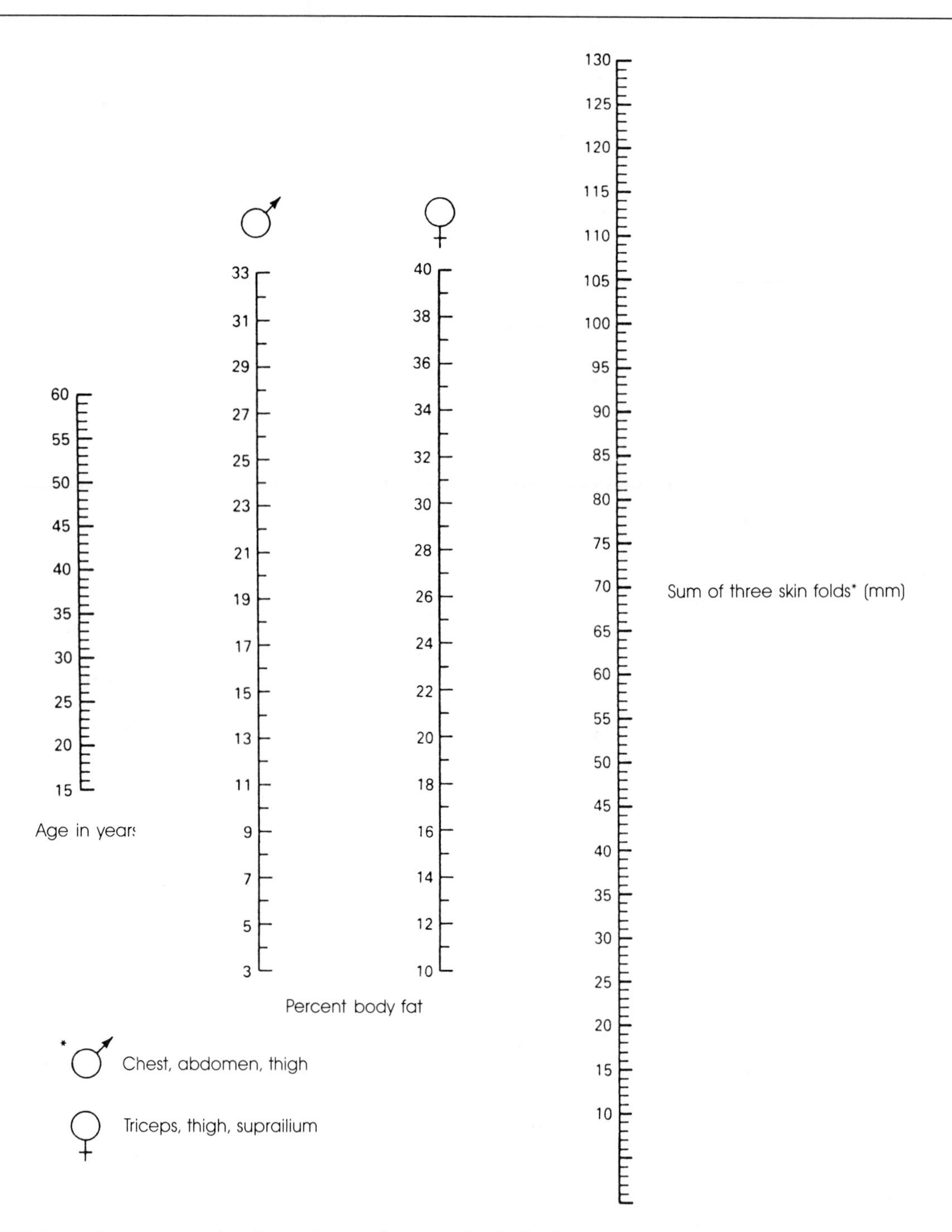

FIGURE 22.2. A nomogram for the estimate of percent body fat from age and the sum of three skin fold thicknesses. Use a straightedge and draw a line from your age to the sum of three skin folds and read your percent fat from the appropriate scale. From B. Sharkey, Physiology of Fitness, 2nd ed. (Champaign, Ill: Human Kinetics, 1984). Used by permission.

CARDIORESPIRATORY ENDURANCE (AEROBIC FITNESS)

Aerobic fitness is really the keystone of any fitness program. It is the ability of the body to use oxygen, an activity that involves the cardiovascular, respiratory, blood, and cellular enzymes systems. Aerobic metabolism is needed if we are involved in any sustained activity that requires a high expenditure of energy. Rhythmic endurance types of exercise such as jogging, swimming, and cycling strengthen these organ systems and increase the ability to use oxygen for energy production.

To achieve and maintain adequate aerobic fitness, an exercise program that requires exercise for 20 to 30 minutes at least three times a week at an intensity that elevates the heart rate to 70% to 80% of the maximum heart rate is recommended. Maximum heart rate declines with age and is estimated as follows:

$$\text{Maximum heart rate} = 220 - \text{age}$$

Thus a 20-year-old college student should exercise at a training heart rate of 140 to 160 beats/min to develop good aerobic fitness.

How can we measure aerobic fitness? The best test is to directly measure the maximum oxygen a person can use. This **maximum O_2 consumption ($\dot{V}O_2$ max)** is determined while the person is exercising at peak load and rate on a treadmill or bicycle ergometer. This test requires some expensive equipment and trained technicians, which make it impractical for testing large groups of people. Also, having an untrained person exercise at maximum load is difficult because of potential for injury and lack of motivation for true peak performance. Other aerobic tests use heart rate instead of O_2 consumption, because above 120 beats/min, the heart rate and O_2 consumption increase at the same rate as the work load increases.

Aerobic fitness can be measured using either exercise heart rates or recovery heart rates, because the heart rate of a trained person is lower at a particular submaximal work load and returns to resting rate faster after exercise than does the heart rate of an untrained individual.

In this exercise, we present two step tests that use recovery heart rate to measure cardiorespiratory fitness. The Harvard Step Test is the classic test from which other tests have been derived. It is, however, quite strenuous and discourages many people who are in the middle to lower fitness categories. For this reason, we have also included the Forest Service Fitness Test, which can be completed by older and less fit individuals.

Caution! *Students who have cardiovascular difficulties, such as cardiac insufficiency or hypertension, should check with their physician before taking part in these exercises.*

Harvard Step Test

This widely used test was developed in the Harvard Fatigue Laboratory in World War II to screen men for physical fitness and evaluate the progress of physical training programs. The Harvard Step Test has been well validated, and measures a general endurance or physical condition that might be held to be desirable for the average citizen.

The test consists of having the subject step up and down on a bench 20 in. high (16 in. for women) and then determining the heart rate during the postexercise recovery period. It should be noted that the test was designed so that only one-third of subjects are able to complete the full 5 minutes of bench stepping. Thus, it is important that the subject stop when he feels he cannot continue. The actual time of stepping is related to the person's endurance and is part of the scoring.

Experimental Procedure

1. The subject stands at attention in front of a bench 20 in. high. An observer stands behind the subject. The subject steps up and down on the bench at the rate of 30 steps (all the way up and down constitutes one step) per minute for as long as possible, up to a maximum of 5 minutes. The observer must be sure that the subject steps fully up on the bench without assuming any crouching position, and that he keeps pace with the counting. If the subject is unable to keep pace for 10 to 15 seconds, the observer stops him.
2. As soon as the subject stops on his own accord, or is stopped by the examiner before or at the end of 5 minutes, he sits down. The observer notes the duration of the exercise in seconds, and records the pulse from 1 to 1 ½ minutes, from 2 to 2 ½ minutes, and from 3 to 3 ½ minutes after the subject has finished the exercise. The actual number of heartbeats during each of these three 30-second periods is recorded, and the three rates are summed.
3. The index of physical fitness is computed from the following formula:

   $$\text{Index} = \frac{\text{Duration of exercise in seconds} \times 100}{2\ (\text{Sum of the 3 pulse counts in recovery})}$$

   The interpretation of scores is as follows:

below 55:	Poor physical condition
55 to 64:	Low average
65 to 79:	High average
80 to 90:	Good
above 90:	Excellent

Modified Harvard Step Test

One consistent criticism of the Harvest test is the use of the same step height for all subjects. This places short persons at a disadvantage because of the larger leverage angle they must use when stepping up on a 20-in. bench. One modification that can be used to equalize the work for persons of different height is to lower the bench height for short subjects. The bench heights in the following list provide a fairer evaluation when using the Harvard Step Test (the adjusted bench heights are for both sexes).

PERSON'S HEIGHT	BENCH HEIGHT
< 5 ft	12 in.
5 ft to 5 ft 3 in.	14 in.
5 ft 3 in. to 5 ft 9 in.	16 in.
5 ft 9 in. to 6 ft	18 in.
> 6 ft	20 in.

Forest Service Fitness Test

This test was developed to screen the physical fitness of potential firefighters for the U.S. Forest Service. It has been compared with laboratory tests of maximum O_2 consumption and found to be a valid and reliable predictor of aerobic fitness. The lower bench height does not seem to discriminate against shorter persons, and the submaximal work load does not place undue stress on the cardiovascular and respiratory systems.

The test consists of stepping up and down on a bench 15¾ in. (40 cm) high for men and 13 in. (33 cm) high for women at a rate of 22½ steps per minute. After 5 minutes the subject sits down and a 15-second recovery pulse count is taken from 15 to 30 seconds after the test.

Experimental Procedure

1. Rest for 5 minutes before taking the test and do not do it after strenuous physical activity,

TABLE 22.3. Men's Fitness Scores.

POSTEXERCISE PULSE COUNT	FITNESS SCORE												
45	33	33	33	33	33	32	32	32	32	32	32	32	32
44	34	34	34	34	33	33	33	33	33	33	33	33	33
43	35	35	35	34	34	34	34	34	34	34	34	34	34
42	36	35	35	35	35	35	35	35	35	35	35	34	34
41	36	36	36	36	36	36	36	36	36	36	36	35	35
40	37	37	37	37	37	37	37	37	36	36	36	36	36
39	38	38	38	38	38	38	38	38	38	38	38	37	37
38	39	39	39	39	39	39	39	39	39	39	39	38	38
37	41	40	40	40	40	40	40	40	40	40	40	39	39
36	42	42	41	41	41	41	41	41	41	41	41	40	40
35	43	43	42	42	42	42	42	42	42	42	42	42	41
34	44	44	43	43	43	43	43	43	43	43	43	43	43
33	46	45	45	45	45	45	44	44	44	44	44	44	44
32	47	47	46	46	46	46	46	46	46	46	46	46	46
31	48	48	48	47	47	47	47	47	47	47	47	47	47
30	50	49	49	49	48	48	48	48	48	48	48	48	48
29	52	51	51	51	50	50	50	50	50	50	50	50	50
28	53	53	53	53	52	52	52	52	52	52	51	51	51
27	55	55	55	54	54	54	54	54	54	53	53	53	52
26	57	57	56	56	56	56	56	56	56	55	55	54	54
25	59	59	58	58	58	58	58	58	58	56	56	55	55
24	60	60	60	60	60	60	60	59	59	58	58	57	
23	62	62	61	61	61	61	61	60	60	60	59		
22	64	64	63	63	63	63	62	62	61	61			
21	66	66	65	65	65	64	64	64	62				
20	68	68	67	67	67	66	66	65					
Body Weight	120	130	140	150	160	170	180	190	200	210	220	230	240

From B. Sharkey, *Physiology of Fitness,* 2nd ed. (Champaign, Ill.: Human Kinetics Publishers, 1984). Used by permission.

after drinking coffee or smoking, in a very warm room (over 78 °F), or when anxious or excited.

2. Set a metronome or other timing device for 90 beats/min. Begin exercising to the beat of the timer with an up-up-down-down cadence of your left and right feet. You must step fully up on the bench without bending your legs and you must keep pace with the timer.

3. After 5 minutes of exercise, sit down and take your pulse count for **exactly** 15 seconds, starting **exactly** at 15 seconds and ending **exactly** at 30 seconds after exercise. The pulse can be felt on the radial artery, just below the base of the thumb, or on the carotid artery in the neck. The test is usually more accurate if a lab partner counts your pulse and checks your stepping cadence. Weigh yourself in the clothes worn during the test and record your weight.

4. Score the test.

   a. Use your body weight and pulse count to find your fitness score in Table 22.3 or 22.4.
   b. Use your fitness score and age in Table 22.5 to find your age-adjusted score.
   c. Use your age-adjusted score to find your physical fitness rating in Table 22.6 or 22.7.

TABLE 22.4. Women's Fitness Scores.

POSTEXERCISE PULSE COUNT	FITNESS SCORE											
45										29	29	29
44								30	30	30	30	30
43							31	31	31	31	31	31
42			32	32	32	32	32	32	32	32	32	32
41			33	33	33	33	33	33	33	33	33	33
40			34	34	34	34	34	34	34	34	34	34
39			35	35	35	35	35	35	35	35	35	35
38			36	36	36	36	36	36	36	36	36	36
37			37	37	37	37	37	37	37	37	37	37
36		37	38	38	38	38	38	38	38	38	38	38
35	38	38	39	39	39	39	39	39	39	39	39	39
34	39	39	40	40	40	40	40	40	40	40	40	40
33	40	40	41	41	41	41	41	41	41	41	41	41
32	41	41	42	42	42	42	42	42	42	42	42	42
31	42	42	43	43	43	43	43	43	43	43	43	43
30	43	43	44	44	44	44	44	44	44	44	44	44
29	44	44	45	45	45	45	45	45	45	45	45	45
28	45	45	46	46	46	47	47	47	47	47	47	47
27	46	46	47	48	48	49	49	49	49	49		
26	47	48	49	50	50	51	51	51	51			
25	49	50	51	52	52	53	53					
24	51	52	53	54	54	55						
23	53	54	55	56	56	57						
Body Weight	80	90	100	110	120	130	140	150	160	170	180	190

From B. Sharkey, *Physiology of Fitness*, 2nd ed. (Champaign, Ill.: Human Kinetics Publishers, 1984). Used by permission.

TABLE 22.5. Age-Adjusted Scores.*

NEAREST AGE	ENTER FITNESS SCORE / AGE-ADJUSTED SCORE																				
	30	31	32	33	34	35	36	37	38	39	40	41	42	43	44	45	46	47	48	49	50
15	32	33	34	35	36	37	38	39	40	41	42	43	44	45	46	47	48	49	50	51	53
20	31	32	33	34	35	36	37	38	39	40	41	42	43	44	45	46	47	48	49	50	51
25	30	31	32	33	34	35	36	37	38	39	40	41	42	43	44	45	46	47	48	49	50
30	29	30	31	32	33	34	35	36	37	38	39	40	41	42	43	44	45	46	47	48	49
35	27	28	29	31	32	33	34	35	36	37	38	39	40	41	42	43	44	45	46	47	48
40	26	27	28	30	31	32	33	34	35	36	37	38	39	40	41	42	43	44	45	46	47
45	25	26	27	29	30	31	32	33	34	35	36	37	38	39	40	41	42	43	44	45	46
50	24	25	26	28	29	30	31	32	33	34	35	36	37	38	39	40	41	42	43	44	45
55	23	24	25	27	28	29	30	31	32	33	34	35	36	37	38	39	40	40	41	42	43
60	22	23	24	25	26	27	28	30	31	32	33	34	35	36	37	37	38	39	40	41	42
65	21	22	23	24	25	26	27	28	29	30	31	32	33	34	35	36	37	38	38	39	40

NEAREST AGE	ENTER FITNESS SCORE / AGE-ADJUSTED SCORE																					
	51	52	53	54	55	56	57	58	59	60	61	62	63	64	65	66	67	68	69	70	71	72
15	54	55	56	57	58	59	60	61	62	63	64	65	66	67	68	69	70	71	72	74	75	76
20	52	53	54	55	56	57	58	59	60	61	62	63	64	65	66	67	68	69	70	71	72	73
25	51	52	53	54	55	56	57	58	59	60	61	62	63	64	65	66	67	68	69	70	71	72
30	50	51	52	53	54	55	56	57	58	59	60	61	62	63	64	65	66	67	68	69	70	71
35	49	50	51	52	53	54	55	56	57	58	59	60	60	61	62	63	64	65	66	67	68	69
40	48	49	50	51	52	53	54	55	55	56	57	58	59	60	61	62	63	64	65	66	67	68
45	47	48	49	50	51	52	52	53	54	55	56	57	58	59	60	61	62	63	64	65	65	66
50	45	46	47	48	49	50	51	52	53	53	54	55	56	57	58	58	59	61	61	62	63	64
55	44	45	46	46	47	48	49	50	51	52	53	53	54	55	56	57	58	59	59	60	61	62
60	42	43	44	45	46	46	47	48	49	50	51	51	52	53	54	55	56	57	57	58	59	60
65	41	42	42	43	44	45	46	46	47	48	49	50	50	51	52	53	54	54	55	56	57	58

From B. Sharkey, *Physiology of Fitness,* 2nd ed. (Champaign, Il.: Human Kinetics Publishers, 1984). Used by permission.

*Example: If your age is 40 years and you score 50 on the step test, your age-adjusted score is 47.

TABLE 22.6. Physical Fitness Rating—Men (Use Age-Adjusted Score).

NEAREST AGE	SUPERIOR	EXCELLENT	VERY GOOD	GOOD	FAIR	POOR	VERY POOR
15	57+	56–52	51–47	46–42	41–37	36–32	31–
20	56+	55–51	50–46	45–41	40–36	35–31	30–
25	55+	54–50	49–45	44–40	39–35	34–30	29–
30	54+	53–49	48–44	43–39	38–34	33–29	28–
35	53+	52–48	47–43	42–38	37–33	32–38	27–
40	52+	51–47	46–42	41–37	36–32	31–27	26–
45	51+	50–46	45–41	40–36	35–31	30–26	25–
50	50+	49–45	44–40	39–35	34–30	29–25	24–
55	49+	48–44	43–39	38–34	33–29	28–24	23–
60	48+	47–43	42–38	37–33	32–28	27–23	22–
65	47+	48–42	41–37	36–32	31–27	26–22	21–

From B. Sharkey, *Physiology of Fitness,* 2nd ed. (Champaign, Ill.: Human Kinetics Publishers, 1984). Used by permission.

TABLE 22.7. Physical Fitness Rating—Women (Use Age-Adjusted Score).

NEAREST AGE	SUPERIOR	EXCELLENT	VERY GOOD	GOOD	FAIR	POOR	VERY POOR
15	54+	53–49	48–44	43–49	38–34	33–29	28–
20	53+	52–48	47–43	42–38	37–33	32–28	27–
25	52+	51–47	46–42	41–37	36–32	31–27	26–
30	51+	50–46	45–41	40–36	35–31	30–26	25–
35	50+	49–45	44–40	39–35	34–30	29–25	24–
40	49+	48–44	43–39	38–34	33–29	28–24	23–
45	48+	47–43	42–38	37–33	32–28	27–23	22–
50	47+	46–42	41–37	36–32	31–27	26–22	21–
55	46+	45–41	40–36	35–31	30–26	25–21	20–
60	45+	44–40	39–35	34–30	29–25	24–20	19–
65	44+	43–39	38–34	33–29	28–24	23–20	19–

From B. Sharkey, *Physiology of Fitness,* 2nd ed. (Champaign, Ill.: Human Kinetics Publishers, 1984). Used by permission.

LABORATORY REPORT

Name ____________________

Date __________ Section __________

Score/Grade ____________________

22. Physical Fitness

Muscular Strength and Endurance

		YOUR OWN	FEMALES MEAN	FEMALES RANGE	MALES MEAN	MALES RANGE
Maximum grip strength	Left: Right:					
Total force 20 contractions	Left: Right:					
Average force 20 contractions	Left: Right:					

Flexibility

Your score = __________

Class mean = __________ Class range = __________

Body Composition

1. Record the skin fold site and average thickness obtained:

Site					
Skin fold thickness (mm)					

Equation for age group __________ Sex __________

Density Equation:

Density = __________

$$\% \text{ fat} = \frac{457}{(\text{Density})} - 414.2 = \_\_\_\_\_\_\_\_$$

Body weight = __________ Lean body wt = __________ Fat wt = __________

Nomogram Method

Sites measured ____________ = ____________ mm

____________ = ____________ mm

____________ = ____________ mm

Sum of skin folds = ____________ mm % fat = ____________

Problem

Laura Wilder weighs 250 lb and has 40% body fat. She wants to lose 60 lb of fat to reduce her weight to 190 lb. What would her new percent fat be at 190 lb? ____________ %

From our study of metabolic rate we know that 2 L of oxygen is required to burn 1 g of fat and that each liter of oxygen equals 4.7 kcal of energy when fat is oxidized. Laura decides to burn off her fat by jogging at a 12-min-mile pace, which requires 10 kcal/mi. How many miles would she need to jog to burn off the 60 lb of fat, not counting any oxygen debt she may accumulate?

____________ mi

Cardiorespiratory Endurance

1. Harvard Step Test

Duration of exercise = ____________ sec

Recovery pulse counts: 1–1½ min = ____________ 2–2½ min = ____________

3–3½ min = ____________

Physical fitness index = $\dfrac{(\qquad\qquad) \times 100}{2(\qquad\qquad)}$ = ____________

Fitness rating = ____________

2. Forest Service Fitness Test

15-second pulse count = ____________ Body weight = ____________

Fitness score = ____________ Age-adjusted score = ____________

Physical fitness rating = ____________

3. Questions

Why is a higher heart rate during the recovery period equated with a lower level of fitness?

What is the difference between aerobic and anaerobic fitness? In which athletic events are each of these types of fitness important?

23 Physiology of Exercise

Of all the environmental stresses to which the body is exposed, that of exercise probably produces the greatest alteration in physiological parameters. The increased metabolic activity of skeletal muscle during exercise (tenfold to twentyfold increase) places heavy demands on the respiratory and circulatory systems and causes profound changes in other systems, such as the digestive and excretory systems. Basically, these changes are brought about to satisfy the increased demand of muscle fibers for more oxygen and energy and for the removal of carbon dioxide and other metabolic waste products.

In this experiment you will demonstrate several of the chief cardiovascular and respiratory responses that allow the body to supply these needs and adjust to the stresses of exercise. Such an adjustment requires integration of nearly all body systems; hence this exercise will give you the opportunity to review many of the physiological principles learned in previous laboratory experiments.

PARAMETERS MODIFIED BY EXERCISE

Oxygen Consumption

Oxygen consumption will be determined using the respirometer, as in the determination of basal metabolic rate (Experiment 13). The gross amount of oxygen consumed is converted to the volume of dry oxygen at standard temperature and pressure by multiplying by the STPD factor. The average individual consumes about 0.250 L of oxygen (250 ml) per minute under basal conditions. At maximal exertion this oxygen consumption may be increased to 2.5–3 L/min (tenfold to twentyfold increase) in an untrained person, and in a trained athlete it may be elevated to 5 L/min (twentyfold increase). The maximal oxygen uptake of a person is probably the best overall index of his or her physical fitness.

Caloric Cost

Caloric cost is the caloric energy expenditure (heat production) of the body. It is calculated by multiplying the corrected oxygen consumption (L/min) times the caloric equivalent of oxygen (4.825 kcal/L O_2 for an average mixed diet). The normal resting caloric cost is around 1.5 kcal/min and may rise to 15–20 kcal/min during exercise.

The severity of work is commonly classified on the basis of the oxygen consumption or energy expenditure, as shown in Table 23.1.

Respiratory Rate

Respiratory rate is determined by counting the deflections per minute on the respirometer record. In the resting state, the respiratory rate varies from 12–16 min; it may increase to as high as 30/min in heavy exercise.

Work

The work performed during an exercise is a measure of force times distance, usually expressed as

TABLE 23.1. Classification of Work on the Basis of Oxygen Consumption or Energy Expenditure.

SEVERITY OF WORK	O_2 CONSUMPTION (L/MIN)	CALORIC COST (KCAL/MIN)
Resting	0.250	1.20
Light Work = 2–3 × resting	0.5 to 0.75	Up to 3.62
Medium Work = 4–7 × resting	1.0 to 1.75	Up to 8.44
Hard Work = 8–12 × resting	2.0 to 3.0	Up to 14.48
Exhaustive Work = 13–20 × resting	3.25 to 5.0	Up to 24.1

foot-pounds per minute or kilogram-meters per minute. In this experiment the exercise will consist of stepping up and down on a small bench of a designated height for a specific period of time. The amount of work performed is calculated as follows:

$$\text{kg-m work/min} = \text{subject's weight (kg)} \times \text{bench height (m)} \times \text{number of steps/min}$$

A moderate work load is considered to be around 600 kg-m/ min, and a load of 1500 kg-m/ min is considered a heavy one. If available, a bicycle ergometer may be used to provide a calibrated work load for the exercise.

Note: 1 kg = 2.2 lb
1 m = 39.37 in.
1 in. = 2.5 cm

Mechanical Efficiency

The efficiency of performing a certain task is simply the ratio of work done to the amount of energy used. It is possible to equate oxygen used, heat produced, and work performed through the following conversion factors:

$$1 \text{ L } O_2 \text{ consumed} = 4.825 \text{ kcal heat} = 2153 \text{ kg-m work}$$

The gross efficiency can be calculated using the following formula:

$$\%\text{ gross efficiency} = \frac{\text{Kg-m work performed}}{\text{L } O_2 \text{ used during work} \times 2153 \text{ kg-m/L } O_2} \times 100$$

Gross efficiency is somewhat misleading, however, because part of the energy used during the work period is being used just to maintain vital body activities and is not being used to perform the work. To allow for this we use the following formula to calculate the net efficiency of the body:

$$\%\text{ net efficiency} = \frac{\text{Kg-m work performed}}{\left(\begin{array}{l}\text{L } O_2 \text{ used} - \text{resting } O_2\\ \text{during work}\end{array}\right) \times 2153 \text{ kg-m/L } O_2} \times 100$$

The gross efficiency will vary from 6% to 25% depending on the kind of work performed and the rate of doing the work. For instance, the efficiency of climbing uphill can be as high as 24%, whereas the efficiency of swimming is as low as 2% to 8%. In climbing uphill the efficiency at a speed of 0.5 mph is around 6%, but at 1.5 mph it increases to 24%. Each type of work has an optimal speed that gives the optimal efficiency for that task.

Oxygen Debt

During the performance of an exercise, a certain amount of the energy used is obtained through anaerobic metabolism. This results in an accumulation of metabolites (e.g., lactic acid) in the tissues and depletion of storage forms of energy (e.g., ATP and creatine phosphate). To remove these metabolites and replenish the energy stores, extra oxygen must be taken in during the recovery period following exercise. This extra oxygen is called the **oxygen debt.** There is a limit to the amount of oxygen debt that a person can tolerate. An untrained person can tolerate an oxygen debt of about 10 L, a highly trained person as much as 17 L. The oxygen debt is calculated using the following formula:

$$\text{Oxygen debt} = \begin{array}{c}\text{Total liters of } O_2\\ \text{consumed in}\\ \text{recovery period}\end{array} - \begin{array}{c}\text{L/min resting}\\ O_2\\ \text{consumption}\end{array} \times \begin{array}{c}\text{Number of min}\\ \text{of recovery}\end{array}$$

For the exercise performed in this experiment, an oxygen debt of 1–8 L may be expected, depending on the severity of the exercise and the physical condition of the individual.

Heart Rate

The heart rate will be determined by palpation of the carotid or radial artery, or via an ECG recording. The heart rate is also a good index of the severity of the work being performed, as is shown by the values given here for work on a bicycle ergometer.

WORK LOAD (KG-M/MIN)	HEART RATE (BEATS/MIN)
Resting	75
277	105
556	132
830	154
1100	177
1380	198

Generally speaking, a heart rate of less than 100 beats/min indicates light work; 100–130 beats/min, moderate work; and greater than 160 beats/min, heavy work. Usually when the heart rate is more than 180 beats/min the subject is near exhaustion, because the efficiency of the heart's pumping action decreases greatly at rates higher than this. Some highly trained individuals can, however, attain rates of 225 beats/min for short periods of time.

Blood Pressure

The systolic and diastolic pressures will be determined by the auscultatory method (using the stethoscope and sphygmomanometer). During exercise the systolic pressure may increase from a normal of 120 mm Hg to around 180–200 mm Hg. The diastolic pressure may increase slightly, remain the same, or even fall slightly. In general, we say that the systolic pressure reflects the force of heart contractility, whereas the diastolic pressure represents the integrity or condition of constriction of the systemic blood vessels.

Cardiac Output

The amount of blood pumped per minute by the heart is commonly measured using the Fick principle; the theory behind this method is explained in most physiology tests. It is based on calculating the volume of blood needed to transport the oxygen taken from the alveoli in a given period of time. Three measurements are required: (1) the oxygen consumption in milliliters per minute, (2) the concentration of oxygen in the arterial blood, and (3) the concentration of oxygen in the venous blood (ml of O_2/100 ml of blood). The arteriovenous (AV) oxygen difference and oxygen consumption are used in the following formula to calculate cardiac output:

$$\text{Cardiac output (ml/min)} = \frac{\text{Oxygen consumed (ml/min)}}{\text{AV oxygen difference (ml/100 ml of blood)}} \times 100$$

In this experiment you will not determine the arterial and venous oxygen concentrations directly, owing to the difficulty of obtaining blood samples. Instead, you will use what is called the modified Fick principle, in which the AV oxygen differences are obtained from the oxygen consumption values as given in the following table. It must be borne in mind that this method is used only for an approximation of cardiac output in the teaching laboratory. It is not to be used in an experimental research situation.

O_2 CONSUMPTION (ML/MIN)	ARTERIOVENOUS O_2 DIFFERENCE (ML/100 ML OF BLOOD)
250	4.5
325	4.8
400	5.0
500	5.5
600	6.0
800	6.5
1000	7.5
1200	8.3
1400	9.0
1600	9.8
1800	10.3
2000	10.9
2200	11.5
2400	12.0
2600	12.5
2800	13.0
3000	13.5
3200	13.9
3400	14.3
3600	14.6
3800	15.0

The normal resting cardiac output is around 5 L/min. During exercise, cardiac output may rise to as high as 22 L/min in an untrained subject and can increase to 30–40 L/min in a highly trained person.

Stroke Volume

The amount of blood forced out of the heart with each systole is around 70 ml in the average subject. During a maximal exertion, the nontrained heart will increase its stroke volume to 100–125 ml/beat, while the trained heart can attain values as high as 150–170 ml/beat. A stroke volume of two times the resting volume is about the maximum that can be expected during exercise.

$$\text{Stroke volume (ml)} = \frac{\text{Cardiac output (ml/min)}}{\text{Heart rate (beats/min)}}$$

Total Peripheral Resistance

Total peripheral resistance (TPR) is a measure of the overall resistance (constriction) of all the systemic blood vessels. It is expressed in resistance units and is calculated from our knowledge of the relationships between pressure, flow, and resistance as given in Ohm's law:

$$\text{Amperes (flow)} = \frac{\text{Volts (pressure)}}{\text{Ohms (resistance)}}$$

$$\text{Cardiac output (blood flow)} = \frac{\text{Mean blood pressure}}{\text{Total peripheral resistance}}$$

$$\text{TPR (units)} = \frac{\text{Mean blood pressure (mm Hg)}}{\text{Cardiac output (ml/sec)}}$$

The mean blood pressure (BP) is a measure of the average blood pressure in the arteries over the entire cardiac cycle.

$$\text{Mean BP} = \text{Diastolic pressure} + \tfrac{1}{3}\ \text{pulse pressure}$$

The normal resting TPR is around 1 unit, and during exercise it sometimes decreases to as low as ¼ unit.

Cardiac Index

To make more accurate comparisons between the cardiac outputs of persons of different size, the cardiac index is often calculated.

$$\text{Cardiac index (L/min/m}^2\text{)} = \frac{\text{Cardiac output (L/min)}}{\text{Body surface area (m}^2\text{)}}$$

For humans, the cardiac index is around 3 L/min/m^2 at rest and can increase to 9–17 L/min/m^2 during exercise.

Modified Tension Time Index

The modified tension time index (MTTI) provides an estimate of the work of the heart. The work performed by myocardial tissue is proportional to the myocardial oxygen consumption, and this has been shown to be closely related to the product of heart rate (HR) and systolic blood pressure (SBP). This index is calculated as follows:

$$\text{MTTI} = \frac{\text{HR} \times \text{SBP}}{100}$$

The MTTI ranges from around 84 at rest to 360 or more during exercise.

Experimental Procedure

1. The subject should come to the laboratory dressed in appropriate work clothes.[1]
2. The class will be divided into teams by the instructor, each team being responsible for obtaining the data on one or two parameters. The four critical measurements are oxygen consumption, heart rate, blood pressure, and work, because all the other parameters are calculated from these values. Record all your data in the Laboratory Report.
3. Resting values will be obtained while the subject is in a standing position next to the respirometer or seated on the bicycle ergometer.
4. The subject will then exercise for 2 minutes by stepping up and down on a bench 12–16 in. high at a rate of 20 steps per minute. The exact number of steps taken during the exercise will be counted by the team assigned to make the calculation. Heart rate and blood pressure will be taken *immediately* after the work and recorded as the work values. Note that oxygen consumption is recorded contin-

[1]These tests cause cardiovascular stress, and students who have cardiovascular difficulties should not take part unless they have permission from their physicians.

uously through the rest, work, and recovery periods.

If a bicycle ergometer is used, the ECG can be recorded during the last 20 seconds of the work period. The work load on the bicycle should be 600–900 kg-m/min depending on the fitness of the subject.

5. After finishing the exercise, the subject will step down from the bench and remain standing (or remain seated on the bicycle) for 8–10 minutes, during which time oxygen consumption, heart rate, and blood pressure will be measured and recorded each minute. **Note:** Breathing 100% oxygen often stimulates mucus production in the respiratory passageways. If the subject begins to accumulate mucus in the mouth he should signal by pointing to his mouth. The operators will then close the intake valve, allow the subject to clear his mouth, place the mouthpiece back in place, and open the valve to the respirometer again. A short piece of the record will be lost, but this will not impair calculations.

6. Draw a separate slope line for each minute on the respirometer tracing and calculate the oxygen consumption for each slope.

7. Place all data on the chalkboard and in the data sheet in the Laboratory Report and discuss the changes occurring in each parameter during exercise and recovery.

LABORATORY REPORT

Name ______________________

Date ____________ Section ____________

Score/Grade ______________________

23. Physiology of Exercise

Data Sheet

PARAMETER MEASURED		RESTING VALUES	WORK VALUES	RECOVERY VALUES (MINUTES IN RECOVERY)						
				1	2	3	4	6	8	10
Oxygen consumption (L/min)										
Caloric cost (kcal/min)										
Respiratory rate (per min)										
Work (kg-m/min)										
Mechanical efficiency (%)	Gross									
	Net									
Oxygen debt (L)										
Heart rate (beats/min)										
Blood pressure (mm Hg), systolic/diastolic										
Cardiac output (L/min)										
Stroke volume (ml)										
Total peripheral resistance (units)										
Cardiac index (L/min/m^2)										
Modified tension time index										

1. The following diagram shows a sample respirometer record obtained during rest, work, and recovery. Record, on the diagram, the values for oxygen consumption, heart rate, and blood pressure obtained with your subject.

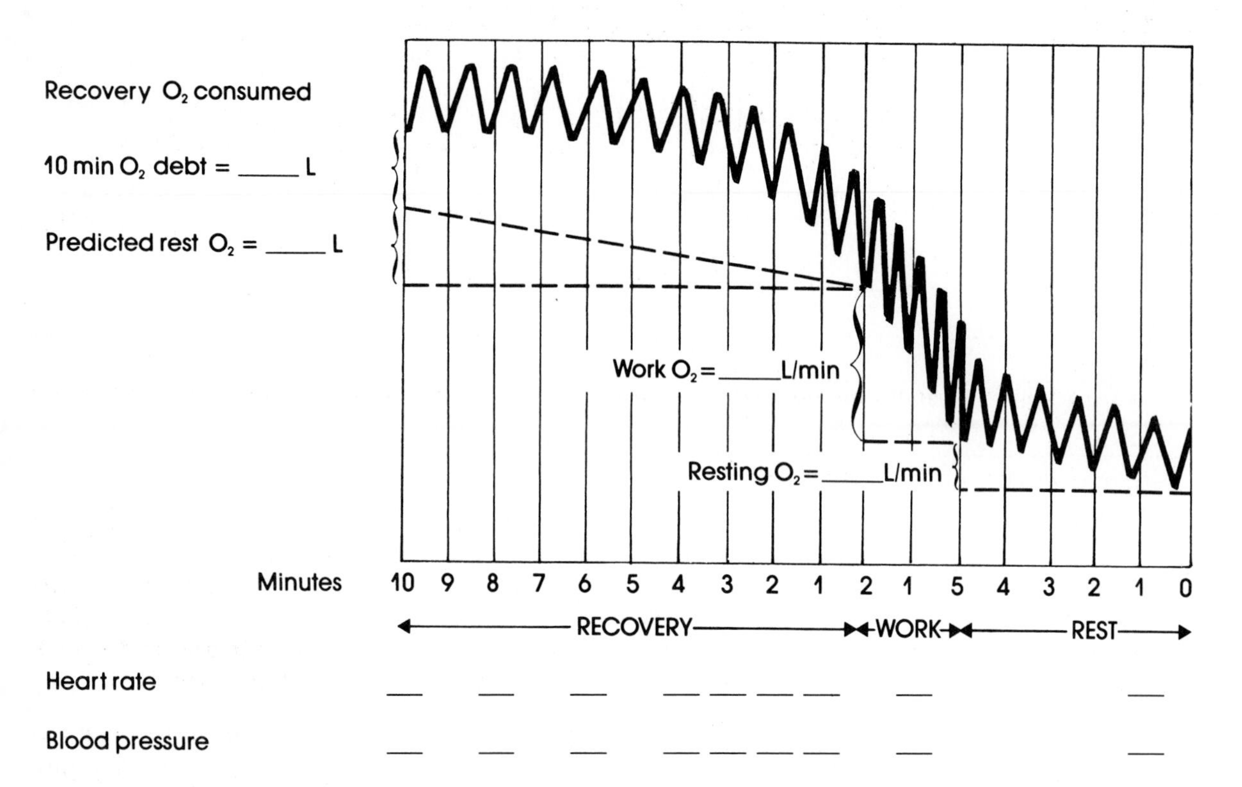

2. What causes the mechanical efficiency of a trained person to be higher than that of an untrained person?

3. What is responsible for the change in total peripheral resistance seen during exercise?

4. Why is the oxygen debt larger in a trained person than in an untrained person?

5. Why is the diastolic blood pressure more critical to watch during exercise and also in the resting state than is the systolic blood pressure?

APPENDIX A

Precautions for Handling Blood

As you are probably aware, Acquired Immune Deficiency Syndrome (AIDS) and other infectious diseases can be transmitted through blood and other body fluids. *All blood and body fluid samples must be treated as potentially infectious.* The following precautions should be observed whenever blood items are used in the lab.

1. Disposable latex gloves must be worn by any person handling blood specimens or supplies that have come in contact with blood (lancets, slides, capillary tubes, cotton, etc.). Wash your hands immediately after removing the gloves.
2. Use only sterile lancets and clean, unused slides, capillary tubes, toothpicks, tissues, and so on. Never reuse these supplies. Used lancets and needles should be placed in a fresh 10% solution of household bleach (sodium hypochlorite) and then placed in a puncture-proof container for disposal. Other items, such as cotton, toothpicks, and the like, that have been in contact with blood are placed in double plastic bags and sealed. Autoclave all items before disposal.
3. Lab coats, masks, and protective eyewear should be worn during procedures that produce blood droplets or splashes.
4. Laboratory surfaces and instruments such as hemocytometers and reusable pipettes should be disinfected with a fresh 10% solution of household bleach and then washed with soap and hot water.
5. Never use mouth pipetting. Mechanical pipetting devices should be used to manipulate liquids in the laboratory.

APPENDIX B

Solutions

1. **Physiological Saline (Mammalian-Homeotherms)**
   0.9% NaCl in distilled water.

2. **Physiological Saline (Amphibian-Poikilotherms)**
   0.7% NaCl in distilled water.

3. **Frog Ringer's Solution**
   NaCl . . . . 6.50 g
   KCl . . . . 0.14 g
   $CaCl_2$ . . . . 0.12 g
   $NaHCO_3$ . . . . 0.20 g
   NaH_2PO_4 . . . . 0.01 g
   Distilled water q.s. to 1 liter.

4. **Mammalian Ringer's Solution**
   NaCl . . . . 9.00 g
   KCl . . . . 0.42 g
   $CaCl_2$ . . . . 0.24 g
   $NaHCO_3$ . . . . 0.20 g
   Distilled water q.s. to 1 liter.

5. **Locke's Solution**
   NaCl . . . . 9.00 g
   KCl . . . . 0.42 g
   $CaCl_2 \cdot 2H_2O$ . . . . 0.24 g
   $NaHCO_3$ . . . . 0.20 g
   Glucose . . . . 1.00 g
   Distilled water q.s. to 1 liter (pH should be 7.6 to 7.8).

6. **Heparinized Mammalian Ringer's Solution**
   Add 100 units (1 mg) of heparin per 500 ml of Ringer's solution.

7. **Anticoagulants for Blood**
   Use 20 mg of sodium oxalate or 200 mg of sodium citrate per 10 ml of blood.

8. **Hayems's Solution**
   Mercury bichloride . . . . 0.5 g
   Sodium chloride . . . . 1.0 g
   Sodium sulfate . . . . 5.0 g
   Distilled water . . . . 200 ml
   Filter the solution.

9. **Gower's Solution**
   Glacial acetic acid . . . . 16.65 ml
   Anhydrous Na_2SO_4 . . . . 6.25 g
   Distilled water . . . . 100.00 ml (exactly)

10. **Turk's Solution**
    Glacial acetic acid . . . . 1.0 ml
    1% aqueous solution of gentian violet . . . . 1.0 ml
    Distilled water . . . . 100.0 ml
    (should be filtered frequently)

11. **Wright's Stain**
    This can be obtained from laboratory supply houses.

12. **Buffer Solution for Use with Wright's Stain**
    KH_2PO_4 . . . . 1.63 g
    Na_2HPO_4 . . . . 3.2 g
    Distilled water . . . . 1000 ml

13. **Benedict's Solution**
    Copper sulfate . . . . 17.3 g
    Sodium citrate . . . . 173.0 g
    Sodium carbonate (anhydrous) . 100.0 g
    Distilled water q.s. to . . . . 1000 ml

    Dissolve the citrate and carbonate in 800 ml of water and filter. Dissolve the copper sulfate in 100 ml of water and then pour it slowly into the first solution while stirring constantly. Cool the solution and add water to make 1 liter.

14. Krebs Manometer Fluid

Anhydrous NaBr . . . . 44 g
Triton X-100 (Rohm & Haas Co.) . . 0.3 g
Evans blue . . . . 0.3 g
Distilled water . . . . 1000 ml

15. Brodie's Manometer Fluid

NaCl . . . . 23 g
Sodium Choleate (Merck) . . . . 5 g
Evans blue . . . . 0.1 g
Distilled water . . . . 500 ml

16. Lugol's Solution

Iodine . . . . 5 g
Potassium iodide . . . . 10 g
Distilled water q.s. to 100 ml

17. Phosphate Buffer

Stock Solutions:

M/15 dibasic sodium phosphate
9.465 g in 1000 ml distilled water solution

M/15 sodium acid phosphate
8.0 g in 1000 ml distilled water solution

For a pH of 7, mix:
60 ml stock dibasic sodium phosphate
40 ml stock sodium acid phosphate

APPENDIX C

Tables and Nomograms

TABLE C.1. Transmission—Optical Density Table.

%T	O.D.	%T	O.D.	%T	O.D.	%T	O.D.
1.0	2.000	26.0	.585	51.0	.292	76.0	.119
1.5	1.824	26.5	.577	51.5	.288	76.5	.116
2.0	1.699	27.0	.569	52.0	.284	77.0	.114
2.5	1.602	27.5	.561	52.5	.280	77.5	.111
3.0	1.523	28.0	.553	53.0	.276	78.0	.108
3.5	1.456	28.5	.545	53.5	.272	78.5	.105
4.0	1.398	29.0	.538	54.0	.268	79.0	.102
4.5	1.347	29.5	.530	54.5	.264	79.5	.100
5.0	1.301	30.0	.523	55.0	.260	80.0	.097
5.5	1.260	30.5	.516	55.5	.256	80.5	.094
6.0	1.222	31.0	.509	56.0	.252	81.0	.092
6.5	1.187	31.5	.502	56.5	.248	81.5	.089
7.0	1.155	32.0	.495	57.0	.244	82.0	.086
7.5	1.126	32.5	.488	57.5	.240	82.5	.084
8.0	1.097	33.0	.482	58.0	.237	83.0	.081
8.5	1.071	33.5	.475	58.5	.233	83.5	.078
9.0	1.046	34.0	.469	59.0	.229	84.0	.076
9.5	1.022	34.5	.462	59.5	.226	84.5	.073
10.0	1.000	35.0	.456	60.0	.222	85.0	.071
10.5	.979	35.5	.450	60.5	.218	85.5	.068
11.0	.959	36.0	.444	61.0	.215	86.0	.066
11.5	.939	36.5	.438	61.5	.211	86.5	.063
12.0	.921	37.0	.432	62.0	.208	87.0	.061
12.5	.903	37.5	.426	62.5	.204	87.5	.058
13.0	.886	38.0	.420	63.0	.201	88.0	.056
13.5	.870	38.5	.414	63.5	.197	88.5	.053
14.0	.854	39.0	.409	64.0	.194	89.0	.051
14.5	.838	39.5	.403	64.5	.191	89.5	.048
15.0	.824	40.0	.398	65.0	.187	90.0	.046
15.5	.810	40.5	.392	65.5	.184	90.5	.043
16.0	.796	41.0	.387	66.0	.181	91.0	.041
16.5	.782	41.5	.382	66.5	.177	91.5	.039
17.0	.770	42.0	.377	67.0	.174	92.0	.036

(continued)

TABLE C.1. *(continued)*

%T	O.D.	%T	O.D.	%T	O.D.	%T	O.D.
17.5	.757	42.5	.372	67.5	.171	92.5	.034
18.0	.745	43.0	.367	68.0	.168	93.0	.032
18.5	.733	43.5	.362	68.5	.164	93.5	.029
19.0	.721	44.0	.357	69.0	.161	94.0	.027
19.5	.710	44.5	.352	69.5	.158	94.5	.025
20.0	.699	45.0	.347	70.0	.155	95.0	.022
20.5	.688	45.5	.342	70.5	.152	95.5	.020
21.0	.678	46.0	.337	71.0	.149	96.0	.018
21.5	.668	46.5	.332	71.5	.146	96.5	.016
22.0	.658	47.0	.328	72.0	.143	97.0	.013
22.5	.648	47.5	.323	72.5	.140	97.5	.011
23.0	.638	48.0	.319	73.0	.137	98.0	.009
23.5	.629	48.5	.314	73.5	.134	98.5	.007
24.0	.620	49.0	.310	74.0	.131	99.0	.004
24.5	.611	49.5	.305	74.5	.128	99.5	.002
25.0	.602	50.0	.301	75.0	.125	100.0	.000
25.5	.594	50.5	.297	75.5	.122		

TABLE C.2. Vapor Pressure of Water (Values Are for Water in Contact with Its Own Vapor).*

TEMPERATURE °C	0	1	2	3	4	5	6	7	8	9
					PH_2O MM HG					
0	4.6	4.9	5.3	5.7	6.1	6.5	7.0	7.5	8.0	8.6
10	9.2	9.8	10.5	11.2	12.0	12.8	13.6	14.5	15.5	16.5
20	17.5	18.7	19.8	21.1	22.4	23.8	25.2	26.7	28.3	30.0
30	31.8	33.7	35.7	37.7	39.9	42.2	44.6	47.1	49.7	52.4
40	55.3	58.3	61.5	64.8	68.3	71.9	75.7	79.6	83.7	88.0
50	92.5	97.2	102	107	113	118	124	130	136	143
60	149	156	164	171	179	188	196	205	214	224
70	234	244	255	266	277	289	301	314	327	341
80	365	370	385	401	417	434	451	469	487	506
90	526	546	567	589	611	634	658	682	707	733
100	760									

*Water vapor pressures obtained from *Handbook of Chemistry and Physics,* 64th ed. (Boca Raton, Fla.: CRC Press, 1983–84). Used by permission.

TABLE C.3. Body Surface Area of Human (M^2).*

BODY SURFACE AREA IN METERS SQUARED (BSA M^2)																	
HEIGHT IN CENTI-	WEIGHT IN KILOGRAMS																
METERS	25	30	35	40	45	50	55	60	65	70	75	80	85	90	95	100	105
200							1.84	1.91	1.97	2.03	2.09	2.15	2.21	2.26	2.31	2.36	2.41
195						1.73	1.80	1.87	1.93	1.99	2.05	2.11	2.17	2.22	2.27	2.32	2.37
190				1.56	1.63	1.70	1.77	1.84	1.90	1.96	2.02	2.08	2.13	2.18	2.22	2.28	2.33
185				1.53	1.60	1.67	1.74	1.80	1.86	1.92	1.98	2.04	2.09	2.14	2.19	2.24	2.29
180				1.49	1.57	1.64	1.71	1.77	1.83	1.89	1.95	2.00	2.05	2.10	2.15	2.20	2.25
175	1.19	1.28	1.36	1.46	1.53	[1.61]	1.67	1.73	1.79	1.85	1.91	1.96	2.01	2.06	2.11	2.16	2.21
170	1.17	1.26	1.34	1.43	1.50	1.57	1.63	1.69	1.75	1.81	1.86	1.91	1.96	2.01	2.06	2.11	
165	1.14	1.23	1.31	1.40	1.47	1.54	1.60	1.66	1.72	1.78	1.83	1.88	1.93	1.98	2.03	2.07	
160	1.12	1.21	1.29	1.37	1.44	1.50	1.56	1.62	1.68	1.73	1.78	1.83	1.88	1.93	1.98		
155	1.09	1.18	1.26	1.33	1.40	1.46	1.52	1.58	1.64	1.69	1.74	1.79	1.84	1.89			
150	1.06	1.15	1.23	1.30	1.36	1.42	1.48	1.54	1.60	1.65	1.70	1.75	1.80				
145	1.03	1.12	1.20	1.27	1.33	1.39	1.45	1.51	1.56	1.61	1.66	1.71					
140	1.00	1.09	1.17	1.24	1.30	1.36	1.42	1.47	1.52	1.57							
135	0.97	1.06	1.14	1.20	1.26	1.32	1.38	1.43	1.48								
130	0.95	1.04	1.11	1.17	1.23	1.29	1.35	1.40									
125	0.93	1.01	1.08	1.14	1.20	1.26	1.31	1.36									
120	0.91	0.98	1.04	1.10	1.16	1.22	1.27										

*From DuBois, E. F., *Basal Metabolism in Health and Disease*, 3rd ed. (Philadelphia: Lea & Febiger, 1936). Used by permission.

TABLE C.4. Kilocalories Per Square Meter of Body Surface Area Per Hour, Mayo Foundation Normal Standards.*

MALES		FEMALES	
AGE AT LAST BIRTHDAY	MEAN	AGE AT LAST BIRTHDAY	MEAN
6	53.00	6	50.62
7	52.45	6½	50.23
8	51.78	7	49.12
8½	51.20	7½	47.84
9	50.54	8	47.00
9½	49.42	8½	46.50
10	48.50	9–10	45.90
10½	47.71	11	45.26
11	47.18	11½	44.80
12	46.75	12	44.28
13–15	46.35	12½	43.58
16	45.72	13	42.90
16½	45.30	13½	42.10
17	44.80	14	41.45
17½	44.03	14½	40.74
18	43.25	15	40.10
18½	42.70	15½	39.40
19	42.32	16	38.85
19½	42.00	16½	38.30
20–21	41.43	17	37.82
22–23	40.82	17½	37.40
24–27	40.24	18–19	36.74
28–29	39.81	20–24	36.18
30–34	39.34	25–44	35.70
35–39	38.68	45–49	34.94
40–44	38.00	50–54	33.96
45–49	37.37	55–59	33.18
50–54	36.73	60–64	32.61
55–59	36.10	65–69	32.30
60–64	35.48		
65–69	34.80		

*From *American Journal of Physiology,* July 1936.

TABLE C.5. Body Surface Area of Rat (M^2).*

WEIGHT (G)	SURFACE AREA (M^2)				
	0	2	4	6	8
50	0.0131	0.0134	0.0137	0.0140	0.0143
60	0.0146	0.0149	0.0152	0.0155	0.0158
70	0.0161	0.0164	0.0167	0.0169	0.0171
80	0.0174	0.0176	0.0179	0.0181	0.0184
90	0.0186	0.0189	0.0192	0.0194	0.0196
100	0.0199	0.0201	0.0204	0.0206	0.0208
110	0.0210	0.0213	0.0215	0.0218	0.0220
120	0.0222	0.0224	0.0227	0.0229	0.0231
130	0.0233	0.0235	0.0237	0.0239	0.0241
140	0.0243	0.0245	0.0247	0.0249	0.0251
150	0.0253	0.0255	0.0257	0.0259	0.0261
160	0.0263	0.0265	0.0267	0.0269	0.0271
170	0.0273	0.0275	0.0277	0.0279	0.0281
180	0.0283	0.0285	0.0287	0.0289	0.0291
190	0.0293	0.0294	0.0295	0.0297	0.0299
200	0.0301	0.0303	0.0304	0.0306	0.0308
210	0.0310	0.0312	0.0313	0.0315	0.0317
220	0.0319	0.0321	0.0322	0.0324	0.0326
230	0.0328	0.0329	0.0331	0.0333	0.0334
240	0.0336	0.0338	0.0339	0.0341	0.0343
250	0.0345	0.0346	0.0348	0.0349	0.0351
260	0.0353	0.0354	0.0356	0.0357	0.0359
270	0.0361	0.0362	0.0364	0.0365	0.0367
280	0.0369	0.0370	0.0372	0.0373	0.0375
290	0.0377	0.0378	0.0380	0.0381	0.0383
300	0.0384	0.0386	0.0387	0.0389	0.0390
310	0.0392	0.0393	0.0395	0.0396	0.0398
320	0.0399	0.0401	0.0402	0.0404	0.0405
330	0.0407	0.0408	0.0410	0.0411	0.0413
340	0.0414	0.0416	0.0417	0.0419	0.0420
350	0.0421	0.0423	0.0424	0.0426	0.0427
360	0.0428	0.0430	0.0431	0.0433	0.0434
370	0.0435	0.0437	0.0438	0.0440	0.0441
380	0.0442	0.0444	0.0445	0.0447	0.0448
390	0.0449	0.0451	0.0452	0.0454	0.0455
400	0.0456	0.0458	0.0459	0.0461	0.0462
410	0.0463	0.0465	0.0466	0.0468	0.0469
420	0.0470	0.0472	0.0473	0.0475	0.0475

*From Taylor, A. B. and F. Sargent II, *Elementary Human Physiology: Laboratory and Demonstration Manual* (Minneapolis: Burgess, 1962). Used by permission.

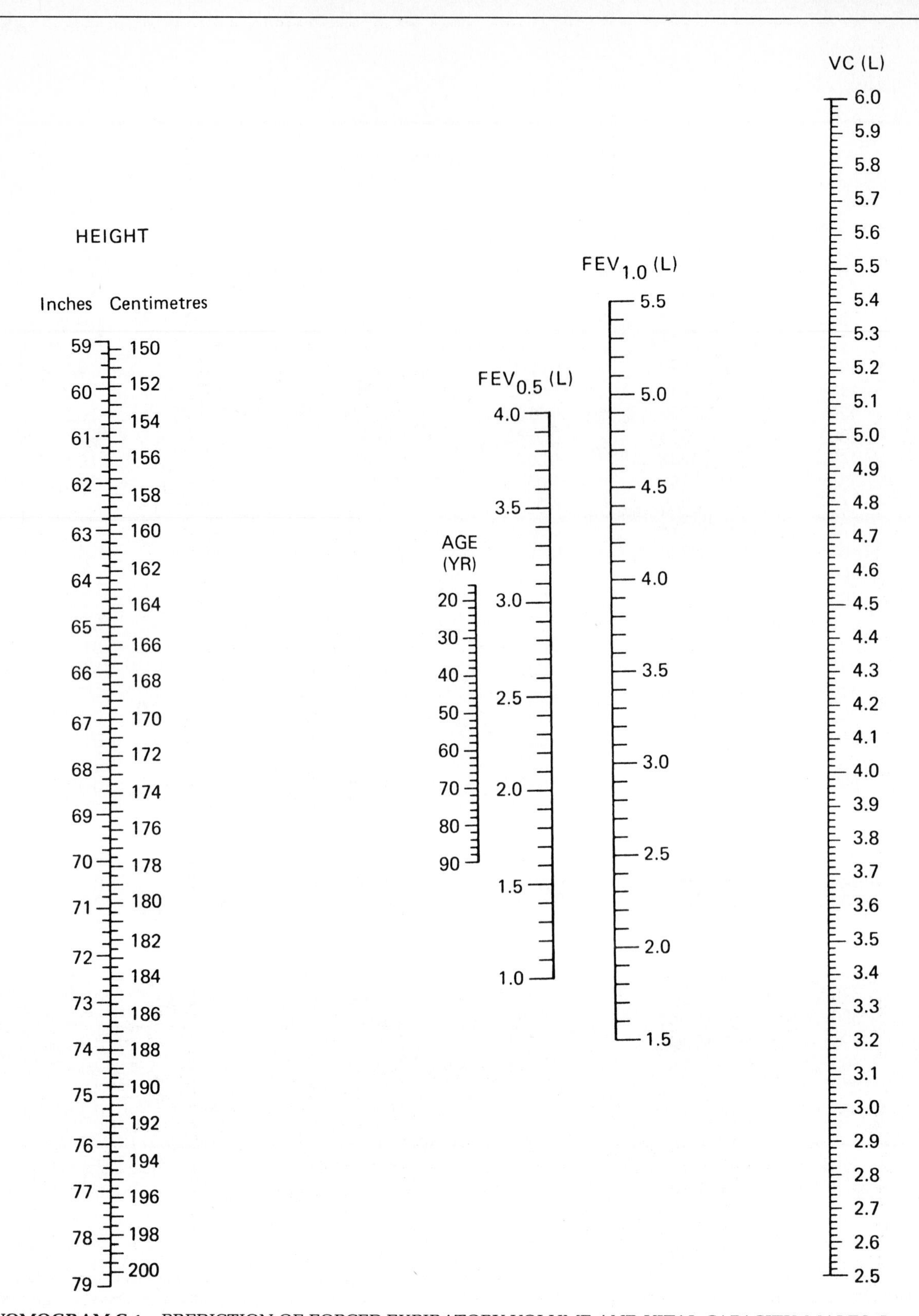

NOMOGRAM C.1. PREDICTION OF FORCED EXPIRATORY VOLUME AND VITAL CAPACITY, MALES. Locate the height in inches (or centimeters) and the age in years. Place a straightedge between these two points; the intersects will give the predicted forced expiratory volume (FEV) and the predicted vital capacity (VC).

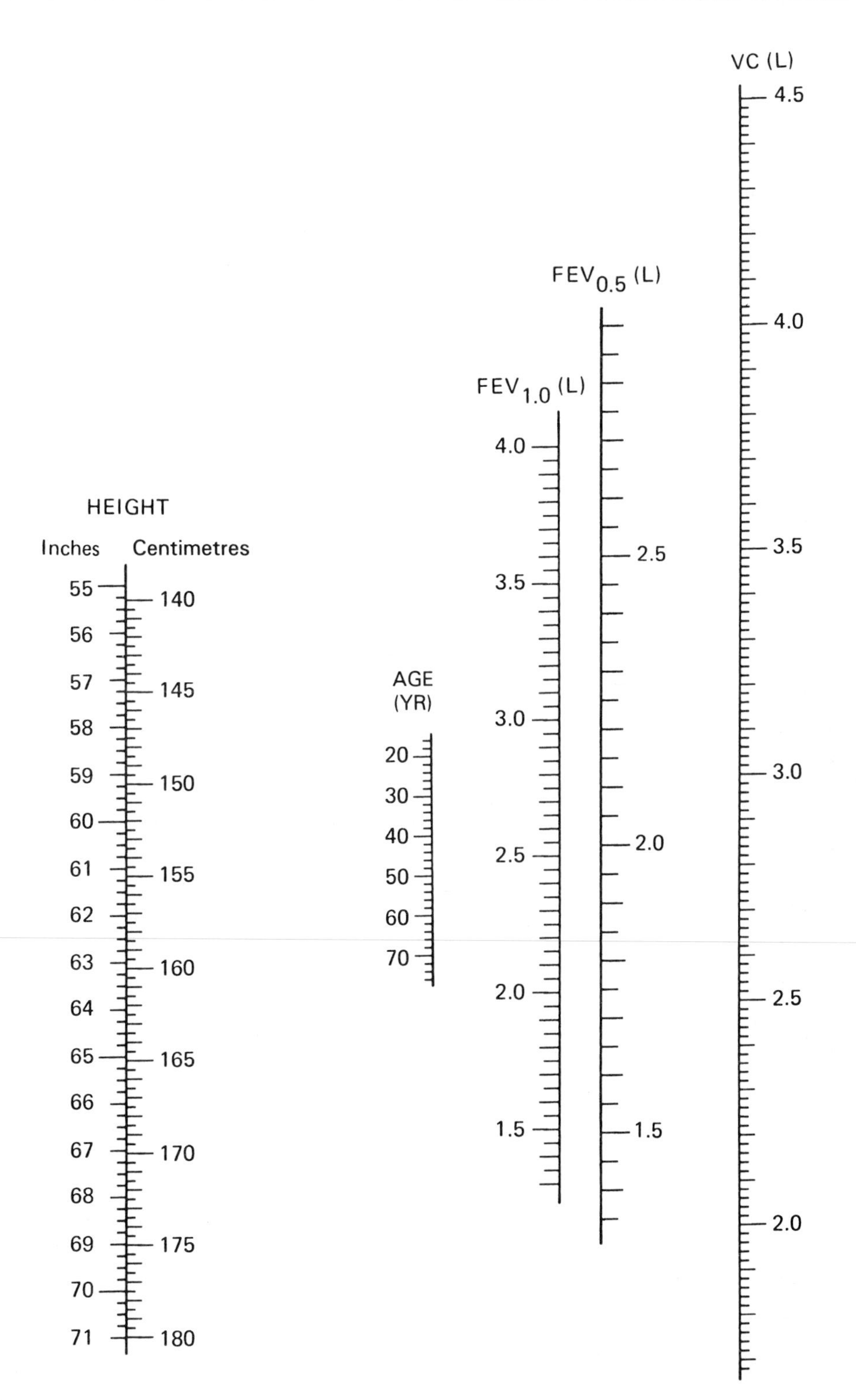

NOMOGRAM C.2. PREDICTION OF FORCED EXPIRATORY VOLUME AND VITAL CAPACITY, FEMALES. Locate the height in inches (or centimeters) and the age in years. Place a straightedge between these two points; the intersects will give the predicted forced expiratory volume (FEV) and the predicted vital capacity (VC).

Index

ABO system, 222–23
Accommodation reflexes, 41, 75
Acid-base balance, 4–6
Action potential, 49, 51–54
Aerobic fitness, 234–39
AIDS (Acquired Immune Deficiency Syndrome), 251
All-or-none law, 168
Amylase, 99
Anemia, 209, 211
Anesthetics for rabbits, 200–202
Ascheim-Zondek pregnancy test, 95
Astigmatism, 76, 79–80
Audiometry, 66
Autonomic neurons, 33

Babinski's reflex, 42
Basal metabolic rate, 122–24
Bile salts, 100–101
Blind spot, 76
Blood, precautions for handling, 251
Blood count, 208–11
Blood pressure, 176–78, 199, 245
Blood typing, 222–24
Body composition, 230–31
Brain anatomy, 36–37
Buffer system, 5–6

Caloric cost, 243
Calorimetry, 121–22
Cardiac index, 246
Cardiac output, 245–46
Castration, 91–92
Cataracts, 80
Chambers-Zweifach capillary unit, 212
Chorionic gonadotropin, 94
Ciliospinal reflex, 42
Cold pressor test, 178
Color blindness, 77–78
Colorimeter, 208
Complementary colors, 77
Corneal reflex, 41–42
Cranial nerves, 35–36
Cutaneous receptors, 63–64
Cyanmethemoglobin, 207–8

Diabetes mellitus, 22, 111
Differential leukocyte count, 220–21
Diffusion, 9
Digestion, 99–101
Diopters, 79

Einthoven's triangle, 181
Electrical axis of the heart, 184
Electrocardiogram, 180–84
Estrus cycle, 92–93
Expiratory reserve volume, 192
Eye, anatomy of, 81–83
Eye reflexes, 41, 75

Fick's law of diffusion, 9
Flexibility, 229, 230
Forced expiratory volume, 193
Forest Service fitness test, 236–37

Glucometer II, 113
Glucose tolerance test, 112

Harvard step test, 235, 236
Hearing impairment, 66–67
Heart block, 167–68
Heart sounds, 175–76
Hematocrit, 205–6
Hemocytometer, 209–10

Hemoglobin, 206
Hemostasis, 224
Heymer test, 193
Holmgren's test, 78
Homeostasis, 4, 9
Hormones, and reproduction, 89–91
Hypermetropia, 79

Immunologic tests for pregnancy, 94–95
Inspiratory reserve volume, 192
Insulin, 111
Ishihara test, 78
Isotonic solutions, 4, 11

Jendrassik's maneuver, 42

Korotkoff sounds, 177
Kymograph system, 149, 190

Labstix test, 22–23
Labyrinthine reflexes, 42–44
Leukocytes, 220
Lipase, 100–101

Maximal breathing capacity, 193
Mean corpuscular hemoglobin concentration, 211
Mean corpuscular volume, 211
Mechanical efficiency, 244
Metabolism, 121, 124, 135
Microcirculation, 212
Modified tension time index, 246–47
Molar solutions, 2–3
Motor points, 149–150
Muscle twitch, 147
Muscular strength and endurance, 229–30
Myopia, 79

Nerve conduction blockade, 54
Nerve-muscle preparation, 143–47
Nervous system, 33
Nystagmus, 43

Obesity, 229, 230
Ophthalmoscope, 83–84
Oscilloscope, 50–51
Osmolar solutions, 3–4, 10
Osmosis, 3, 10
Osmotic pressure, 3, 10
Ovariectomy, 92–93
Oxygen debt, 244–45

Pepsin, 100
Percentage (%) solutions, 2
Perimetry, 80–81
Permeability of cell membrane, 12–13
Physiograph system, 146, 148, 190
Plantar reflex, 42
Pneumograph, 189–90
Pregnancy tests, 94–95
Proprioception, 44
Pulse wave, 178–80
Pupillary reflex, 41

Reaction times, 44
Referred pain, 64
Reflex arc, 42
Refractory period, 53
Residual volume, 192
Respirometer, 122, 243
Resting potential, 49
Rh system, 223–24
Rinne test, 65–66

Sensory receptors, 63
Smooth muscle, 107
Snellen test, 76
Solutions, 253–54
Somatic neurons, 33
Spatial orientation, 44
Sphygmomanometer, 176
Spinal cord, 33–35
Spinal nerves, 33–35
Spinal reflexes, 41, 42
Spirometer, 122, 192
Starling's law, 147, 166
Stethoscope, 175
Stimulation of tissues, 49–50, 147
Strength-duration curve, 53
Stroke volume, 246
Summation of subliminal stimuli, 53
Surface area law, 124

Tendon reflexes, 42
Testicular and gonadotropic hormones, 91–92
Tetanization, 166
Threshold stimuli, 52, 147
Thyroid, 135–37
Tidal volume, 191
Total peripheral resistance, 246

Urinalysis, 22
Urogenital system, 90

Vaginal smears, 93
Vital capacity, 192

Watch tick test, 65
Weber test, 65
Weber's law, 64